T0327567

WELDING METALLURGY AND WELDABILITY

WELDING METALLURGY AND WELDABILITY

JOHN C. LIPPOLD
Ohio State University

Published by John Wiley & Sons, Inc., Hoboken, New Jersey
Published simultaneously in Canada

For general information on our other products and services or for technical support, please contact our Customer Care Department within the United States at (800) 762-2974, outside the United States at (317) 572-3993 or fax (317) 572-4002.

Wiley also publishes its books in a variety of electronic formats. Some content that appears in print may not be available in electronic formats. For more information about Wiley products, visit our web site at www.wiley.com.

Library of Congress Cataloging-in-Publication Data:

Lippold, John C., author.
  Welding metallurgy and weldability / John C. Lippold.
    pages   cm
  Includes bibliographical references and index.
  ISBN 978-1-118-23070-1 (hardback)
  1. Stainless steel–Weldability.   2. Stainless steel–Metallurgy.   3. Nickel–Weldability.
4. Nickel–Metallurgy.   5. Nickel alloys–Weldability.   6. Nickel alloys–Metallurgy.   I. Title.
  TS227.L657 2009
  669'.142–dc23
                                                                                    2014033412

Dedicated to four giants of the international welding metallurgy community who have preceded me,

Henri Granjon (France)
Warren F. "Doc" Savage (USA)
Trevor Gooch (United Kingdom)
Fukuhisa Matsuda (Japan)

CONTENTS

PREFACE

This textbook is a companion to previous books on welding metallurgy and weldability of stainless steels and Ni-based alloys published in 2005 and 2009, respectively. In retrospect, this book should have been published first because it lays the groundwork for many of the metallurgical and weldability concepts that are described in those texts. The subject matter in this book is based on a course that I have taught in the Welding Engineering Program at Ohio State University since 1986. The content is designed to provide engineers with the necessary background to understand the basic concepts of welding metallurgy and to interpret failures in welded components.

The main topic of this book is material "weldability." As described in Chapter 1, there is considerable disagreement over the meaning of the term weldability and the subjects it encompasses. In this book, it is meant to describe weld failure mechanisms associated with either fabrication or service. In addition, the failure mechanisms described are related to the microstructure of the weldment and do not address non-metallurgical defects, such as lack-of-fusion, porosity, design deficiencies, or other design- or process-related issues. For readers who are not proficient in welding metallurgy, Chapter 2 reviews basic principles that will be helpful in understanding the concepts presented in subsequent chapters.

Weldability issues are divided into fabrication- and service-related failures. Chapters 3–5 address hot cracking, warm (solid-state) cracking, and cold cracking that occur during initial fabrication, or repair. In each of these chapters, the basic mechanisms for cracking are described and preventive measures recommended. Chapter 6 addresses corrosion-related failures, and Chapter 7 discusses some general concepts with respect to fracture and fatigue. Chapter 8 provides guidance on failure analysis and includes examples of SEM fractography that will aid in determining failure mechanisms. Finally, Chapter 9 describes a number of weldability testing techniques that can be used to quantify susceptibility to various forms of weld cracking. Appendices are included which list compositions of base and filler metals referenced in the text and etching techniques that are used for metallurgical analysis.

This book has been dedicated to four "giants" of the international welding metallurgy community. *Henri Granjon* was the director of the welding school at Institut de Soudure (French Welding Institute) where he introduced many of the early concepts of welding metallurgy, particularly with regard to steels. He is also credited with developing the implant test for assessing hydrogen-induced cracking that is still in use today. I never had the pleasure of meeting Henri Granjon but have been inspired by his work and reputation.

Trevor Gooch of The Welding Institute in the United Kingdom was a talented metallurgist but also well versed in corrosion and hydrogen embrittlement. My first encounter with Trevor was at an AWS annual meeting in 1978. I attended a session where a tall, bespectacled man in the back of the room rose to ask a question of a speaker bellowing "Gooch, TWI." He proceeded to take the speaker to task on a number of issues and I learned to be extra vigilant (and prepared) if I expected Dr. Gooch to be in attendance at one of my presentations. In later years, I came to know Trevor quite well and found him to be a humble and gentle man who loved bluegrass music and had an incredible passion for welding metallurgy.

Warren F. "Doc" Savage is arguably the Father of modern Welding Metallurgy. From the 1950s through the 1970s, he and his students defined the field of welding metallurgy and established many of the principles that exist today. I met **"Doc"** when I was an undergraduate at Rensselaer Polytechnic Institute and, after a little maneuvering on my part, he took me on as a graduate student. It turned out to be a lifelong relationship. He served as my advisor and mentor, and taught me a little bit about life along the way including the pleasures of drinking scotch whiskey. I have adopted his hands off style of advising graduate students, allowing student to think creatively. Doc has had a tremendous impact on my professional career.

Fukuhisa Matsuda was a long-time faculty member at Osaka University, but spent 1 year in the welding metallurgy laboratory at RPI. His research extended the basic concepts proposed by Savage by making use of weldability testing and advanced characterization. His single greatest accomplishment was the development of the fundamental understanding of weld solidification cracking that is still widely accepted today. His body of research is impressive and has greatly influenced my thinking. There are many references to his work in this book.

I would also like to recognize the many students who have taken the course on which this book is based. As part of that class, students are required to prepare a review paper on a topic relevant to the various aspects of weldability. These papers were a very valuable resource for me as I prepared this book, and I am sincerely grateful to all those students for being so diligent.

I would also like to thank many of my current graduate students and postdocs who have assisted me in the preparation of this text, particularly in providing photomicrographs and other figures that are used to illustrate welding metallurgy and weldability principles. In particular, I would like to acknowledge Xiuli Feng, Xin Yue, Adam Hope, and David Tung.

I would also like to acknowledge a number of my current and former students who took time out of their busy careers to review individual chapters in the book. These individuals include Adam Hope and David Tung (both OSU PhD students), Jeff Sowards (NIST), Jeremy Caron (Haynes International), Seth Norton (BP), Xin Yue (ExxonMobil), Morgan Gallagher (Shell), and Mikal Balmforth (Materials and Engineering Group LLC). Their thoughtful, and occasionally critical, reviews have made this a much improved book.

Finally, I would like to thank Ohio State University for providing the opportunity and support to prepare this textbook. Much of the work on this book was completed during a professional leave period (a.k.a. sabbatical) for which faculty are eligible every 7 years. Without this opportunity, it would have been difficult to complete this book in the normal course of teaching and research. I am particularly grateful to my department chair, Rudy Buchheit, who has been very supportive of my scholarly endeavors.

College of Engineering Distinguished Professor   JOHN C. LIPPOLD
Department of Materials Science and Engineering
The Ohio State University

AUTHOR BIOGRAPHY

Dr. Lippold is a Professor in the Welding Engineering Program and Head of the Welding and Joining Metallurgy Group at Ohio State University. He is also a College of Engineering Distinguished Faculty at OSU.

He received his B.S., M.S., and Ph.D. degrees in Materials Engineering from Rensselaer Polytechnic Institute where he was a student of Dr. Warren F. Savage. Upon completion of his formal education, he worked for 7 years at Sandia National Laboratories, Livermore, CA, as a member of the technical staff, specializing in the area of stainless steel and high alloy weldability. From 1985 to 1995, Dr. Lippold worked for Edison Welding Institute. In 1995, he joined the faculty of the Welding Engineering Program at OSU.

Over the past 35 years, Dr. Lippold has been involved in research programs designed to gain a better understanding of the welding metallurgy and properties of engineering materials. His research has involved both fundamental and applied topics with a high degree of industrial relevance. He has been actively involved in the evaluation of weldability test techniques and the development of testing approaches that provide useful engineering information. Based on this research, Dr. Lippold has published over 300 technical papers and reports. He is recognized internationally in the field of stainless steel and high alloy welding metallurgy, and weldability testing. Since joining OSU in 1995, he has advised over 60 graduate students. In 2005, he coauthored the textbook *Welding Metallurgy and Weldability of Stainless Steels* and in 2009 coauthored a companion textbook entitled *Welding Metallurgy and Weldability of Nickel-Base Alloys*.

He has previously won the Charles H. Jennings Memorial Award (1977, 1980, 2005), the William Spraragen Memorial Award (1979, 1992, 2012), the Warren F. Savage Memorial Award (1993, 1999, 2010, 2011, 2012, 2014), the McKay-Helm

Award (1994, 2011), the Lincoln Gold Medal (1983), the A.F. Davis Silver Medal (2001), the Hobart Memorial Award (2011), the Irrgang Award (2002), and the Plummer Memorial Educational Lecture Award (2002) from the American Welding Society (AWS). He has also been the recipient of the Buehler Technical Paper Merit Award (1985, 1989) from the International Metallographic Society. In 1997, he presented the Comfort A. Adam's Memorial Lecture at the AWS Annual Convention in Los Angeles. In 2008, he won the Jaeger Lecture Award and in 2009 the Yoshiaki Arata Award from the International Institute of Welding. Dr. Lippold is a Fellow of both ASM International (1994) and the American Welding Society (1996). He is currently coeditor of the international journal *Welding in the World* published by IIW.

1

INTRODUCTION

This textbook addresses the topics of welding metallurgy and weldability. The two topics are inextricably intertwined since the weldability of a material is closely related to its microstructure. While the term *welding metallurgy* is universally accepted as a subset of physical metallurgy principles, the term *weldability* has been subject to a wide range of definitions and interpretations. In its broadest context, weldability considers aspects of design, fabrication, fitness for service, and, in some cases, repair. This broad treatment is reflected in the definitions for weldability that are provided by both the American Welding Society and the ISO Standard 581:1980. Thus, weldability can be used to describe both the ability to successfully fabricate a component using welding and the capacity for that component to perform adequately in its intended service environment.

AWS Definition of Weldability

The capacity of a material to be welded under fabrication conditions imposed into a specific, suitably designed structure and to perform satisfactorily in the intended service.

In a *Welding Journal* article published in 1946 entitled "This Elusive Character Called Weldability," W.L. Warren from the Watertown Arsenal in the United States stated, "That word (weldability)…has grown to mansize in stature and importance in respect to its significance in modern welding application. This term has been and is used with such a variety of shades of meaning that one may rightly conclude weldability to possess a value as changeable as a chameleon" [1].

Welding Metallurgy and Weldability, First Edition. John C. Lippold.
© 2015 John Wiley & Sons, Inc. Published 2015 by John Wiley & Sons, Inc.

FIGURE 1.1   Henri Granjon, Institut de Soudure.

ISO 581:1980 Definition of Weldability

Metallic material is considered to be susceptible to welding to an established extent with given processes and for given purposes when welding provides metal integrity by a corresponding technological process for welded parts to meet technical requirements as to their own qualities as well as to their influence on a structure they form.

Henri Granjon (Fig. 1.1) in his text *Fundamentals of Welding Metallurgy* defined weldability as "…the behavior of (those) joints and the constructions containing them, during welding and in service…" [2] R.D. Stout in *Weldability of Steels* states that "the term weldability has no universally accepted meaning and the interpretation placed upon the term varies widely according to individual viewpoint" [3]. At a conference held at The Welding Institute (TWI) in 1988 entitled *Quantifying Weldability* [4], Trevor Gooch from TWI (Fig. 1.2) stated that "…the concept of weldability of a material is complex." At the same conference, A.D. Batte of British Gas Corporation is quoted as saying that "…it is incongruous to find that the definition of weldability is still an active area of debate," and W.G. Welland from BP International stated that "the concept of weldability is of little interest to the builders and users of most welded fabrications." Despite the many papers published by Warren F. Savage (Fig. 1.3) and

FIGURE 1.2   Trevor Gooch, The Welding Institute, 1992.

FIGURE 1.3   Warren F. "Doc" Savage, Rensselaer Polytechnic Institute, 1986.

FIGURE 1.4 Fukuhisa Matsuda, Osaka University, 1988 (W.A. "Bud" Baeslack III in the background).

his students at Rensselaer Polytechnic Institute and Fukuhisa Matsuda (Fig. 1.4) and his students at Osaka University, there are no definitions of weldability attributed to them (perhaps for good reason).

In this text, weldability will be considered from the standpoint of materials' resistance or susceptibility to failure. From a fabrication standpoint, this relates to the ability to produce welds that are defect-free. There are multiple weld defects that can occur during fabrication, as described in Section 1.1, and these can be separated into those that are related to the welding process and procedures and those associated with the material. For example, defects such as lack of fusion, undercut, and slag inclusions are related primarily to the welding process and can usually be avoided by changes in process conditions. Defects such as solidification cracks and hydrogen-induced cracks are primarily related to the metallurgical characteristics of the material and are usually difficult to eliminate by changes in process conditions alone.

The term weldability also describes the behavior of welded structures after they are put into service. There are many examples of welded structures that are free of fabrication defects that later fail in service. These include failure modes involving corrosion, fatigue, stress rupture (creep), or complex combinations of these and other failure mechanisms. The service-related failure modes are perhaps the most serious

FIGURE 1.5 Liberty ship failure.

of the weldability issues discussed here, since failure by these mechanisms can often be unexpected and catastrophic. As an example of this, consider the catastrophic Liberty ship failures (see Fig. 1.5) during World War II that led to the sinking of many transport ships and the loss of many lives.

This text will focus primarily on the aspects of weldability that are influenced by the metallurgical properties of a welded structure. As such, chapters addressing various fabrication cracking mechanisms are included. These chapters are designed to not only describe the underlying mechanisms for cracking but to provide insight into how such forms of cracking can be avoided. Similarly, the various forms of service cracking are described, particularly those associated with corrosion, brittle fracture, and fatigue. In order to provide the reader with sufficient metallurgical background to interpret the contents of these chapters, a chapter on welding metallurgy principles has been included.

1.1 FABRICATION-RELATED DEFECTS

Fabrication-related defects include cracking phenomena that are associated with the metallurgical nature of the weldment and process- and/or procedure-related defects. A list of common fabrication defects is provided in Table 1.1. The defects associated with the metallurgical behavior of the material can be broadly grouped by the temperature range in which they occur.

Hot cracking includes those cracking phenomena associated with the presence of liquid in the microstructure and occurs in the fusion zone and PMZ region of

TABLE 1.1   Fabrication-related defects

"Hot" cracking
Weld solidification
HAZ liquation
Weld metal liquation

"Warm" cracking
Ductility dip
Reheat/PWHT
Strain age
Liquid metal embrittlement (LME)

"Cold" cracking
Hydrogen-induced (or hydrogen-assisted) cracking
Delayed cracking

Process control
Lack of fusion
Weld undercut
Excessive overbead
Incomplete penetration
Slag inclusions

Others
Geometric defects (design or fit-up)
Metallurgical anomalies (e.g., local softening or embrittlement)
Porosity

---

the HAZ. Liquid films along grain boundaries are usually associated with this form of cracking.

Warm cracking occurs at elevated temperature in the solid state, that is, no liquid is present in the microstructure. These defects may occur in both the fusion zone and HAZ. All of the warm cracking phenomena are associated with grain boundaries.

Cold cracking occurs at or near room temperature and is usually synonymous with hydrogen-induced cracking. This form of cracking can be either intergranular or transgranular.

A number of nonmetallurgical defects that can occur during fabrication are also listed in Table 1.1. These are generally associated with poor process/procedure control and include lack of fusion, undercut, incomplete joint penetration, and geometric defects. Such defects can usually be remedied by careful attention to process conditions, joint design, material preparation (cleaning), etc. This text will not address the nature or remediation of these types of defects.

1.2   SERVICE-RELATED DEFECTS

Welds are subject to a wide range of service-related defects. Since welds are metallurgically distinct from the surrounding base metal and may contain residual stresses, they are often susceptible to failure well in advance of the base metal.

TABLE 1.2　Service-related defects

Hydrogen induced
Environmentally induced (i.e., corrosion)
Relaxation cracking
Fatigue
Stress rupture
Creep and creep fatigue
Corrosion fatigue
Mechanical overload

These defects are usually manifested as cracks that form under specific environmental and/or mechanical conditions. A list of service-related defects is provided in Table 1.2.

Corrosion of welds is often a problem due to both the microstructural and local mechanical conditions of welded structures. The presence of fabrication-related defects can often accelerate service failures, particularly by fatigue. Welds in many engineering materials may contain softened regions that can promote mechanical overload failures. Conversely, local hard zones can result in reduced ductility and possible brittle failure.

1.3　DEFECT PREVENTION AND CONTROL

Although the understanding of the mechanisms leading to various forms of cracking is important, developing a methodology to prevent cracking is the ultimate goal of the welding engineer. Preventative measures can usually not be developed until the nature of the failure is understood. In some cases, changes in welding technique, or procedure, may be effective. For example, simple changes in heat input and bead shape can sometimes prevent weld solidification cracking. Another example is the use of preheat and interpass temperature control to prevent hydrogen-induced cracking.

Before such preventative measures can be implemented, the nature of failure must be determined in order that the "cure" does not lead to other weldability problems. Many Ni-base weld metals are susceptible to both solidification cracking and ductility-dip cracking, but the remedy for each defect type is different.

This textbook provides the necessary background to understand and interpret weld failures and recommends possible remedies for such failures. It should be noted, however, that the solution for many weldability problems will require a change in material rather than a "tweaking" of composition or process parameters. For example, reheat and strain-age cracking are significant problems when welding thick-section or highly restrained Cr–Mo steels and Ni-base superalloys, respectively. Again, knowledge of the precise mechanism of failure is required before remedial measures can be implemented.

REFERENCES

[1] Warner WL. This elusive character called weldability. Weld J 1946;25 (3):185s–188s.

[2] Granjon H. *Fundamentals of Welding Metallurgy*. Cambridge, UK: Woodhead Publishing Ltd.; 1991 (Translated from French edition published in 1989).

[3] Stout RD. *Weldability of Steels*. 4th ed. New York: Welding Research Council; 1987.

[4] Pargeter RJ, editor. *Quantifying Weldability*. Cambridge, UK: The Welding Institute; 1988.

2

WELDING METALLURGY PRINCIPLES

2.1 INTRODUCTION

The purpose of this chapter is to review the basic principles that govern microstructure evolution in welds. Since the cracking susceptibility of welded structures is a function of microstructure, environment, and applied stress, it is essential to understand the basic principles that govern the evolution of microstructure during welding. This chapter will focus primarily on fusion welds, but a section specific to solid-state welds is also included. This is not meant to be an exhaustive review of welding metallurgy principles. For a more detailed treatment of this topic, the reader is referred to textbooks by Kou entitled *Welding Metallurgy* [1] and Easterling entitled *Introduction to the Physical Metallurgy of Welding* [2].

There are a number of metallurgical processes that control the microstructure and properties of welds. Melting and solidification are important processes, since they are the key to achieving acceptable joints in all fusion welding processes. Coupled with solidification are segregation and diffusion processes resulting in local compositional variations that influence both weldability and service performance.

Many metallurgical processes occur in the solid state, including phase transformations, precipitation reactions, recrystallization, grain growth, etc. The extent of these reactions may significantly alter the microstructure and properties of the weldment (weld metal and heat-affected zone (HAZ)) relative to the base metal. Many of these reactions, or complex combinations of reactions, can result in embrittlement, or

Welding Metallurgy and Weldability, First Edition. John C. Lippold.
© 2015 John Wiley & Sons, Inc. Published 2015 by John Wiley & Sons, Inc.

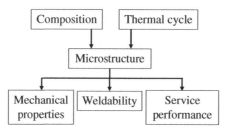

FIGURE 2.1   Block diagram for weld microstructure evolution and performance.

cracking, of welds. This embrittlement can occur due to liquation, the presence of liquid films in an otherwise solid matrix, or in the solid state due to a loss in ductility.

Thermal expansion during heating and contraction during cooling can result in complex stress patterns in and around welds. These stresses can subsequently affect the microstructure and properties of the weldment and may promote cracking in regions where the tensile strain resulting from these stresses exceeds the ductility of the material.

The nature of the weld microstructure for a given material results from the combination of the weld thermal cycle and the material composition. In general, the heating and cooling rates associated with welding are quite high (10–1000°C/s) and usually prevent prediction of microstructure based on equilibrium thermodynamic principles. All of the metallurgical processes that influence the weld microstructure are temperature and heating/cooling rate dependent, and thus, the welding thermal cycle plays a key role in the evolution of microstructure and, ultimately, the weldability of the material, as shown schematically in Figure 2.1.

2.2   REGIONS OF A FUSION WELD

Examination of a welded joint reveals distinct microstructural regions. The fusion zone is associated with melting. The HAZ, though not melted, is affected by the heat from the joining process. Beyond the HAZ is the unaffected base metal. The fusion zone and HAZ can be further subdivided, as described in this section.

The fusion zone is described as such because it is the region where melting and solidification occur to form the joint, or weld. Since all metals are crystalline in nature, many possessing cubic crystal lattices, there are general solidification phenomena common to all metals. In many materials, solidification behavior is very sensitive to composition. For example, the addition of small amounts of carbon and nitrogen to some steels can change their solidification behavior from ferritic (bcc) to austenitic (fcc). Minute additions of sulfur to steels can promote severe solidification cracking in the fusion zone. Aluminum alloys that are otherwise crack susceptible can be welded with a filler material containing more than 6% of silicon in order to avoid cracking.

The microstructure and properties of the HAZ are solely controlled by the thermal conditions experienced during welding and postweld heat treatment (PWHT). Aluminum alloys are routinely precipitation hardened or work hardened to increase

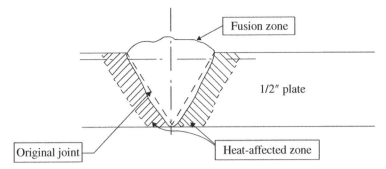

FIGURE 2.2   Early schematic of regions of a fusion weld (From Ref. [3]. © AWS).

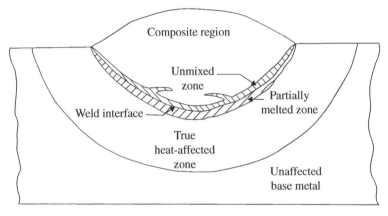

FIGURE 2.3   Regions of a fusion weld (From Ref. [4]. © AWS).

strength; welding can completely eliminate these strengthening effects in the HAZ. Steel undergoes a phase transformation, which can result in a HAZ that has a radically different microstructure and properties than either the base metal or the fusion zone.

The understanding of regions of a weld has evolved tremendously since the 1960s. Prior to that time, a fusion weld was thought to consist of only two regions, the fusion zone and a surrounding HAZ, as shown in Figure 2.2 from a lecture by E.F. Nippes in 1959 [3]. Considerable research conducted by W.F. Savage and his students at RPI in the 1960s and 1970s revealed that other distinct regions of a fusion weld existed [4, 5].

In 1976, Savage *et al.* [4] proposed several changes to the terminology used to describe fusion weld microstructure regions, as shown in Figure 2.3. The fusion zone was considered to consist of two regions. The composite region represented the portion of the fusion zone where base metal and filler metal were mixed in a "composite" composition. Surrounding this region along the fusion boundary, they defined a region called the unmixed zone (UMZ). The UMZ consists of melted and resolidified base metal that does not mix with the filler metal. In some alloy systems, the UMZ can exhibit microstructures and properties very different from those of the composite region, particularly when dissimilar filler metals are used.

The HAZ was subdivided into two regions, the partially melted zone (PMZ) and the "true" heat-affected zone (T-HAZ). The PMZ exists in all fusion welds made in alloys since a transition from 100% liquid to 100% solid must occur across the fusion boundary. In addition, other mechanisms were identified that resulted in local melting (or liquation) in a narrow region surrounding the fusion zone. These include grain boundary melting due to segregation and a phenomenon described as "constitutional liquation" that results from local melting associated with a constituent particle. The designation of a T-HAZ was used to differentiate that region of the HAZ within which all metallurgical reactions occur in the solid state, that is, no melting, or liquation, occurs.

Little has changed since 1976 regarding terminology for describing regions of a fusion weld, although considerable research has been conducted on a variety of alloy systems to verify that these regions actually exist in these material systems. Additional refinements have been made to this original terminology. For example, the T-HAZ in steels has been subdivided into various subregions, such as the coarse-grained HAZ (CGHAZ), the fine-grained HAZ (FGHAZ), and the intercritical HAZ (ICHAZ) regions.

The only potential addition to the terminology in Figure 2.3 is a transition region within the fusion zone. In heterogeneous welds, where the filler metal is of different composition from the base metal, this would represent a composition transition from the composite region to the UMZ. In some alloy systems, this transition zone (TZ) can exhibit a microstructure distinctly different from the surrounding regions. For example, in welds between stainless steels and low-alloy steels, a martensitic structure may form in the transition region that does not occur elsewhere in the weld.

A new schematic of the regions of a fusion weld is provided in Figure 2.4 for a heterogeneous weld. It is similar to the illustration in Figure 2.3 but contains a composition TZ that may be present in some systems. The following sections will review the various regions defined earlier in considerable detail and will describe the mechanisms involved in their formation.

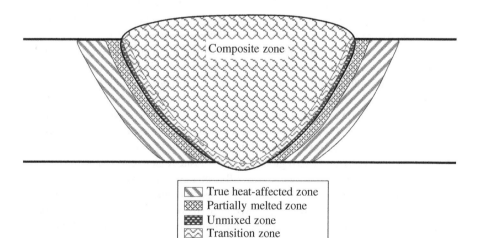

Composite zone

▨	True heat-affected zone
▨	Partially melted zone
▨	Unmixed zone
▨	Transition zone

FIGURE 2.4 Modern schematic showing regions of a fusion weld.

2.3   FUSION ZONE

The fusion zone represents that region of a fusion weld where there are complete melting and resolidification during the welding process. The microstructure in the fusion zone is a function of composition and solidification conditions. Small differences in composition often result in large variations in microstructure and properties. In some systems, changing the solidification and cooling rates can also alter the microstructure, sometimes dramatically.

The fusion zone is normally very distinct from the surrounding HAZ and base metal when samples are prepared metallographically. This is due to both macroscopic and microscopic fluctuations in composition resulting from the solidification process.

In welds where the filler metal is of a different composition from the base metal, three regions theoretically exist. The largest of these is the composite zone (CZ), consisting of filler metal uniformly diluted with base metal. Adjacent to the fusion boundary, two additional regions may exist. The unmixed zone (UMZ) consists of melted and resolidified base metal where negligible mixing with filler metal has occurred. Between the UMZ and CZ, a transition zone (TZ) must exist where a composition gradient from the base metal to the CZ is present.

Three types of fusion zones have been defined: autogenous, homogenous, and heterogeneous. The classifications are based on whether or not a filler metal is used and the composition of the filler metal with respect to the base material. All three types of fusion zones are commonly encountered.

Autogenous welds are those where no filler metal is added and the fusion zone is formed by the melting and resolidification of the base metal. These are common in situations where section thicknesses are minimal and penetration can easily be achieved by the process selected. In thin sections, autogenous welding can often be applied at high speeds, and normally, a minimum amount of joint preparation is required, that is, butt joints can be used. Welding processes that are, or can be, adapted to autogenous welding include GTAW, EBW, LBW, PAW, and resistance welding. The fusion zone is essentially the same composition as the base metal, except for possible losses due to evaporation or pickup of gases from the shielding atmosphere. Not all materials can be joined autogenously because of weldability issues.

Homogenous welds involve the use of a filler metal that closely matches the base metal composition. This type of fusion zone is used when the application requires that filler and base metal properties must be closely matched. Properties such as heat treatment response or corrosion resistance are examples of such properties. Some common examples include the use of Type 316L base metal joined with 316L filler for matching corrosion properties and the use of E10016-D2 filler metal on AISI 4130 Cr–Mo steel, which is usually given a full PWHT to provide uniform strength.

Heterogeneous welds are fusion welds made with filler metals whose composition is different from that of the base metal. In many situations, matching filler metals may not exist or the weld properties desired may not be achievable with a matching composition. It should also be recognized that many base metal compositions may have inherently poor weldability and that dissimilar filler metals are required to achieve acceptable properties or service performance. Some considerations that would require the use of a dissimilar composition filler metal include strength, weld defect formation (e.g., porosity), weldability/solidification cracking resistance, heat treatment response, corrosion resistance, filler metal cost, and operating characteristics of the consumable.

When using a filler metal that has a composition different from the base metal, dilution effects must be carefully considered or the desired outcome may not be as expected. Common examples of heterogeneous welds include the use of Type 308L filler metal on Type 304L base metal for weldability and corrosion resistance and the use of a 4043 aluminum filler metal with 6061 aluminum base metal for solidification cracking resistance.

As noted earlier, the use of heterogeneous welds often requires close attention to dilution effects. Dilution can be defined as a change in composition of a filler metal due to its mixing with the base metal during the melting process. In many cases, dilution is not desirable and must be carefully controlled. Alteration of the deposited weld metal composition by dilution can negate or lessen the desired weld metal properties that would be achieved by a filler metal in its undiluted condition. One case where dilution is particularly undesirable is in surfacing operations where filler metals are significantly different from the base material and chosen to produce very specific properties such as abrasion resistance, corrosion resistance, or impact properties. For example, if stainless steels are used as cladding on carbon steels for corrosion resistance, significant dilution (~40%) can reduce the chromium content to a level where the clad layer is no longer corrosion resistant.

Dilution is expressed in terms of dilution of the filler metal by the base metal and is shown schematically in Figure 2.5. Mathematically, dilution is the ratio of the amount of melted base metal to the total amount of fused metal. For example, a weld with 10% dilution will contain 10% base metal and 90% filler metal. For most welding processes, dilution is normally controlled below 50%. Cross sections of welds, as shown in Figure 2.5, can be used to estimate dilution based on the original joint geometry, or the actual composition of the weld metal can be determined by analysis, and the dilution calculated if the compositions of the base and filler metals are known.

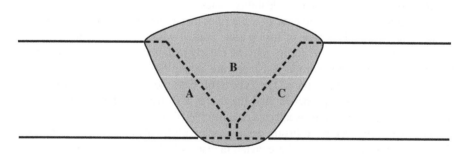

$$\text{Dilution (\%)} = \frac{A + C}{A + B + C} \times 100$$

FIGURE 2.5   Schematic illustration of the determination of dilution in a heterogeneous weld.

2.3.1 Solidification of Metals

Melting and solidification are primary metalworking phenomena that allow for mixing of various elements to form an alloy that can then be solidified, or cast, into a form that will be used as an as-cast part or subsequently thermomechanically processed into other useful shapes (bar, plate, pipe, etc.). These phenomena are also the basis of the fusion welding processes, and a general knowledge of the solidification of metals is required to understand the metallurgical nature of a fusion weld.

There are several requirements for solidification. First, it is necessary to nucleate, or form, solid species within the liquid phase. Once the initial solid forms and the liquid-to-solid transformation proceeds, it is required that heat of fusion generated by the transformation be removed or dissipated. This normally occurs by conduction through the solid away from the solidification front. During the solidification of an alloy, it is also necessary to redistribute solute between liquid and solid, since the composition of the liquid and solid in contact at the solidification front changes continuously as the temperature decreases within the solidification range. This redistribution will result in local variation in composition in the solidified structure if the solid does not have time to reach its equilibrium composition, which is common in most casting and welding processes.

Most pure metals and alloys undergo a negative volume change when they solidify. This "shrinkage" phenomenon requires special precautions during casting to prevent shrinkage voids from forming. Solidification shrinkage also imparts stresses upon the as-solidified structure that may lead to solidification cracking. This shrinkage also contributes to the residual stress that is associated with fusion welds.

Using a simple phase diagram (Fig. 2.6), the equilibrium solidification behavior of a two-component alloy can be reviewed. For Alloy 1, solidification to solid A begins when the liquid temperature drops below the liquidus and ends when the alloy cools below the solidus. Within the solidification temperature range, the composition of liquid and solid in contact with each other at the solidification front is dictated by the isothermal tie line connecting the liquidus and solidus at a given temperature. At the end of solidification, Alloy 1 is 100% A.

For Alloy 2, solidification proceeds as described earlier until the alloy reaches the eutectic temperature (T_e). At this point, the remaining liquid, which is of eutectic composition, undergoes a eutectic reaction $(L \rightarrow A + B)$. The final structure will then be a mixture of A and eutectic $(A + B)$. The relative proportions can be determined using the lever rule.

For Alloy 3, solidification will not proceed until the system reaches the eutectic temperature. At this temperature, the liquid will completely transform to a eutectic structure with the composition of the A and B phase determined by the maximum solid solubility (C_{Smax}) of B in A and A in B at T_e.

2.3.1.1 Solidification Parameters
A number of parameters are useful in describing microstructure development and solute redistribution during solidification. These are defined as follows:

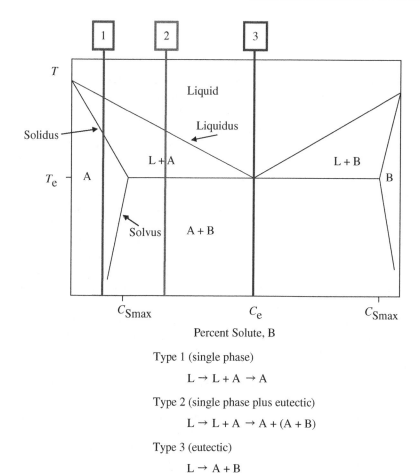

Percent Solute, B

Type 1 (single phase)

$$L \rightarrow L + A \rightarrow A$$

Type 2 (single phase plus eutectic)

$$L \rightarrow L + A \rightarrow A + (A + B)$$

Type 3 (eutectic)

$$L \rightarrow A + B$$

FIGURE 2.6   Examples of different solidification paths in a simple eutectic system.

- Partition coefficient: $k = C_S/C_L$
- Liquid temperature gradient: $G_L = dT_L/dx$
- Solidification rate: $R = dx/dt$
- Cooling rate: $G_L \cdot R = dT/dt$

The partition coefficient, k, sometimes called the solute redistribution coefficient, is simply the ratio of the solid and liquid composition in contact with each other at a given temperature within the solidification range. For most alloy systems, k is not a constant and varies as a function of temperature. It can only be constant in systems where the liquidus and solidus lines are straight, which is uncommon. When considering solute segregation during solidification, it is typical to assume an average value of k. When the value of k is less than 1, solute will partition to the liquid. When k is greater than one, solute will be depleted in the liquid. As the value of k approaches 1, solute redistribution during solidification is reduced.

The temperature gradient in the liquid (G_L) is also an important parameter since it dictates the nature of the temperature field in advance of the solid–liquid (S–L) interface. In situations where some undercooling of the liquid has occurred prior to solidification, this gradient will be negative. This would be the typical situation for the solidification of a casting. During weld solidification, however, this gradient is normally positive since the weld pool is superheated by the welding heat source.

Solidification growth rate (R) is dictated by how fast the S–L interface is moving during the solidification process. When coupled with the temperature gradient in the liquid, the local cooling rate at the S–L interface can be determined. This latter value ($G_L \cdot R$) will have an influence on the dimensions of the solidification substructure, such as dendrite arm spacing.

2.3.1.2 Solidification Nucleation

In order for the solidification process to begin, it is necessary to nucleate solid within the liquid phase. This can occur either homogeneously or heterogeneously when a nucleating particle or solid substrate is present. Homogeneous nucleation requires that solid of a critical, or threshold, size form within the liquid. The size of this spherical nucleant can be defined by a critical radius size, r^*, where

$$r^* = 2\gamma_{SL} \frac{T_m}{\Delta H_M \Delta T}$$

γ_{SL} is the S–L interfacial energy, T_m is the melting temperature, ΔH_M is the latent heat of melting, and ΔT is the amount of liquid undercooling. Note that as the amount of undercooling increases, the critical radius size decreases. Solid spheres less than r^* will simply remelt, while those exceeding r^* will grow.

In many solidification processes, heterogeneous nucleation may accompany homogeneous nucleation, or completely dominate it (as in the case of welding). Heterogeneous nucleation occurs from a foreign particle (such as an oxide, nitride, sulfide, etc.) or an existing solid substrate. Since these heterogeneous sites are stable at or above the melting temperature of the alloy, little or no undercooling is required for nucleation to occur. For example, single-crystal Ni-base turbine blades are manufactured using a "seed" crystal of a given orientation as a heterogeneous nucleation site.

In fusion welds, a number of heterogeneous nucleation events are possible, as illustrated schematically in Figure 2.7 [1]. Convective fluid flow during solidification can result in the tips of dendrites at the solidification front or grains in the surrounding solid metal to detach and be swept into the liquid. Depending on the liquid undercooling and size of the detached solid species, this solid can act as a nucleation site. In some casting operations, the casting mold walls are vibrated to induce dendrite fragmentation and subsequent refinement of the solidification structure. These heterogeneous nucleation events are only possible if the adjacent liquid is undercooled (negative G_L). In the case where the liquid is superheated (positive G_L), these nucleants will remelt. This is generally the case with weld solidification, and thus, this form of heterogeneous nucleation is generally not possible during fusion welding.

Higher melting point particles added to or formed within the liquid can also serve as nucleation sites. Sometimes called inoculants, these particles can substitute for

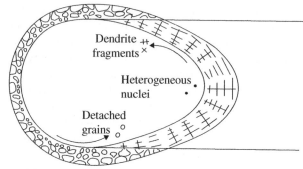

Mechanism 1: Dendrite fragmentation
Mechanism 2: Grain detachment
Mechanism 3: Heterogeneous nucleation

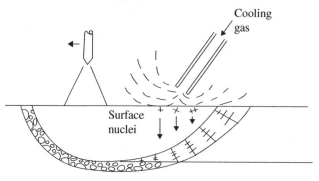

Mechanism 4: Surface nucleation

FIGURE 2.7   Various forms of heterogeneous nucleation associated with a molten weld pool (From Ref. [1]).

the homogeneous nuclei described previously. This type of nucleation can occur on the surface of the liquid, particularly if an oxide surface layer forms. In some cases, it may be possible to add nucleants directly to the molten pool, but this is usually not practical.

Heterogeneous nucleation off a solid substrate is called epitaxial nucleation, coming from the Greek word *epitaxis*, which means "to grow from." As noted earlier, the use of "seed crystals" in some casting or crystal growth applications is a form of epitaxial nucleation. Epitaxial nucleation is the dominant form of heterogeneous nucleation during weld solidification.

Epitaxial nucleation requires essentially no undercooling or other driving forces. As a result, solidification begins immediately upon cooling below the liquidus temperature. When the compositions of the base metal substrate and liquid are similar, the solidification front that grows from a given grain on that substrate will retain the same crystallographic orientation. Since grain orientation of the substrate is generally random, this results in a continuation of the crystallographic misorientation of the

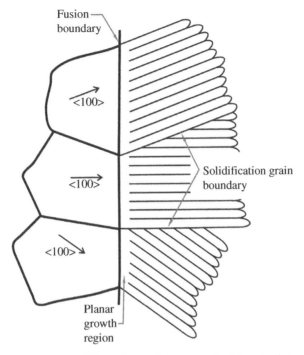

FIGURE 2.8   Schematic illustration of epitaxial nucleation.

base metal grains across the fusion boundary into the solidifying solid, as illustrated in Figure 2.8. That is to say, grain boundaries are continuous across the original fusion boundary where epitaxial nucleation occurred.

In fcc and bcc metals, solidification growth occurs preferentially along the cube edge, or <100> crystallographic directions. These are sometimes called "easy growth" directions, since solidification is most efficient in this crystallographic orientation. In hcp metals, growth occurs parallel to the basal plane in the <1010> direction. Growth is most favorable when these easy growth directions are parallel to the heat flow vector through the S–L interface. This results in growth that is roughly perpendicular to the S–L interface.

2.3.1.3   Solidification Modes   Multiple solidification modes can occur in metals. These modes describe the different morphological forms that can exist at the S–L interface and in many cases are still apparent when cooled to room temperature. Under conditions of low solidification rates, steep temperature gradients, or both, plane front solidification can occur. In most practical cases, the plane front breaks down into other modes described by either cellular or dendritic morphologies. The solidification mode that is most stable is dictated by the combined effect of composition, temperature gradient, and solidification rate.

The range of solidification modes that are observed in metals are illustrated in Figure 2.9 [6]. A planar solidification front first breaks down into a cellular front and

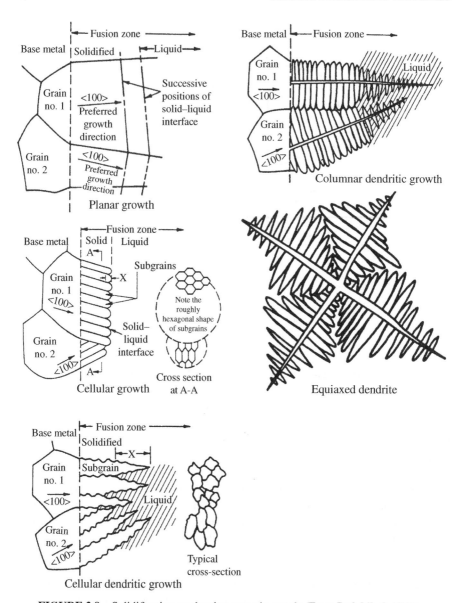

FIGURE 2.9 Solidification modes that occur in metals (From Ref. [6]. © AWS).

then into more complex dendritic morphologies, depending on solidification conditions. Most alloys solidify in either cellular, cellular dendritic, columnar dendritic modes, or a combination of these. According to Chalmers [7], the solidification mode is determined by the degree of *constitutional supercooling*, or undercooling, that exists in the liquid immediately in front of the S–L interface. As the extent of supercooling increases, more dendritic modes of solidification are favored. This concept will be described later in this chapter.

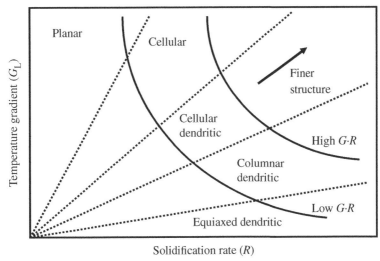

FIGURE 2.10 Effect of temperature gradient in the liquid, G_L, and solidification growth rate, R, on solidification mode.

The two most common solidification modes found in fusion welds and castings are the cellular and cellular dendritic modes. A planar solidification mode is normally not stable in actual practice. In laboratory experiments, it can be maintained at very slow growth rates in pure materials. Under normal solidification conditions, the planar interface quickly breaks down into a cellular or cellular dendritic growth front.

Equiaxed dendritic growth is also not normally observed in fusion welds due to the large constitutional supercooling required. It is sometimes observed in the terminal weld crater of fusion welds of some materials where the temperature gradient is very low due to the extinction of the arc (or other heat sources).

In metal alloys, the combination of temperature gradient, G_L, and solidification growth rate, R, influences solidification mode. As shown in Figure 2.10, a planar growth mode is favorable when the gradient is high and/or the solidification rate is very low. As solidification rate increases for all but very high temperature gradients, the solidification mode shifts to cellular and then dendritic. If the temperature gradient is extremely low, equiaxed dendritic growth is possible. As the product $G \cdot R$ increases, reflecting an increase in cooling rate, the structures that form become finer. This results in cellular or dendritic structures that are spaced much more closely relative to the distance between adjacent cell and dendrite arms.

Composition also has an effect on solidification mode. As shown in Figure 2.11, a pure metal will solidify with a planar front under most combinations of G_L and R. When solute or impurity elements are added, planar growth is only favorable when the gradient is very high, the solidification rate is very low, or a combination of these conditions. Most alloys will solidify in dendritic or cellular modes, as illustrated by the shaded region in the diagram. As described previously, equiaxed dendritic growth is only favored when the gradient is very shallow, a condition that usually does not exist in fusion welds.

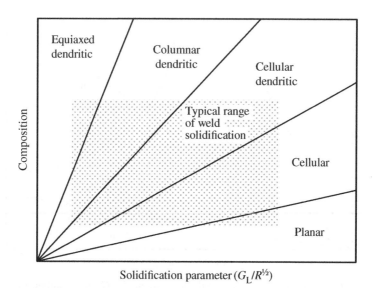

FIGURE 2.11 Effect of composition and solidification parameter on solidification mode (From Ref. [7]).

The vector product of temperature gradient and solidification rate represents the local cooling rate at the S–L interface, where

$$\frac{dT}{dx} \cdot \frac{dx}{dt} = \frac{dT}{dt}$$

The local cooling rate has a significant influence on the size (or scale) of the solidification structure that forms, with high values of $G_L \cdot R$ promoting very fine structures. This structural aspect of solidification is normally quantified by measuring the distance between the axial centers of the cells or dendrites, sometimes called the cell or dendrite core. This distance may range from several millimeters in very large castings to a few microns in laser or electron beam welds. The terms primary dendrite arm spacing (PDAS) and secondary dendrite arm spacing (SDAS) are often used to define the effect of cooling rate on solidification substructure size.

2.3.1.4 Interface Stability   Plane front solidification in metals only occurs in special cases. This includes the solidification of ultrapure metals that are not significantly undercooled and conditions that involve very low solidification rates, steep temperature gradients, or a combination of both. In almost all practical situations, the plane front is unstable and breaks down into a cellular or dendritic mode. Two theories are most prominent in describing this interfacial breakdown, both involving the solute redistribution that must occur at the S–L interface during alloy solidification.

The constitutional supercooling theory proposed by Chalmers [7] involves an *effective* undercooling of the liquid at the S–L interface that promotes planar interface

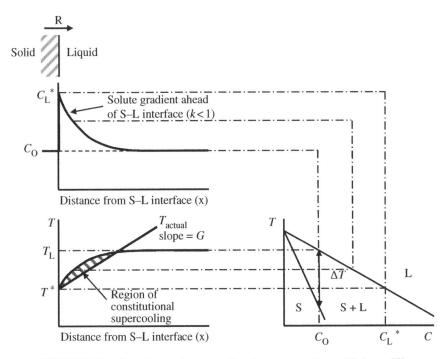

FIGURE 2.12 Constitutional supercooling theory according to Chalmers [7].

instability. Mullins and Sekerka [8] have used perturbation theory along the planar interface to mathematically predict a similar instability. Both theories are consistent with experimental observations. The Mullins and Sekerka approach is a mathematical treatment of the phenomenon and is not discussed here.

The constitutional supercooling theory of Chalmers is based upon the premise that solute partitioning occurs in advance of the S–L interface. Assuming a plane front, a solute gradient exists perpendicular to the front to a certain distance into the liquid. In Figure 2.12, for the case of the partition coefficient $k < 1$, the solute concentration decreases with distance. This solute concentration can be translated to temperature by using the phase diagram.

Simplifying Figure 2.12 by removing the phase diagram, it is possible to more easily describe the concept of constitutional supercooling that leads to the breakdown of a planar solidification front. Assuming the solute profile in advance of the S–L interface (for $k < 1$), an effective temperature profile can be constructed that *increases* as a function of distance from the interface, as shown in Figure 2.13. This is an *effective* temperature profile since an actual temperature gradient exists in the liquid that has previously been defined as G_L. If the actual gradient (G_{actual}) is less than the slope of the line tangent to the effective temperature profile at the S–L interface ($G_{critical}$), a region of *constitutional supercooling* will exist and the planar interface will be unstable. If the temperature gradient exceeds the slope of the tangent, the planar front is stable. Based on this theory, plane front solidification of alloys is only possible

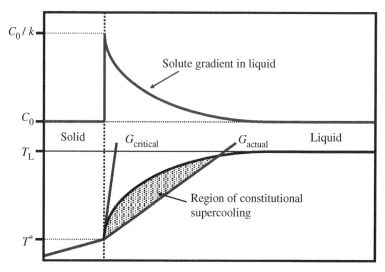

FIGURE 2.13 Simplified schematic of the constitutional supercooling theory for the case of $k < 1$.

when the temperature gradient (G_L) is very steep. In fusion welds, this condition is only satisfied at the fusion boundary, as discussed later in this chapter.

Sections 2.3.2 and 2.3.3 summarize the macroscopic and microscopic aspects of weld solidification and are not intended to be a comprehensive review of these topics. For more detailed coverage of weld solidification, the reader is referred to review papers by Davies and Garland [9], David and Vitek [10], and Katayama [11, 12].

2.3.2 Macroscopic Aspects of Weld Solidification

Solidification of welds occurs under nonequilibrium conditions and must be studied from both a macroscopic and microscopic viewpoint. Macroscopic solidification will be considered at both the trailing edge of the weld pool and along solidification grain boundaries (SGBs). From a solute redistribution standpoint, the macroscopic approach considers the solidification front as a plane front even though microscopically this front is usually cellular or dendritic. The macroscopic shape of the weld pool will be shown to be strongly influenced by welding conditions, particularly heat flow, heat input, and travel speed.

Microscopic solidification will be used to describe the formation and solute redistribution of solidification subgrains, such as cells and dendrites. Solidification parameters k, G_L, R, and $G \cdot R$ affect the nature of microscopic solidification and segregation.

As noted previously and illustrated in Figure 2.8, nucleation in fusion welds is dominated by epitaxial growth from the surrounding base metal. The thermodynamic driving force required for epitaxial nucleation is very low, and essentially, no undercooling is required for nucleation to occur. The newly formed grains maintain the same crystallographic orientation as the base metal grains from which they nucleate. As a result, grain boundaries are continuous across the fusion boundary.

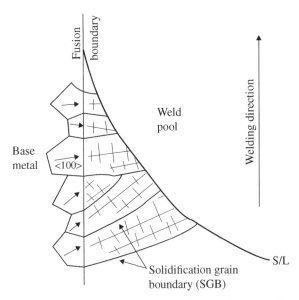

FIGURE 2.14   Illustration of epitaxial nucleation and competitive growth (From Ref. [16]. © AWS).

In fcc and bcc metals, which constitute the bulk of the engineering alloys that are commonly welded, solidification occurs preferentially along the cube edge, or <100> directions. These are sometimes called "easy growth" directions because solidification is most efficient in these orthogonal directions. This growth direction is maintained as long as the solidifying grain remains in contact with the S–L interface or until it is grown out of existence by adjacent weld metal grains that are more favorably oriented. This latter phenomenon is called "competitive" growth. The boundaries between these grains are defined as SGBs. The concepts of epitaxial nucleation and competitive growth are illustrated in Figure 2.14.

Occasionally, nonepitaxial nucleation and growth may occur in the fusion zone. The introduction of heterogeneous nuclei directly into the weld pool may cause nucleation to occur in advance of the S–L interface. For example, the addition of titanium oxide powder into titanium and aluminum welds has been shown to promote nucleation and grain refinement [13, 14].

In some systems, heterogeneous nuclei may actually form in the liquid ahead of the advancing interface. This is most likely to occur along the weld centerline where the temperature gradient is shallow and these nuclei cannot be swept into hotter regions of the weld pool. Nonepitaxial nucleation has also been observed along the fusion boundary in some systems. For example, in lithium-bearing aluminum alloys, small equiaxed grains have been observed at the fusion boundary that nucleate from $Al(Li, Zr)_3$ particles from the base metal [15].

Nonepitaxial nucleation has also been identified in systems where the base metal and weld metal have different crystal structures. For example, when austenitic stainless steel or Ni-base alloys (fcc) are deposited onto a ferritic steel (bcc), there is no evidence

(a)

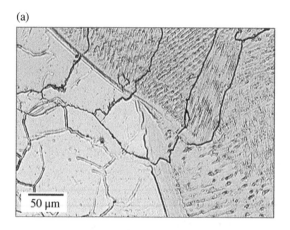

(b)

FIGURE 2.15 Examples of (a) epitaxial nucleation in austenitic stainless steel and (b) non-epitaxial nucleation of fcc weld metal (Monel) deposited on bcc base metal (Type 409 SS).

of epitaxy at the fusion boundary. The bcc substrate effectively acts as a "mold wall," and nucleation of fcc crystals occurs heterogeneously [16, 17]. Examples of epitaxial and nonepitaxial nucleation at the fusion boundary are provided in Figure 2.15.

Because of epitaxial nucleation and growth, grains will solidify along easy growth directions at the trailing edge of the weld pool. Base metal grains in polycrystalline metals are normally randomly oriented, and the resulting fusion zone grains will adopt the same degree of misorientation. Growth is most favorable along the heat flow direction or, conversely, perpendicular to the temperature isotherms at the S–L interface. These isotherms run roughly parallel to the S–L interface. Grains are most favored whose growth direction is most nearly perpendicular to the S–L interface.

The macroscopic weld pool shape is determined by a combination of material physical properties, process parameters, and heat flow conditions. Two general types of pool shape, teardrop and elliptical, are normally encountered as illustrated in Figure 2.16 [1]. Elliptical pools are usually associated with high heat input, low travel speeds, and 3-D heat flow conditions. Materials with high thermal conductivity, such as aluminum and copper, form elliptical weld pools over a wide range of conditions.

Teardrop shape

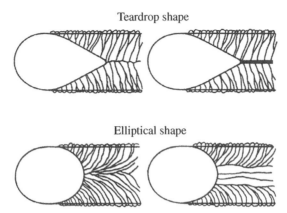

Elliptical shape

FIGURE 2.16   Illustration of the elliptical and teardrop-shaped weld pools (From Ref. [18]. © Wiley).

Teardrop pool shapes are most favored when travel speeds are rapid, thermal conductivity is low, and heat flow is 2-D. For example, austenitic stainless steels and nickel-base alloys often exhibit teardrop shape pools when welded in thin-sheet form at high travel speeds. This shape is generally to be avoided in high-restraint conditions, since centerline cracking can be a problem.

Fluid flow in the weld is affected by a number of forces including buoyancy, electromagnetic, and surface tension. Surface tension-induced fluid flow may dominate these other forces in some cases, often resulting in significant heat-to-heat variations in weld pool shape and penetration characteristics. This behavior is shown in Figure 2.17.

In systems where surface tension decreases as temperature increases, the hot fluid under the arc flows along the surface to the periphery of the weld and causes melting at the weld edge, or toe. If the surface temperature gradient is positive, a strong downward flow occurs and melting is most efficient at the root of the weld. The latter case provides the best weld penetration.

Small changes in composition can promote large changes in penetration due to the so-called "Marangoni" effect [18]. For example, reducing sulfur content in stainless steels from 0.010 to 0.003 wt% can result in a 50% reduction in weld penetration. Other elements, including oxygen, titanium, and aluminum, may have similar effects.

2.3.2.1   Effect of Travel Speed and Temperature Gradient   Weld pool shape is usually controlled by adjusting the weld travel speed. At high travel speeds in materials with low thermal conductivity, heat extraction from the weld becomes more difficult and the pool tends to elongate. This results in a gradual evolution from an oval, or elliptical, shape to a teardrop, as illustrated in Figure 2.18. In the elliptical pool, the angular relation of the velocity vector and principal heat flow direction at the S–L interface relative to the fusion boundary gradually changes from perpendicular to parallel upon moving toward the centerline. This results in considerable competitive growth along the solidification front.

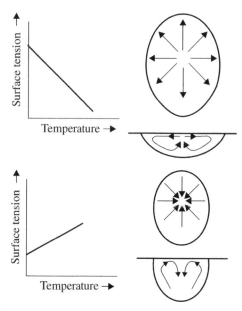

FIGURE 2.17 Surface tension-induced fluid flow (From Ref. [19]. © AWS).

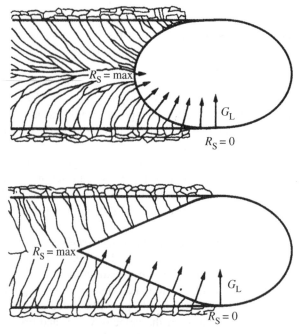

FIGURE 2.18 Effect of weld pool shape on the solidification parameters G_L and R and the macroscopic grain structure (From Ref. [6]. © AWS).

In the teardrop pool, the S–L interface is essentially straight (constant angle). As a result, grains that are favorably oriented relative to this interface can grow continuously from the fusion boundary to the centerline. Because of this, competitive growth is restricted, and these grains grow until they impinge on those from the other side of the weld along the weld centerline. Under extreme cases, this centerline can be extremely sharp, representing an SGB parallel to the fusion boundary. As noted previously, the orientation and compositional nature of this centerline make it susceptible to weld solidification cracking.

Solidification mode often varies significantly within a given weld, depending on composition, local temperature gradient, and solidification rate. At the fusion boundary, the gradient (G_L) is the steepest of anywhere along the S–L interface since heat flow into the surrounding base metal is most efficient at this point. The local solidification rate is very low here since the angular relationship to the travel speed vector, V_W, is approximately 90°. The combination of high gradient and low solidification rate can promote the establishment of a planar solidification front.

A short distance from the fusion boundary, the planar front breaks down into cellular and dendritic modes. This occurs due to both a decrease in G_L and an increase in R, as shown in Figure 2.18. Normally, this transition occurs within only a few microns of the fusion boundary. The centerline represents the location along the solidification front where R is the highest (equal to V_W) and G_L the lowest. This combination will promote more dendritic structures, potentially even equiaxed dendrites if the gradient is shallow enough.

Figure 2.19 demonstrates the change in solidification growth rate, R, as a function of the location along the macroscopic S–L interface. Note that the welding velocity and local growth rate are always equivalent at the weld centerline. The growth rate progressively decreases upon moving along the S–L interface toward the fusion boundary.

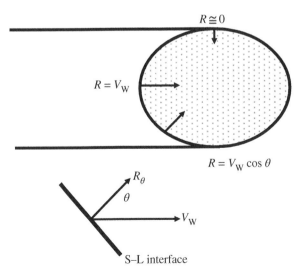

FIGURE 2.19   Relationship between weld travel speed, V_W, and local solidification rate, R.

The growth rate at the fusion boundary is extremely low relative to the weld centerline and approaches zero at the point that epitaxial nucleation and solidification begin at the fusion boundary. As noted previously, this low growth rate (in addition to higher temperature gradients in the liquid) can support planar solidification at the fusion boundary. As R increases, the planar front breaks down and cellular and dendritic growth modes dominate.

These shifts in solidification mode are illustrated schematically in Figure 2.20 from Easterling [2]. In actuality, most of the fusion zone solidifies in the same mode, usually cellular or cellular dendritic. In materials that solidify as fcc (austenite), such as austenitic stainless steels and Ni-base alloys, it is common to see evidence of planar growth at the fusion boundary, as shown in Figure 2.21a. The remainder of the fusion zone normally solidifies in a cellular/cellular dendritic mode, as shown in Figure 2.21b. Equiaxed dendritic solidification is rarely observed, except in the terminal weld crater as shown by the cross section and SEM micrograph in Figure 2.22.

2.3.3 Microscopic Aspects of Weld Solidification

On a microscopic scale, the fusion zone consists of a solidification microstructure exhibiting various types of interfaces or boundaries. It is important to understand the nature of boundaries in the fusion zone, since many of the defects associated with this region, both during fabrication and service, are associated with these boundaries. At least three different boundary types can be observed metallographically, as shown schematically in Figure 2.23.

Solidification subgrain boundaries (SSGBs) are the finest resolvable boundaries in the microstructure. These result from the formation of cells and dendrites during the solidification process. These boundaries form under microscopic solidification conditions and solute redistribution according to the boundary conditions described later in this section. Crystallographic misorientation across these boundaries is small, that is, they represent low-angle boundaries.

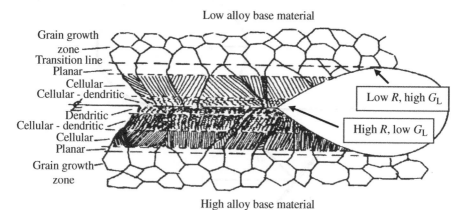

FIGURE 2.20 Change in solidification growth mode as a function of location in the weld (From Ref. [2]. © Wiley).

(a)

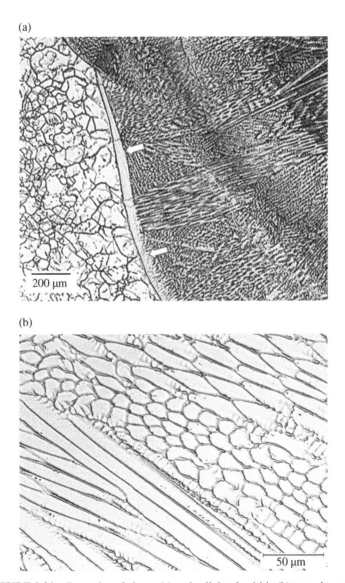

(b)

FIGURE 2.21   Examples of planar (a) and cellular dendritic (b) growth modes.

Solidification grain boundaries (SGBs) arise from the intersection of packets of subgrains, resulting in a crystallographic misorientation across the boundary. Solute and impurity segregation to these boundaries during solidification is defined under the macroscopic solute redistribution conditions described in Section 2.3.4.1.

Migrated grain boundaries (MGBs) represent true crystallographic grain boundaries in the fusion zone. These boundaries maintain the misorientation of the parent SGBs that they migrated from following solidification.

(a)

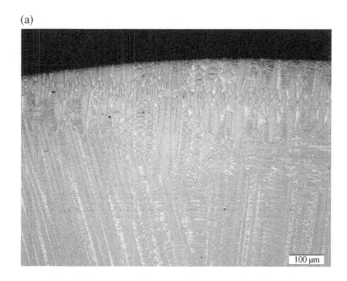

(b)

FIGURE 2.22 Equiaxed dendritic growth in the terminal crater of a weld in Alloy 690.
(a) Metallographic cross section and (b) SEM micrograph (Courtesy of Adam Hope).

2.3.3.1 Solidification Subgrain Boundaries (SSGB)   The solidification subgrains represent the finest structure that can be resolved in the optical microscope.
These subgrains are normally present as cells or dendrites, and the boundary separating adjacent subgrains is known as an SSGB. These boundaries are evident in
the microstructure because their composition is different from that of the bulk
microstructure. Solute redistribution that creates this compositional gradient at the
SSGB is dictated by microscopic solute redistribution, as described in the next
section.

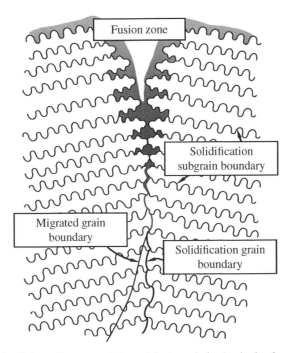

FIGURE 2.23 Schematic representation of the boundaries in single-phase weld metals.

There is virtually no crystallographic misorientation across the SSGB, and these boundaries are characterized as "low-angle" boundaries. The low misorientation (typically <5°) results from the fact that subgrain growth occurs along preferred crystallographic directions (or easy growth directions). In fcc and bcc metals, these are <100> directions. Because of this, the dislocation density along the SSGB is generally low since there is not a large misorientation to accommodate. Examples of solidification subgrains having cellular and cellular dendritic character are shown in Figure 2.24.

2.3.3.2   Solidification Grain Boundaries (SGB)   The SGB results from the intersection of packets, or groups, of subgrains. Thus, SGBs are the direct result of competitive growth that occurs along the trailing edge of the weld pool. Because each of these packets of subgrains has a different growth direction and orientation, their intersection results in a boundary with high angular misorientation. These are often called "high-angle" grain boundaries. This misorientation results in the development of a dislocation network along the SGB.

The SGB also exhibits a compositional component resulting from solute redistribution during solidification. This redistribution can be modeled using macroscopic boundary conditions and often results in high concentrations of solute and impurity elements at the SGBs, as described in the next section. This compositional partitioning may lead to the formation of low-melting liquid films along the SGBs at the conclusion of solidification that can promote weld solidification cracking. An example of an SGB is shown in Figure 2.24.

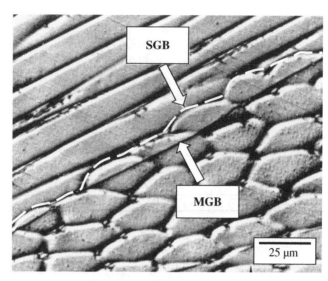

FIGURE 2.24   Examples of boundaries in the fusion zone of a fully austenitic (fcc) stainless steel.

2.3.3.3   Migrated Grain Boundaries (MGB)   The SGB that forms at the end of solidification has both a compositional and crystallographic component. In some situations, it is possible for the crystallographic component of the SGB to migrate away from the compositional component. This new boundary that carries with it the high-angle misorientation of the "parent" SGB is called an MGB.

The driving force for migration is the same as for simple grain growth in base metals, a lowering of boundary energy. The original SGB is quite tortuous since it forms from the intersection of opposing cells and dendrites. The crystallographic boundary can lower its energy by straightening and pulling away from the original SGB. Further migration of the boundary is possible during reheating, such as during multipass welding. An example of an MGB associated an SGB is also indicated in Figure 2.24.

Because it carries the crystallographic misorientation of the SGB with it, the MGB represents a high-angle boundary with misorientations typically greater than 30°. The composition of the boundary varies locally, depending on the composition of the microstructure where it has migrated. It is also possible that some segregation can occur along MGBs, possibly by a "sweeping" or diffusion mechanism.

MGBs are most prevalent in single-phase weld metals, particularly austenitic stainless steels and Ni-base alloys. In weld metals that contain a second phase or form a eutectic constituent at the end of solidification, the crystallographic component of the SGB is often "pinned" and is not free to move, thus preventing the formation of an MGB.

2.3.4   Solute Redistribution

The solidification of alloys requires that solute be redistributed between liquid and solid as dictated by the phase diagram. In alloy systems where the partition coefficient, k, is less than one, the liquid and solid become increasingly enriched in solute as the temperature decreases within the solidification temperature range, and the liquid phase is more highly enriched in solute than the solid phase. This redistribution requires mass transport of solute atoms between liquid and solid to maintain equilibrium at the interface.

This mass transport can occur both through the liquid via mixing and in the solid by diffusion. These two methods of mass transport (diffusion) establish the boundary conditions by which solute redistribution during solidification can be described. Three cases can be used to describe solidification based on mass transport boundary conditions. If mass transport is unrestricted in both the liquid and solid, solute redistribution can occur according to the phase diagram and solidification proceeds under equilibrium conditions.

If mass transport is restricted in the solid, liquid, or both, then solidification proceeds in a nonequilibrium fashion. In the case that approximates microscopic weld solidification (formation of cellular and dendritic subgrains), solid diffusion is considered negligible and liquid mixing is complete. This mixing may actually be considered to occur by diffusion within the liquid, since distances are so small (a few microns) and diffusion in the liquid is rapid as will be discussed later in this chapter.

If mass transport is restricted in both the solid and liquid, the situation approximates macroscopic weld solidification. The actual model does allow for some short-range diffusion in a liquid boundary layer at the S–L interface, but no long-range diffusion or mixing. The macroscopic and microscopic modes of weld solidification were shown schematically in Figure 2.23.

2.3.4.1 *Macroscopic Solidification*   Solute redistribution during macroscopic weld solidification can be approximated using plane front solidification of a small volume of liquid. Based on the mathematical approach of Smith *et al.* [19], macroscopic solidification can be defined by three distinct regions: (i) an initial transient, (ii) a steady-state region, and (iii) a final transient. The solute profiles shown in Figure 2.25 represent an alloy with $k<1$.

The initial transient represents the start of the solidification process, such as that which would occur at the fusion boundary. The initial solid to form must be of composition kC_0 since it is in contact with liquid of nominal composition (C_0). As solidification proceeds, the composition of the solid phase increases (for $k<1$). The liquid composition at the interface also increases since microscopic equilibrium (as dictated by the phase diagram) must be maintained. Away from the interface, the liquid composition is of composition C_0. The initial transient stage ends when the solid composition reaches C_0. Rigorous solutions for solute redistribution have been developed by Smith *et al.* [19] Use of these solutions is very cumbersome, and simplified relationships for solute profiles for the solid in the initial transient and the liquid in the steady-state region have been given by Flemings as follows [20]:

$$C_S = C_0 \left[1 - (1-k)\exp\left(\frac{-kRx_C}{D_L} \right) \right] \tag{2.1}$$

$$C_L = C_0 \left[1 + \left(\frac{1-k}{k} \right)\exp\left(\frac{-Rx_C{'}}{D_L} \right) \right] \tag{2.2}$$

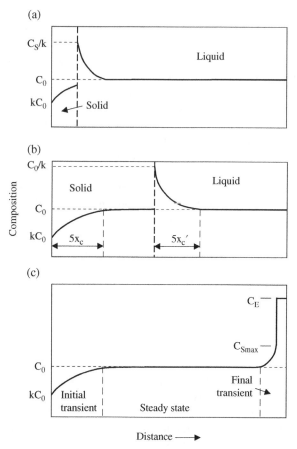

FIGURE 2.25 Solute profiles for macroscopic weld solidification showing the (a) initial transient, (b) steady state region, and (c) final transient.

The width of the initial transient is a function of D_L, R, and k, where the term x_C represents a "characteristic distance." Based on Equation 2.1, the width of the initial transient is approximated by $5x_C = 5D_L/kR$. Note that as k or R decreases, the width of the transient increases.

Under macroscopic solidification conditions, steady-state solidification occupies most of the solidification process. In this stage, solid of composition C_0 forms from liquid of C_0/k, based on microscopic equilibrium at the S–L interface. For macroscopic solidification, C_0 represents the average solid composition. In advance of the macroscopic interface, a solute gradient is established in the liquid of finite width, as described in Equation 2.2. The width of this gradient is a function of D_L and R and is approximately equal to $5x_C' = 5D_L/R$. Note that as the solidification rate (R) increases, the width of the solute gradient decreases.

As the final liquid is consumed at the end of the solidification process, the solid composition again rises (for $k < 1$). This "dumping" of solute occurs over a very

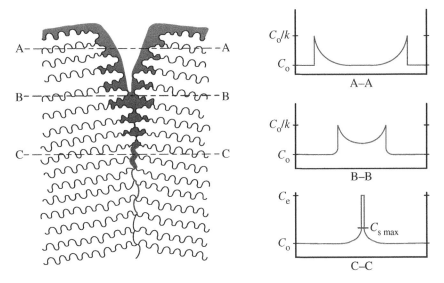

FIGURE 2.26   Solute profiles during formation of a solidification grain boundary, assuming $k<1$.

narrow region, typically on the order of a few microns or less. The solute enrichment in the final transient must equal the depletion in the initial transient. Because of this enrichment, the solidification temperature decreases relative to the bulk alloy, and in systems that exhibit a eutectic reaction, some eutectic constituents are formed. This solute/impurity "dumping" effect gives rise to the compositional component of an SGB. Upon reheating, this would be the first region in the microstructure to melt.

Macroscopic weld solidification models solute redistribution at the trailing edge of the weld pool and along SGBs. The schematic in Figure 2.26 shows how solute builds up along the grain boundary as solidification proceeds (again for the case of $k<1$). This results in the formation of low-melting liquid films along these boundaries that can potentially promote weld solidification cracking. This type of solute segregation also results in the formation of low-melting constituents in the terminal weld pool and may promote the formation of "crater" cracks in some materials.

This model for solute segregation along the SGB is greatly simplified since it considers the advancing interfaces that form the boundary as a plane front rather than as an array of cells or dendrites.

2.3.4.2   Microscopic Solidification

Solute redistribution under microscopic conditions is quite different than those during macroscopic solidification, since complete mixing in the liquid is considered. This requires that the liquid composition remains constant throughout the process, while no diffusion is allowed in the solid. The composition of solid and liquid in contact at the interface is dictated by microscopic equilibrium and is determined by the phase diagram.

This form of solute redistribution is representative of subgrain (cells and dendrites) solidification and predicts solute profiles across SSGBs. The consideration of complete

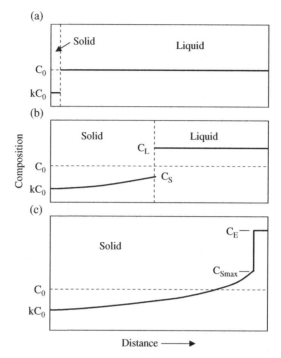

FIGURE 2.27 Microscopic solidification considering no solid diffusion and complete mixing in the liquid.

mixing is valid in this case because distances are short and rapid diffusion in the liquid eliminates any solute gradients. A schematic for solute redistribution during microscopic solidification is shown in Figure 2.27.

The Scheil equation, sometimes referred to as the nonequilibrium lever law, is very effective in modeling microscopic solute redistribution [21, 22]. This model considers that the solidification front is planar on a microscopic scale and that the solid, once formed, does not change its composition by diffusion, as shown in Figure 2.28. The Scheil equation for describing the composition of the solid as a function of the solute distribution coefficient, k, and the fraction solidified, f_s, is as follows:

$$C_S = kC_0 \left(1 - f_S\right)^{k-1} \tag{2.3}$$

Solidification begins at the tip of the cell or dendrite and proceeds until solidification is complete at the cell/dendrite boundary.

Solute redistribution across a cell boundary is shown in Figure 2.29 at several stages for the case of $k < 1$. Note that in all cases, the liquid composition is uniform. The initial solid to form is of composition kC_0. The solid composition gradually increases outward from the cell core and then rises rapidly at the conclusion of solidification. In systems that exhibit a eutectic reaction, this may result in the formation of eutectic constituents along the subgrain boundary.

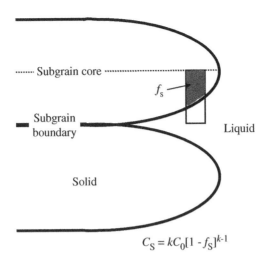

FIGURE 2.28 Scheil equation for determining solute segregation during microscopic solidification.

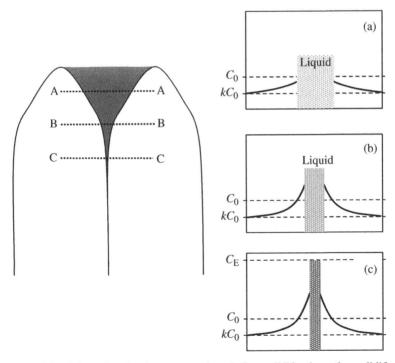

FIGURE 2.29 Schematic of solute segregation during solidification of a solidification subgrain.

In systems that contain a eutectic reaction, the Scheil equation can be rearranged to allow the prediction of the fraction eutectic. This is done by setting C_S equal to C_{Smax}, the solid composition in equilibrium with the liquid eutectic composition, C_E. By setting $(1-f_S)$ equal to the fraction eutectic, f_E, the equation can be rearranged in the following form:

$$f_E = \left(\frac{C_E}{C_0} \right)^{1/(k-1)} \tag{2.4}$$

This relationship can be useful in determining the fraction eutectic in eutectic systems where the solute level is less than C_{Smax}. Since the amount of eutectic constituent can influence solidification cracking resistance, this relationship can be used to predict the weldability of some alloy systems.

Although solid diffusion is not considered in Equation 2.3, it can, in fact, be considered under microscopic solidification conditions by the addition of an α-factor to the equation, as shown in the following:

$$C_S = kC_0 \left[1 - \frac{f_S}{(1+\alpha k)} \right]^{k-1} \tag{2.5}$$

where $\alpha = 4D_s t_f / L^2$ and includes the diffusivity of solute in the solid (D_S), the local solidification time (t_f), and the spacing between the cell and dendrite cores (L). When the value of α exceeds approximately 0.1, diffusion can have a significant influence on solute redistribution since there will be back diffusion into the solid that will alter the solute profile and result in less solute partitioning to the final liquid. In practice, it is difficult to determine the α-factor since values for D_S are not well known at elevated temperature and local solidification time must be estimated based on solidification rate. In systems that contain fast diffusing elements such as carbon and nitrogen, diffusion in the solid must be considered in order to accurately approximate the solute gradients and microstructure evolution during solidification.

2.3.5 Examples of Fusion Zone Microstructures

The fusion zone microstructure can vary widely based on alloy type and composition and welding process and conditions. This section is included to demonstrate the range of fusion zone microstructures that are observed in a number of common material systems, including steels, stainless steels, aluminum alloys, Ni-base alloys, and Ti alloys.

Most steels, including plain-carbon and low-alloy steels, solidify as bcc ferrite (delta ferrite) and transform to austenite almost immediately upon cooling below the solidification temperature range. The combination of solidification as ferrite and transformation to austenite on cooling tends to eliminate any evidence of the solidification substructure. In addition, these weld metals transform to lower-temperature products (ferrite, bainite, and martensite) upon cooling below the upper critical

(a)

(b)

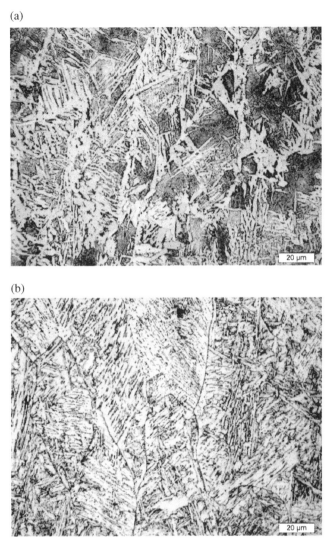

FIGURE 2.30   Fusion zone microstructure of (a) plain-carbon steel, (b) low-alloy steel.

temperature (A_3). The resultant fusion zone microstructures show evidence of a columnar solidification pattern, but SGBs and SSGBs (as shown in Fig. 2.24) are not observed. Examples of the fusion zone microstructure in two such steels are shown in Figure 2.30.

When steels solidify as fcc austenite, the solidification substructure becomes more apparent, as already shown in the austenitic stainless steel fusion zone in Figure 2.24. Other alloy systems, such as Ni-base, Cu-base, and Al-base alloys, also solidify as an fcc phase and exhibit distinct solidification substructure, as shown in Figure 2.31. This occurs since diffusion is relatively sluggish in the fcc phase relative to bcc, and the original solute segregation patterns established during solidification are preserved. Note that in some aluminum alloys a twinning phenomenon can occur following solidification, giving rise to what are described as "feather crystals" in the fusion zone microstructure.

Ti alloys solidify as a bcc beta phase and exhibit little or no evidence of solidification substructure. Since the beta phase is stable over a relatively wide temperature range prior to transformation to the hcp alpha phase (at the beta transus), solid-state diffusion is effective in eliminating solute segregation resulting from solidification. In addition, and as with plain-carbon and low-alloy steels, the transformation to

(a)

(b)

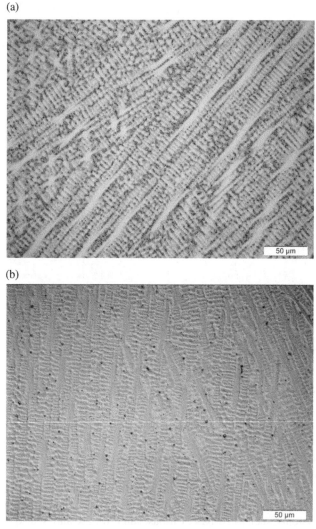

FIGURE 2.31 Representative fusion zone microstructure of different alloy systems: (a) Ni base, (b) Cu base,

(c)

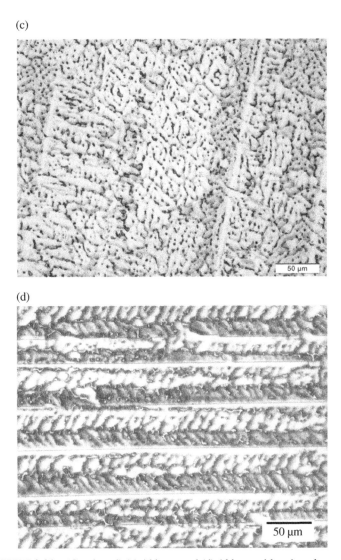

(d)

FIGURE 2.31 (Continued) (c) Al base, and (d) Al base with twinned crystals.

alpha below the beta transus tends to eliminate evidence of the preexisting solidification structure. An example of the fusion zone in Ti–6Al–4V is shown in Figure 2.32.

2.3.6 Transition Zone (TZ)

In heterogeneous welds, a composition TZ must exist between the fully mixed (diluted) weld metal and the base metal. If the composition difference between the base and filler metal is not large, this TZ may be undetectable, particularly if the TZ and composite fusion zone have the same microstructure. When the composition

FIGURE 2.32   Fusion zone microstructure of Ti–6Al–4V.

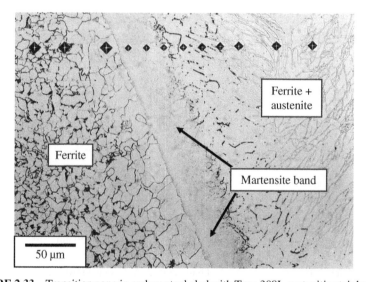

FIGURE 2.33   Transition zone in carbon steel clad with Type 308L austenitic stainless steel.

difference is large, such as when austenitic stainless steel or Ni-base alloy cladding is applied to carbon steels, the TZ is very apparent and may exhibit different micro-structure and properties relative to the base and filler metals. Two examples are provided here.

When carbon steels are clad with austenitic stainless steels (such as Type 308L) for corrosion protection, the TZ reaches a composition where the austenite that forms during solidification transforms to martensite on cooling to room temperature. This results in a narrow band of martensite close to the fusion boundary that exhibits

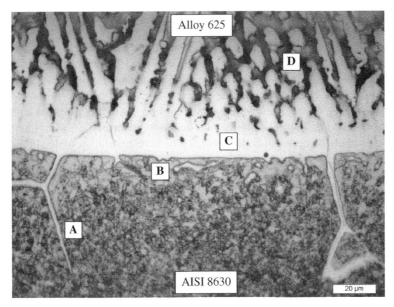

FIGURE 2.34 Transition zone between AISI 8630 steel clad with Ni-base Alloy 625 after PWHT: (a) penetration of weld metal down the grain boundary, (b) carbon-depleted zone, (c) planar growth region, and (d) cellular growth region. (From Ref. [23]. © Springer)

as-welded hardness much higher than either the base metal or composite fusion zone, as shown in Figure 2.33. In some situations, this may require a PWHT to temper this hardened region.

When Ni-base alloys are used to join or clad steels, a similar TZ structure can form, as shown in Figure 2.34. In this case, an apparent planar growth region exists at the fusion boundary within the TZ. This planar growth region quickly breaks down into cellular and cellular dendritic solidification. Because of the high Ni content, the TZ in this combination tends to be austenitic (fcc) rather than martensitic, since the austenite that forms at elevated temperature is quite stable and resists transformation to martensite. When such a combination is subjected to PWHT to temper the HAZ of the steel, carbon migration from the steel to the cladding occurs. This results in depletion of carbon in the steel and subsequent buildup in the cladding at the interface. For the 8630/Alloy 625 combination shown in Figure 2.34, this leads to the transformation to ferrite in the carbon-depleted zone and the formation of a very hard austenite band in the planar growth region of the cladding [23].

2.4 UNMIXED ZONE (UMZ)

The UMZ represents the region of the fusion zone immediately adjacent to the fusion boundary, as shown in Figure 2.4. It is normally very narrow relative to the other regions of a weld, which can lead to the notion that it is insignificant. In many systems, it may be difficult to even distinguish the UMZ. For some dissimilar combinations, the mechanical properties or corrosion properties of the UMZ can be significantly different from those of the base and filler metals. For example, the UMZs in some combinations are subject to cracking or localized corrosive attack.

Theoretically, a UMZ is present in every fusion weld. Because the fluid velocity in the weld must go to zero at the fusion boundary, a stagnant liquid layer of some finite thickness will exist. Even in autogenous and homogenous welds, a UMZ would exist since the adjacent weld metal would be of slightly different composition due to evaporation or contamination effects. In practice, the UMZ is normally indistinguishable in these welds since its microstructure is similar to that of the bulk fusion zone.

UMZs are almost always associated with heterogeneous welds, particularly where the relative compositions and physical properties of the base and filler metals are quite different. The size and nature of the UMZ can vary tremendously depending on a number of material and process variables.

There are many factors that affect the formation of the UMZ. Large relative differences in the composition of base and filler metals may result in important differences in melting temperature and fluid properties. Base metals with a higher melting temperature than the weld metal are probably more prone to UMZ formation, although this effect is complicated by other factors. Differences in fluid viscosity may also be important. For example, if the molten base metal is more viscous than the weld metal, the UMZ is less likely to be disturbed. The miscibility, or ability of the two fluids to mix, may also influence UMZ formation.

Welding process has a significant effect on UMZ formation. High-energy-density processes, such as EBW and LBW, almost never exhibit a UMZ, probably because of the low heat inputs, steep fusion boundary temperature gradients, and vigorous weld pool stirring associated with these processes. Arc welding processes, such as GTAW, GMAW, and PAW, are much more likely to exhibit a UMZ in certain alloy systems.

Both fluid flow in the weld pool and the temperature gradient along the fusion boundary have an important influence on UMZ formation. If fluid flow is vigorous, the UMZ will be "stirred" into the weld metal. Regions of arc welds where fluid flow is sluggish often exhibit distinct UMZs. The temperature gradient at the fusion boundary can influence the width of the UMZ, since it affects the distance over which the base metal is molten.

UMZs have been observed in a number of systems. In their original paper proposing the expanded regions of a fusion weld, Savage *et al.* [4] studied the UMZ in a high-strength low-alloy (HSLA) steel, HY-80, designed for submarine hulls. Baeslack and Lippold determined that a UMZ formed in Type 304L stainless steel when welded with GMAW using either Type 312 or 310 filler metal was more susceptible to corrosion than either the weld metal or base metal [24]. Similar corrosion issues have been found with the UMZ of the high-Mo, high-N "superaustenitic" stainless steels when they are welded with Ni-base filler metals [25].

Other alloy systems in which UMZ formation has been reported in the literature include aluminum alloys [26], Ni-base alloys [27], and some dissimilar combinations, particularly those involving steels and Ni-base alloys. Many other

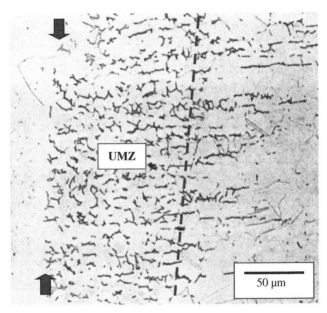

FIGURE 2.35 Unmixed zone that forms between Type 304L base metal and Type 310 filler metal (Arrows indicate the fusion boundary.) (From Ref. [24]. © AWS).

systems undoubtedly exhibit a UMZ, but because this region is so narrow, it may go undetected unless there is some particular weldability or service-related issue associated with the UMZ.

A UMZ that forms between Type 304L base metal and Type 310 filler metal is shown in Figure 2.35. Both the base metal and composite region of the weld are fully austenitic. A distinct UMZ consisting of a two-phase mixture of austenite + ferrite is formed between these two regions. The UMZ microstructure is consistent with that of melted and resolidified Type 304L, thus indicating no mixing with the filler metal [24].

When exposed to certain corrosive environments, localized attack is possible within the UMZ. In the earlier example, a constant extension rate, stress corrosion cracking test was conducted on this weldment in a 5N H_2SO_4+0.5N NaCl solution. Under these conditions, corrosive attack was localized in the UMZ due to the presence of the two-phase microstructure. Similar attack is possible in certain service environments.

The Li-bearing aluminum alloys may form an equiaxed grain zone (EQZ) along the fusion boundary, as shown in Figure 2.36 [15]. It is postulated that this region represents a UMZ in which nonepitaxial nucleation occurs. This results in small equiaxed grains along the fusion boundary that are morphologically distinct from the composite region of the fusion zone. This region has proven to be susceptible to liquation cracking during repair welding and may have lower toughness and ductility than the base metal and weld metal.

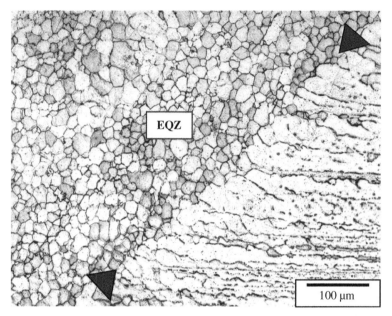

FIGURE 2.36 Unmixed zone in an Al–Li alloy (Arrows indicate the fusion boundary) (From Ref. [26]. © AWS).

2.5 PARTIALLY MELTED ZONE (PMZ)

The PMZ represents a transition region between 100% melting in the fusion zone (or the UMZ at the fusion boundary) and the 100% solid region of the weld (the T-HAZ). In a pure metal, a PMZ will not exist because there is no liquid–solid temperature range. In an isotropic alloy (one in which no segregation or local variations in composition exist), the PMZ represents the temperature range between the alloy liquidus and solidus temperature. For most alloys, this range is typically narrow (25–100°C in most iron- and nickel-base alloys) and would predict a narrow PMZ.

In most engineering alloys, segregation of alloying and impurity elements increases the "effective" melting temperature range of the base material. The temperature range between the liquidus and "effective" solidus temperatures is generally used to describe the extent of the PMZ. There are a number of phenomena that influence the magnitude of this range by promoting liquation reactions under nonequilibrium thermal conditions.

As indicated earlier, the PMZ in isotropic alloys would simply represent the temperature range between the solidus and liquidus on the phase diagram. In actuality, solute and impurity elements are not distributed uniformly in the base metal, and further segregation may occur during the weld thermal cycle. The net effect is that local variations in composition in the HAZ adjacent to the fusion boundary will promote melting at temperatures below that of the bulk microstructure.

Grain boundaries typically have a higher concentration of alloy and impurity elements than the grain interiors. This concentration gradient lowers the melting

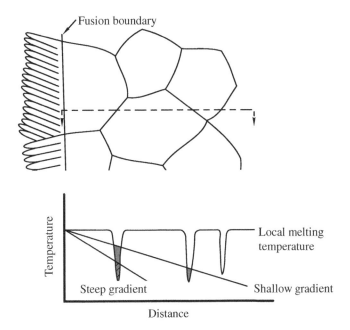

FIGURE 2.37   Illustration of local melting along grain boundaries in the PMZ associated with the temperature gradient in the solid.

temperature of the grain boundaries compared to that of the bulk material, and consequently, these boundaries will generally melt at lower temperatures than the bulk microstructure during heating of the base metal surrounding the weld. The amount of melting depends on the nature and degree of segregation, while the extent, that is, distance from the fusion boundary, depends on the temperature gradient. This grain boundary melting phenomenon is illustrated in Figure 2.37.

In addition, specific particles or precipitates may undergo a phenomenon called "constitutional liquation" whereby the constituent particle reacts with the surrounding matrix resulting in interfacial melting. This will be described in more detail later.

Two basic liquation mechanisms can be described. The ***penetration mechanism*** requires that both a localized, or discrete, liquation phenomenon occur and a mobile grain boundary interact with this liquated region. In most systems, the liquid will then "penetrate" and spread along the grain boundary. Localized liquation may also occur in the grain interiors, but this is generally not as deleterious as liquid spreading along the grain boundary.

In the absence of a discrete liquation phenomenon, local melting can occur along grain boundaries via a ***segregation mechanism***. The segregation of solute or impurity elements to grain boundaries can also lead to liquation in an appropriate temperature field, that is, above a certain threshold temperature. Since most solutes and impurities tend to depress the melting temperature of the solvent metal, high concentrations along grain boundaries can reduce the melting temperature relative to the bulk composition, as illustrated in Figure 2.37. There are at least three segregation mechanisms, or combinations of these mechanisms, that can lead to grain boundary liquation in the HAZ. These include Gibbsian segregation, grain boundary sweeping, and pipeline diffusion. Each will be described in the following sections.

2.5.1 Penetration Mechanism

The penetration mechanism for grain boundary liquation requires both a liquation phenomenon and grain boundary motion. As shown in Figure 2.38, liquation (in this case constitutional liquation) and boundary motion must be simultaneous. When the boundary encounters the liquated region surrounding the particle, it will be "pinned" and further motion inhibited. Depending on the wetting characteristics of the liquid/boundary combination, the liquid may then penetrate along the boundary. This gives rise to grain boundary liquid films. The degree of penetration depends on the temperature field, wetting characteristics, and the amount of liquid. In some alloys, continuous grain boundary liquid films may form in the PMZ.

Localized melting in the PMZ may occur via a number of other mechanisms. At temperatures very close to the solidus, all polycrystalline materials will undergo incipient melting. This type of melting normally occurs at grain boundaries, since these are high-energy sites, that is, the boundary energy combined with the thermal energy contributes to allow melting at a temperature below bulk melting. In most materials, incipient melting occurs within a few degrees of the solidus, although the presence of impurity elements may allow melting at much lower temperatures. In general, this form of localized melting is not very important in the formation of the PMZ.

Cast materials will naturally contain local regions (interdendritic) that melt at a lower temperature than the matrix. Thus, welds made on cast materials will contain a PMZ dictated by the melting temperature of the interdendritic region. Materials that

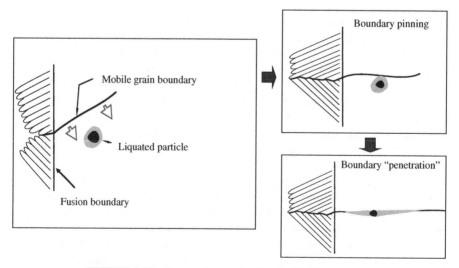

FIGURE 2.38 Penetration mechanism for PMZ formation.

are thermomechanically processed following casting often exhibit some local compositional banding. This may be particularly true in plate materials that experience a series of directional rolling operations. Because of these local fluctuations in composition, the melting temperature will also fluctuate. In the PMZ, it is possible that some regions may melt at a lower temperature than an adjacent region due to this phenomenon.

A number of materials undergo a phenomenon described as "constitutional liquation." This form of liquation is manifested by localized melting at a particle/matrix interface. It is somewhat unique to welding since it occurs under transient thermal conditions, normally requiring relatively rapid heating rates.

The constitutional liquation mechanism was first proposed by Savage on a theoretical basis in the late 1950s and was demonstrated experimentally by Savage and his students in the 1960s [28, 29]. The basis of this mechanism requires the reaction between a "constituent" particle and the surrounding matrix such that local melting occurs at the constituent/matrix interface, hence the term "constitutional" liquation.

It is important to note that the particle itself does not melt. Most of the particles that undergo constitutional liquation have melting temperatures far exceeding that of the base metal (e.g., NbC and TiC). Rather, it is the intermediate composition in the reaction zone between the particle and the matrix that melts. In order for this mechanism to be operative, (i) the particle must react with the matrix to create a composition gradient around the particle, and (ii) the reaction zone composition must undergo melting below the melting temperature of the surrounding matrix.

A simple binary phase diagram that portrays a eutectic reaction between a particle phase, A_xB_y, and the matrix, α, can be used to describe constitutional liquation, as shown in Figure 2.39. When an alloy of composition C_0 is heated, a reaction between

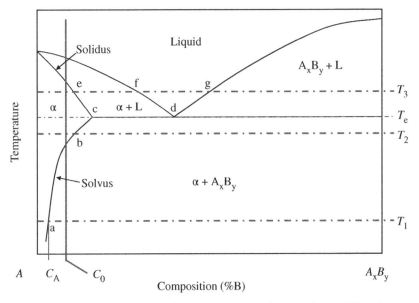

FIGURE 2.39 Phase diagram for a system undergoing constitutional liquation.

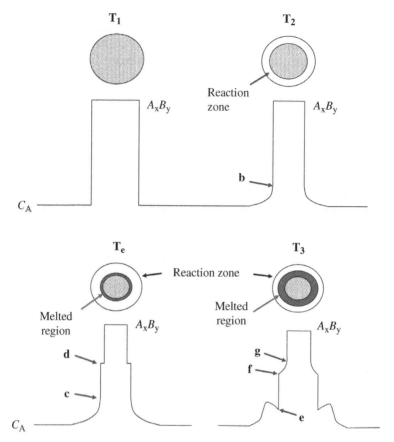

FIGURE 2.40 Reaction at the particle/matrix interface for temperatures T_1, T_2, T_e, and T_3 from Figure 2.39.

the matrix and particle will occur. The nature of that reaction is described at temperatures T_1, T_2, T_e, and T_3 in Figure 2.40.

For constitutional liquation to occur, the particle must partially, but not completely, dissolve upon heating to the eutectic temperature (T_e). If complete dissolution occurs, as predicted by the phase diagram, no constitutional liquation can occur. When only partial dissolution occurs, microscopic equilibrium is maintained at the particle/matrix interface, and melting occurs along this interface when the temperature exceeds the eutectic temperature, T_e. Again, the particle itself does not melt since its melting temperature is well above that of the matrix.

According to the phase diagram, an alloy of composition C_0 consists of particles of composition A_xB_y in a matrix of composition C_A. At low temperatures, such as T_1 shown in Figure 2.39, the particle remains in equilibrium with the matrix of composition C_A. Note that at this temperature, there is no reaction between the particle and surrounding matrix, that is, no solid diffusion gradient has developed. As the alloy is rapidly heated to T_2, the particle starts to react with the surrounding matrix. Since equilibrium must be maintained at the particle/matrix interface at all times, the particle of composition A_xB_y is in contact with a matrix composition indicated by the point "b" on the solvus line. A composition gradient resulting from partial particle dissolution now exists in the matrix. This occurs in a reaction zone surrounding the particle where B atoms are diffusing into the surrounding matrix.

As the alloy is heated from T_2 to T_e, the interface composition continues along the solvus line of the phase diagram. When this composition reaches point "c" at the eutectic temperature, equilibrium dictates that a liquid of composition "d" must be in contact with the solid. As a result, liquid is formed in the system within the reaction zone. This liquid completely surrounds the particle and represents the onset of "constitutional liquation."

As the alloy is heated to T_3, additional liquid forms in the system. The composition of this liquid ranges from "g" at the particle interface to "f" at the matrix interface. The composition of the matrix in contact with the liquid is now represented by point "e" on the solidus line. Again, all interface compositions must obey microscopic equilibrium as dictated by the phase diagram. Note that a solute "hump" exists in the solid matrix within the reaction zone. This occurs because the solid composition achieved at T_e (point "c") has now decreased to "e" since the solid composition must lie along the solidus line. Above T_3, additional constitutional liquation will occur around the particle until the alloy exceeds the solidus temperature and bulk melting of the matrix begins.

The SEM photomicrograph in Figure 2.41 shows constitutional liquation of a TiC particle in Waspaloy (Ni–20Cr–14Co–4Mo–3Ti–1.5Al) [30]. Note that most of the particle is intact and a distinct reaction zone surrounds the particle. This reaction zone was liquid at elevated temperature and has resolidified as a gamma/Laves eutectic. The large degree of constitutional liquation in Waspaloy makes it susceptible to HAZ liquation cracking.

Constitutional liquation has been observed in a number of engineering alloy systems, as listed in Table 2.1. In all the systems listed, the matrix is austenitic (fcc). The structural steel, while ferritic (or martensitic) at room temperature, is austenitic at the elevated temperatures where constitutional liquation occurs. This suggests that the slower diffusion rates in austenite better support the formation of a critical reaction zone than a ferritic (bcc) matrix. For example, NbC and TiC in ferritic stainless steels do not undergo constitutional liquation.

Constitutional liquation has not been reported in other alloy systems, such as aluminum-, titanium-, or copper-based alloys. Although it is theoretically possible, the particle/matrix reactions in these alloy systems do not seem to support this mechanism.

2.5.2 Segregation Mechanism

Grain boundary melting is also observed in the PMZ of alloys that do not undergo constitutional liquation or otherwise form grain boundary liquid films via a penetration mechanism. This suggests that another mechanism, or mechanisms, must be operative to allow boundary melting. Grain boundaries are quite complex, consisting of arrays of dislocations that accommodate the crystallographic misorientation of the

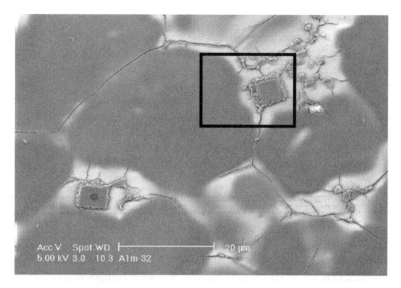

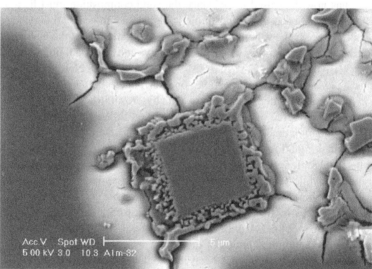

FIGURE 2.41 Constitutional liquation associated with Ti-rich MC carbides in a Waspaloy hot ductility sample heated to a peak temperature of 1300°C (2370°F) (From Ref. [30]. © AWS).

grains. For more information on the nature of grain boundaries, the reader is referred to an authoritative text, such as that recently published by Priester [31]. The "segregation mechanism" is proposed to explain this. Via the segregation mechanism, grain boundaries are enriched in melting point depressant elements that diffuse or segregate to the boundary at elevated temperature. A number of mechanisms have been used to explain this segregation, including "Gibbsian" segregation, grain boundary sweeping, and pipeline diffusion. A schematic that illustrates the segregation mechanism is shown in Figure 2.42.

TABLE 2.1   Systems reported to undergo constitutional liquation

Alloy system	Susceptible alloys	Constituent	References
	Alloy 718	NbC	1,2
Ni-base	Waspaloy	TiC	3
	Udimet 700	TiC, M_3B_2	4
	Alloy 903	NbC	5
	Type 347	NbC	6
Stainless steel	A-286	TiC	7
	Alloy 800	TiC	8
High-strength steel	18Ni maraging	TiS	9

[1]Thompson RG, Genculu S. Microstructural evolution in the HAZ of Inconel 718 and correlation with the hot ductility test. Weld J 1983;62(12):337s–345s.

[2]Qian M, Lippold JC. Liquation phenomena in the simulated HAZ of alloy 718 after multiple postweld heat treatment cycles. Weld J 2003;82(6):145s–150s.

[3]Qian M, Lippold JC. Effect of multiple postweld heat treatment cycles on the weldability of Waspaloy. Weld J 2002;81(11):233s–238s.

[4]Owczarski WA, Duvall DS, Sullivan CP. A model for heat affected zone cracking in nickel base superalloys. Weld J 1966;46(4):145s–155s.

[5]Baeslack WA, Ernst SC, Lippold JC. Weldability of high-strength, low-expansion superalloys. Weld J 1989;68(10):418s–430s.

[6]Nippes EF, Savage WF, Bastian BJ, Mason HF, Curran RM. An investigation of the hot ductility of high temperature alloys. Weld J 1955;34(4):183s–196s.

[7]Brooks JA. Effect of alloy modification on HAZ cracking of A-286 stainless steel. Weld J 1974;53(11):324s–329s.

[8]Lippold JC. An investigation of heat-affected zone hot cracking in alloy 800. Weld J 1983;62(1):1s–11s.

[9]Pepe JJ, Savage WF. Effects of constitutional liquation in18Ni maraging steel weldments. Weld J 1967;46(9):411s–422s.

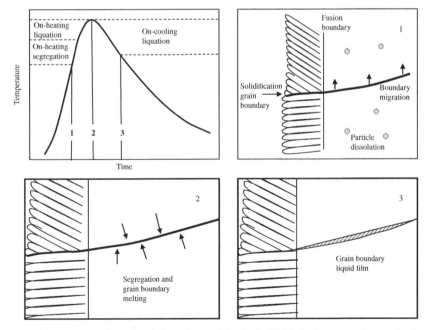

FIGURE 2.42   Illustration of grain boundary melting in the PMZ via the segregation mechanism.

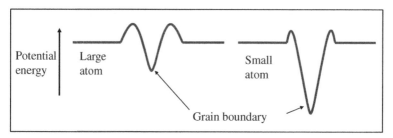

FIGURE 2.43   Driving force for grain boundary segregation.

2.5.2.1 Gibbsian Segregation   There is a natural tendency for atoms to segregate, or diffuse, to a grain boundary. Since a grain boundary consists of any array of dislocations, there are numerous sites in the boundary that can capture atoms that diffuse into it. The lowering of system free energy is the primary driving force for the type of segregation known as "Gibbsian" segregation [32, 33].

Depending on the nature of the atomic species and boundary, two general situations can be described, as shown in Figure 2.43. In the case of a large substitutional atom (such as Cr or Ni in steel), there is a large potential barrier required to get the atoms into the boundary. This barrier is primarily a function of the diffusivity of the atom in the matrix. Once in the boundary, the potential well to keep it there is not so deep, that is, the boundary binding force is not high. When the boundary moves (grain growth), these large atoms are left behind and may contribute to the formation of "ghost" grain boundaries in some materials [34]. Ghost grain boundaries indicate the previous location of a mobile grain boundary by the residual solute left behind when the boundary migrates.

For small interstitial atoms (such as S, P, or O in steel), the situation can be quite different. The potential barrier for the small atom to enter the boundary is quite low, since the diffusivity of these elements is normally high. Once in the boundary, these atoms may be quite tightly bound (deep potential well) since there are many interstitial sites where they can reside. As a result, it may take considerable thermal energy to remove these atoms from the boundary once they have segregated there. Also, these small atoms may be able to move with the boundary if the boundary migrates.

2.5.2.2 Grain Boundary Sweeping   In the HAZ, grain growth normally occurs at temperatures above about $0.5T_m$. In some alloy systems, this requires the dissolution of second phases or particles that can pin the boundaries. As the boundary moves, it is possible for it to "sweep up" atoms in the matrix. The elements most easily swept up by the moving boundary are those that have a high affinity for boundaries or surfaces. These include interstitial impurity elements such as S, P, O, and B and certain solute elements such as Ti and Si [35, 36]. Larger substitutional elements are not able to migrate with the boundary and are not swept along as the boundary moves.

Once they are swept into the boundary, they can be bound there via the Gibbsian energy mechanism described previously. If they are to remain in the moving boundary, however, they must be able to migrate with the boundary. This requires a high diffusivity in the matrix at elevated temperatures.

The solute and impurity elements that are swept up may impose a drag effect on the boundary [37, 38]. As the solute/impurity concentration increases, it becomes more difficult for the boundary to "drag" these atoms along. The boundary either slows down due to this drag effect, or it may break away, leaving a solute-/impurity-rich region behind. This latter effect may result in the formation of a "ghost" grain boundary.

When the solute/impurity concentration reaches some critical level in a given temperature field, the boundary will melt. The extent of this melting will define the bounds of the PMZ.

2.5.2.3 Pipeline Diffusion   Another mechanism for grain boundary segregation in the HAZ is so-called "pipeline" diffusion because the grain boundary represents a fast diffusion path [39, 40]. Because grain boundaries are continuous across the fusion boundary due to epitaxial nucleation and growth, a natural grain boundary pipeline from the fusion zone into the HAZ is created, as illustrated in Figure 2.44.

In the fusion zone, solute redistribution along the SGBs results in a high concentration of alloying and impurity elements (for $k < 1$) in close proximity to the HAZ. Diffusion of these elements along the grain boundary pipeline into the HAZ can result in significant enrichment of the HAZ boundaries and promote grain boundary melting.

Grain boundary diffusion can be quite rapid, perhaps 100–1000 times faster than in the matrix [41]. As a result, solute gradients created by solidification in the fusion zone can promote segregation along the grain boundary pipeline into the HAZ.

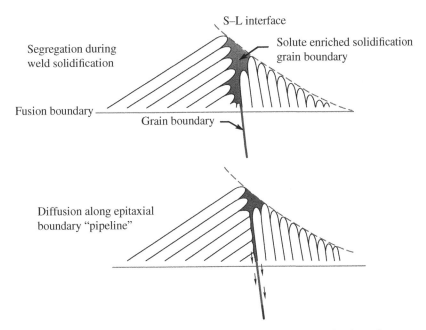

FIGURE 2.44   Schematic of pipeline diffusion across the fusion boundary.

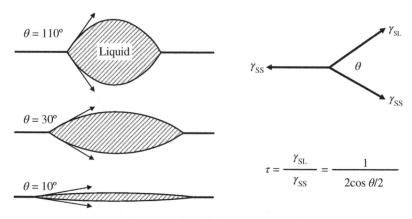

FIGURE 2.45   Grain boundary wetting.

2.5.2.4   Grain Boundary Wetting   The presence of liquid along grain boundaries in the PMZ is not necessarily a problem, unless the liquid forms a continuous liquid film along the grain boundary. The wetting characteristics of the liquid at the boundary can be estimated if the relative solid–solid and S–L interfacial energies are known. Using the relationship in Figure 2.45, the wetting angle (θ) can be determined. If θ is greater than about 30°, wetting is not very effective and continuous boundary films are not likely to form. As θ approaches 0, wetting becomes much more effective and the liquid will spread out along the boundary. Under these conditions, the boundary becomes susceptible to cracking since solid–solid contact on the boundary is reduced.

In practice, elevated temperature interfacial energy data is not readily available, particularly for complex systems. It has been shown, however, that impurity elements such as P and S tend to reduce the liquid–solid interfacial energies, thereby promoting boundary wetting by the liquid. These elements also reduce the melting temperature of the grain boundary, so they tend to play an important role in the formation of liquid films along grain boundaries.

2.5.3   Examples of PMZ formation

Some examples of PMZ formation in various materials are provided in Figure 2.46. Alloy 907 is an Fe–Ni low-expansion alloy with additions of Nb (~4.5 wt%). Constitutional liquation of NbC in the PMZ has led to the presence of continuous liquid films along the grain boundaries [42]. Austenitic stainless steels can also form a PMZ, as shown in Figure 2.46b. The arrows in this figure indicate the presence of a liquid film that was present along the boundary at elevated temperature. This melting occurs due to the segregation of impurity elements (S and P) to the boundary. The formation of ferrite along these boundaries will suppress liquation in these alloys, since it is difficult for liquid films to wet austenite–ferrite boundaries. PMZ formation is also common in many aluminum alloys. An example of grain boundary liquation and cracking in a 6000-series alloy is shown in Figure 2.46c. In these alloys, the segregation of Mg and Si to the boundary depresses the local melting temperature and promotes melting.

(a)

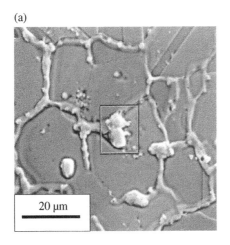

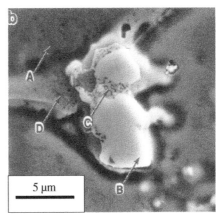

Location	Structure	Composition (wt%)					
		Fe	Ni	Co	Nb	Ti	Si
A	Gamma matrix	Bal	37.0	11.8	5.8	1.5	0.7
B	Nb-rich carbide	Bal	0.2	–	85.3	13.4	0.3
C	Laves phase	Bal	33.4	10.9	25.9	3.3	1.0
D	Grain boundary gamma	Bal	38.5	11.3	10.2	2.6	0.9

(b)

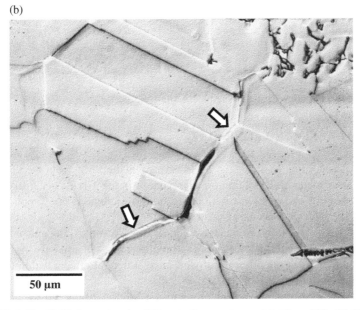

FIGURE 2.46 PMZ formation in different alloy systems: (a) Alloy 907, (b) Type 304 stainless steel,

(c)

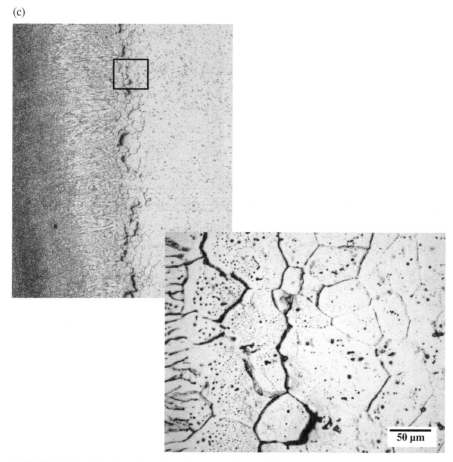

FIGURE 2.46   (Continued) and (c) aluminum Alloy 6022.

2.6   HEAT AFFECTED ZONE (HAZ)

The T-HAZ according to Savage *et al.* [4] separates the PMZ from the unaffected parent, or base material. By definition, all reactions in the T-HAZ occur in the solid state. No melting or liquation reactions occur in this region. For simplicity, this region will be referred to as the HAZ in this text.

Microstructure evolution in the HAZ can be quite complex, depending on both composition and thermal factors. Heating and cooling rates will influence the reactions in this region and can often have profound microstructural effects within the same alloy or alloy system.

The HAZ microstructure surrounding a fusion weld is influenced by many variables. Welding process and heat input, material type, and material condition prior to welding are all important variables that can affect the HAZ thermal history and hence the resultant microstructure. Because the HAZ experiences a spectrum of thermal cycles (peak temperatures and cooling rates), a wide range of microstructure is possible within the same weld and local variations may be large.

The reactions that occur in the HAZ can be quite complex. There are many possible metallurgical reactions, and any given area in the HAZ may undergo one or more of the following possible reactions:

- Recrystallization
- Grain growth
- Phase transformations
- Dissolution/overaging of precipitates
- Precipitate formation
- Residual stress and stress relaxation

The equilibrium phase diagram is a guideline for determining possible reactions in most material systems. During welding, rapid heating and cooling may suppress these equilibrium reactions and/or promote others not predicted by the phase diagram. For example, continuous cooling transformation (CCT) diagrams have been developed for steels that predict transformations to metastable phases that are not present on the equilibrium phase diagram, such as austenite transformation to bainite and martensite.

Heat input and heat flow conditions can influence the dimensions and nature of the HAZ. These dimensions are controlled by the temperature gradient from the fusion boundary into the surrounding base metal and the nature of the metallurgical reactions that occur over that temperature range. In situations where heat input is low and/or heat flow is effective (high thermal diffusivity), a narrow HAZ will result. This is the case in copper and aluminum alloys that possess high thermal conductivity allowing heat to readily flow away from the weld region. The use of low heat input processes and procedures will also minimize the HAZ, since less heat is introduced into the weld. When heat input is high and/or heat flow away from the weld is restricted, a wider HAZ will result. In stainless steels and nickel-base alloys, which exhibit low thermal conductivity, heat builds up around the weld resulting in a shallow temperature gradient. The net effect is to form a wide HAZ. Similarly, a wide HAZ will exist in welds in thin-sheet materials, since heat flow is controlled by section thickness.

2.6.1 Recrystallization and Grain Growth

When welding is conducted on materials that have been strengthened by cold work, the HAZ will be softened by recrystallization and grain growth of the cold-worked microstructure. This process actually occurs in three stages as shown in Figure 2.47.

During the *recovery stage*, internal energy is reduced by the rearrangement of dislocations. This often results in what is described as a cellular dislocation structure that produces strain-free regions in the structure. It is these strain-free regions that act as nuclei for newly formed grains. Since the mechanism involved results in

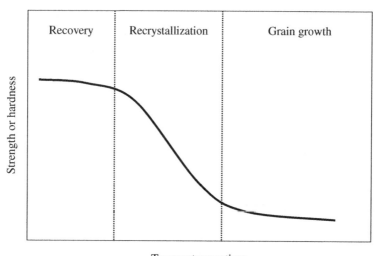

FIGURE 2.47 Change in strength (hardness) as a function of recrystallization and grain growth of cold-worked materials.

dislocation rearrangement rather than annihilation, there is little effect on the strength or ductility of the material.

In the *recrystallization stage*, dislocations are annihilated as the strain-free nuclei grow. These nuclei become new grains and continue to grow and consume the previous, highly dislocated microstructure. This results in a dramatic decrease in strength and hardness, with a corresponding increase in ductility.

With additional time or temperature, *grain growth* continues. The driving force for this reaction is the reduction of overall grain boundary energy by reducing boundary area. This process gradually slows down as the grains get larger. The increase in grain size beyond the recrystallization stage results in a further decrease in strength. This is often referred to as the Hall–Petch effect, where yield strength (σ_y) is related to grain size by the following relationship [43–45]:

$$\sigma_y = \sigma_0 + k_y / d^{1/2} \tag{2.6}$$

where σ_0 is a material constant for the starting stress for dislocation movement (or the resistance of the lattice to dislocation motion), k_y is the strengthening coefficient (a constant unique to each material), and d is the average grain diameter.

The threshold temperature at which recrystallization occurs is a function of the amount of stored energy in the form of cold work. As the percent cold work increases, the amount of additional thermal energy required for recrystallization decreases. As shown in Figure 2.48, the recrystallization temperature of iron drops by 400°C as cold work increases from 5 to 50% [46]. This effect has important implications in the HAZ, since the width of the softened HAZ will increase as the percent cold work in the base metal increases. Other materials, such as aluminum, copper, nickel, and stainless steels, behave in a similar fashion.

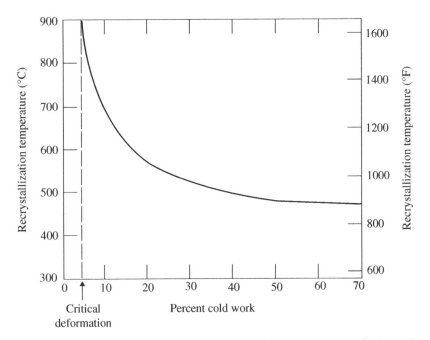

FIGURE 2.48 Effect of cold work on the recrystallization temperature for iron (From Ref. [46]. © Wiley).

The schematic in Figure 2.49 illustrates how the strength and ductility of cold-worked materials are influenced by temperature [46]. Note that a threshold recrystallization temperature exists above which strength decreases and ductility increases. This temperature effect is analogous to the behavior of the HAZ upon moving from the base metal (left) toward the fusion boundary (right).

The starting base metal microstructure can significantly influence the microstructure and properties in the HAZ. In cold-worked materials, recrystallization and grain growth in the HAZ will result in significant softening relative to the base metal. Even in annealed materials, grain growth in the HAZ can result in some softening, as illustrated in Figure 2.50. Although the strength (hardness) plot in Figure 2.50 is somewhat exaggerated, the fusion zone is typically slightly harder (stronger) than the HAZ in single-pass welds due to the presence of a solidification substructure (cells and dendrites) resulting from the solidification process. Transverse tensile tests of welds in cold-worked materials normally fail in the HAZ rather than the weld metal.

2.6.2 Allotropic Phase Transformations

A number of metals undergo allotropic transformations. The term "allotropy" means that the metal can take on different crystallographic forms as a function of temperature. Most prominent among these metals are iron and titanium. Allotropic transformations in steels can be used to great advantage for optimizing mechanical

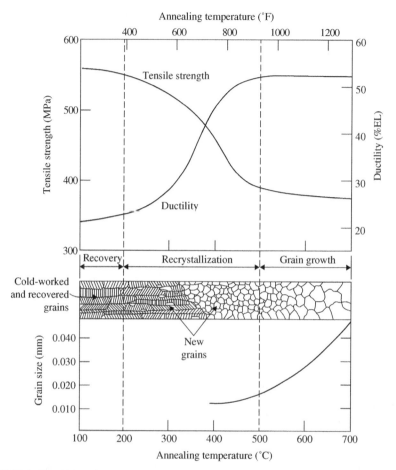

FIGURE 2.49   Effect of temperature on microstructure, strength, and ductility of cold-worked materials (From Ref. [46]. © Wiley).

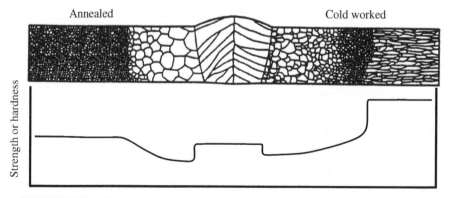

FIGURE 2.50   Change in hardness as a function of recrystallization and grain growth.

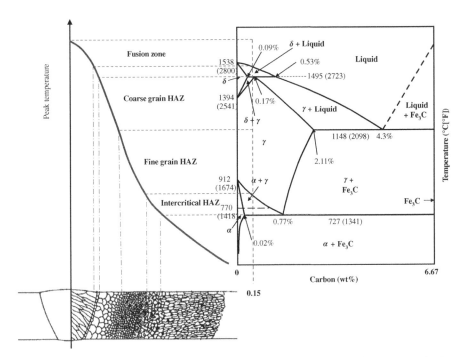

FIGURE 2.51 Relationship between Fe–Fe$_3$C phase diagram and the microstructure in the HAZ of plain-carbon steels.

properties, particularly strength. For steels, because the HAZ spans a very large temperature range from the solidus to the lower critical temperature (A_1), a variety of microstructures can result in these materials due to allotropic transformations. For titanium alloys, the region of the HAZ heated above the alpha (hcp) to beta (bcc) transformation temperature (beta transus) will exhibit a microstructure very distinct from the base metal. Other systems that undergo allotropic transformations include the duplex stainless steels and some copper alloys. Nickel and aluminum alloys are not allotropic.

The microstructure of the HAZ in steels can be predicted, to a first approximation, using the Fe–Fe$_3$C phase diagram, as shown in Figure 2.51. The base metal starts to transform to austenite upon heating above the lower critical temperature (A_1) and is fully austenitic above the upper critical temperature (A_3)[i]. The narrow region of the HAZ heated between these two temperatures is known as the ICHAZ. Above the A_3, the HAZ is fully austenitic, and increasing the peak temperature results in austenite grain growth. This gives rise to the FGHAZ and CGHAZ, as shown in Figure 2.51.

The formation of austenite at elevated temperatures in the HAZ of steels results in a wide variety of transformation products upon cooling. These include ferrite, pearlite

---

[i]Note that the heating and cooling A_1 and A_3 temperatures are different. The heating temperatures are indicated as A_{c1} and A_{c3} and the cooling temperatures as A_{r1} and A_{r3}, where "c" and "r" come from the French *chauffage* (heating) and *refroidissement* (cooling), respectively.

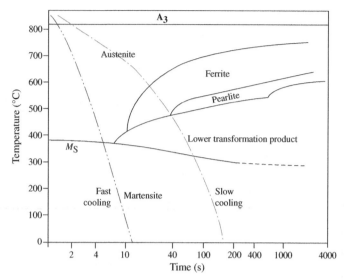

FIGURE 2.52   Schematic illustration of a continuous cooling transformation diagram for steels.

(mixture of ferrite and Fe$_3$C), bainite (both upper and lower), and martensite. The nature of the transformation and the resulting HAZ microstructure are controlled by both composition and cooling rate. CCT diagrams, such as the schematic shown in Figure 2.52, have been developed for many steels in order to predict their microstructure as a function of cooling conditions. These diagrams are very useful in predicting HAZ microstructure in plain-carbon, low-alloy, and high-strength steels if the cooling rate of the HAZ can be measured or approximated.

In multipass welds in steels, the evolution of microstructure in the HAZ becomes very complicated because of the reheating of the underlying microstructure and the overlap of the CGHAZ, FGHAZ, and ICHAZ regions. This is shown schematically in Figure 2.53. A detailed discussion of these phenomena is beyond the scope of this text. The reader is referred to other resources [47, 48].

2.6.3   Precipitation Reactions

Many engineering alloys (such as Al alloys and Ni-base superalloys) are strengthened by precipitation reactions. During welding, the strengthening precipitates can be dissolved or otherwise modified in the HAZ. Dissolution will typically occur above some critical temperature, often called a solvus temperature, as shown in Figure 2.54 for HAZ cycles A and B. Below this temperature, the precipitate will not dissolve but it may actually grow (cycle C). This can produce an overaging reaction that can result in softening in the HAZ.

Upon cooling, precipitation may occur in areas of the HAZ where dissolution has occurred on heating. The degree of precipitation will depend on the composition and cooling rate. In order to recover the strength lost in the HAZ during welding, welds are often given a PWHT. This can consist of a full solution heat treatment to dissolve

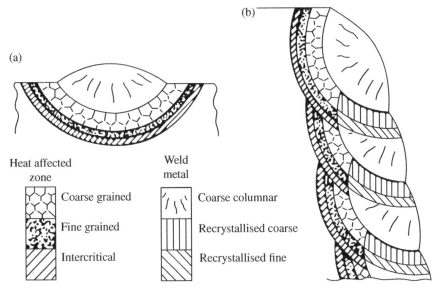

FIGURE 2.53   HAZ microstructure evolution during multipass welding of steels.

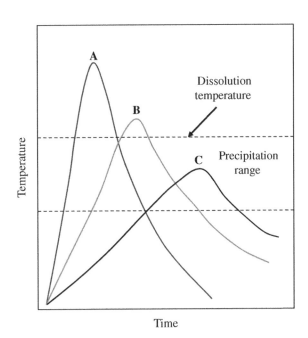

FIGURE 2.54   Effect of HAZ thermal cycle on precipitate dissolution.

all the precipitates and homogenize the structure, followed by aging, or it may consist of a simple aging treatment. The latter is most common, since high-temperature solution treatments of actual structures are often impractical. However, the aging treatment alone will not recover strength in the regions of the HAZ that have been overaged and full strength cannot be achieved by simple aging. In situations where full strength must be restored (as is the case for Ni-base superalloys used in turbine engines), a full solution heat treatment and aging cycle must be applied.

Many precipitation reactions exhibit a "C-curve" behavior with temperature and time, as shown in Figure 2.55. As the nose of this curve moves to shorter times, the precipitation reaction occurs more rapidly. The position of this C curve in temperature–time space has important implications during weld cooling, since it will dictate to what degree precipitation will occur. In most cases, precipitation during cooling is desirable since it will allow some strength recovery. This is true in aluminum- and nickel-base alloys. In some situations, these reactions are undesirable, such as when Cr-rich carbides form in the HAZ of stainless steels. This phenomenon, called "sensitization," can result in localized corrosion attack along grain boundaries.

Aging following welding can take two forms. In most alloy systems strengthened by precipitation, "artificial aging" is required. This consists of heating the material into the temperature range where precipitates can nucleate and grow. This may be preceded by a solution heat treatment. As shown in Figure 2.56, the aging temperature can have an important influence on the peak strength achievable, with lower aging temperatures achieving higher strengths at long aging times. There must be some compromise between temperature and time to achieve a desired strength. For example, some aluminum alloys may require hundreds of hours at a low aging temperature to achieve high peak strength. This is usually not practical, and higher temperatures and shorter times are used with the sacrifice of some strength. At some point, the strength

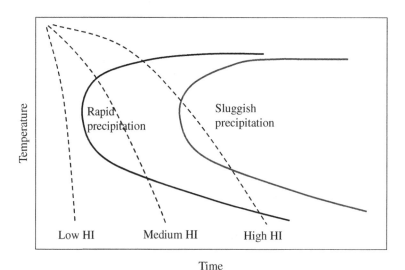

FIGURE 2.55 C-curve precipitation behavior during cooling.

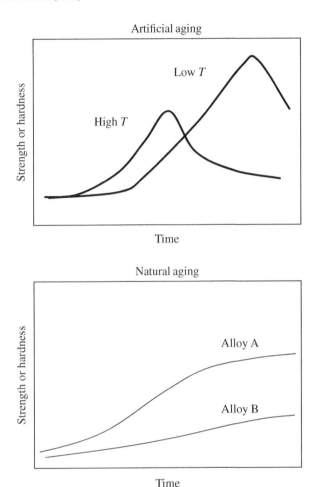

FIGURE 2.56   Artificial and natural aging behavior.

of the material will start to decrease due to the growth of the precipitates beyond an optimum size. This phenomenon is known as "overaging."

"Natural aging," or room temperature aging, is possible in some systems, most notably aluminum and magnesium alloys. This will occur in regions of the HAZ that are solutionized (precipitates dissolved) during welding. Natural aging is typically a very slow process, often requiring weeks or months to achieve significant increases in strength. It is also very alloy dependent, that is, some alloys naturally age much faster to higher strengths than others due to their precipitation kinetics (such as Alloy A in Figure 2.56).

2.6.4   Examples of HAZ Microstructure

Two examples of HAZ microstructure in steels are provided to demonstrate how the HAZ can differ significantly from the base metal. The HAZ in a GTA weld on a fine-grained Type 304L stainless steel is shown in Figure 2.57. The fusion boundary

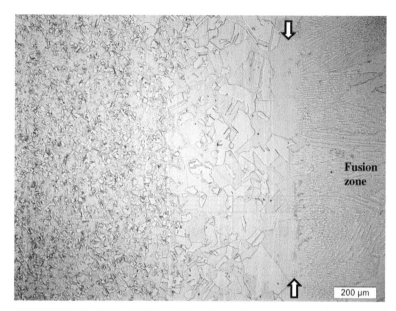

FIGURE 2.57   HAZ of GTA weld in a fine-grained austenitic stainless steel.

is indicated by the arrows. In this case, the HAZ thermal cycle results in significant grain growth with HAZ grains an order of magnitude larger than those in the base metal. Due to the Hall–Petch effect, the HAZ will be softer than the base metal and weld metal.

An even more dramatic change in microstructure occurs in the HAZ of a carbon steel, as shown in Figure 2.58. The base metal consists of a mixture of ferrite and pearlite. The HAZ adjacent to the fusion boundary (boxed area in Fig. 2.58) represents the CGHAZ and exhibits large prior austenite grain size and a microstructure consisting of a mixture of martensite and bainite. In the as-welded condition, this microstructure would exhibit high hardness and would generally require a tempering heat treatment to restore ductility and toughness properties.

2.7   SOLID-STATE WELDING

By definition, melting and solidification do not occur during solid-state welding. These processes depend on the formation of metallic bonds at the atomistic level to produce a joint. There are a number of nonfusion welding processes. They are normally distinguished by the way heat and deformation are generated at the interface. These processes include frictional heating by translation of the pieces relative to each other (friction welding) or by a third member (friction stir welding (FSW)), high-velocity collision (explosion welding), or simple heating to accelerate interdiffusion (diffusion welding).

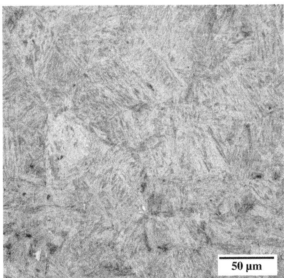

FIGURE 2.58   HAZ of a carbon steel in as-welded condition.

In all cases, a combination of heat and deformation is required to produce a sound weld. As a result, a HAZ will also exist in a solid-state weld, the properties of which may be significantly different from the base metal. In general, two distinct regions can be identified: the *heat and deformation zone* and the ***T-HAZ***. In general, solid-state welding processes do not result in melting at the interface.

Within the heat and deformation zone, significant forging action may occur that promotes continuous (dynamic) recrystallization resulting in extremely fine grain size. Often, the strength of this region exceeds that of the T-HAZ and the base metal.

A few examples of microstructure evolution during solid-state welding are included in the following sections.

2.7.1   Friction Stir Welding

Friction stir welding (FSW) is a novel solid-state process that was introduced in the early 1990s and is shown schematically in Figure 2.59. Bonding relies on the frictional heat of a tool rotating between the two pieces to be welded. The friction heats the material to a temperature where it flows easily and the abutting pieces are joined by this metallic stirring action. No melting takes place, and a high integrity, solid-state joint is formed.

Three distinct regions can be identified in a friction stir weld. The *stir zone* is sometimes referred to as the weld "nugget" and represents the region that consumes the original joint. Metal is heated under the shoulder of the tool and is moved around the tool from front to back by the rotation of the tool. The end of the tool, called the pin (or probe), will sometimes have features machined into it that facilitate material flow in the stir zone.

The *thermomechanically affected zone (TMAZ)* represents the region surrounding the stir zone where some metal flow occurs. Minor recrystallization is often observed in this region. In some materials (steels, Ti alloys), it may be difficult to distinguish a TMAZ.

The *HAZ* of a FSW is analogous to that in a fusion weld—only the heat source is different. The HAZ of FSWs often exhibits the same metallurgical reactions as fusion

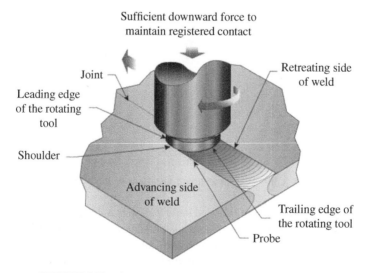

FIGURE 2.59   Schematic illustration of friction stir welding.

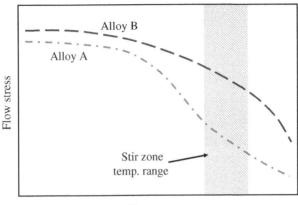

FIGURE 2.60   Effect of temperature on flow stress.

welds. For example, the softening of the HAZ in aluminum due to overaging occurs in both FSWs and fusion welds.

In order to successfully weld a material using FSW, it is necessary for the material to flow at elevated temperature. Flow stress as a function of temperature can be used to determine the stir characteristics of a material. In general, the more rapidly the flow stress drops with increasing temperature, the easier it can be welded by FSW. In the example shown in Figure 2.60, Alloy A would be much easier to weld than Alloy B because it exhibits a lower flow stress in the stir zone temperature range. Higher stirring temperatures also affect tool wear. For materials such as steels, stainless steels, and Ni-base alloys, the very high temperatures required to achieve material flow result in high tool wear and significant microstructural changes in the stir zone, TMAZ, and HAZ relative to the base metal.

Some examples of microstructure evolution during FSW or friction stir processing (FSP) are provided in Figures 2.61–2.63. FSP is mechanistically the same as FSW, except that a joint is not being created. FSP is used to alter the original base metal microstructure through dynamic recrystallization. Aluminum alloys can be easily friction stir welded/processed (FSW/P) since the material can be plastically deformed at relatively low temperatures (400–500°C), allowing the use of steel tool materials. As shown in Figure 2.61, the stir zone undergoes dynamic recrystallization under the action of the tool resulting in extremely fine grains in this region.

Titanium has also been successfully FSW/P, but the temperatures required are typically in the range from 800 to 1000°C (1470–1830°F). The transformation from the low-temperature alpha phase to the higher-temperature beta phase (beta transus) is in this same temperature range, so it is possible to process the material in either the two-phase alpha/beta or single-phase beta regime. The effect on stir zone micro-structure can be quite dramatic, as shown in Figure 2.62 for the near-alpha alloy Ti–5Al–1V–1Sn–1Zr–0.8Mo (commonly known as Ti-5111). When processing above the beta transus, the prior beta grain size is reduced from nearly 500 µm in the base metal to 10 µm in the stir zone. When processed below the beta transus, an equiaxed

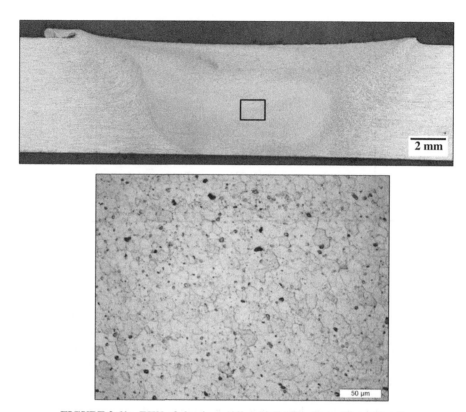

FIGURE 2.61   FSW of aluminum Alloy 6061 (Courtesy of Peter Ditzel).

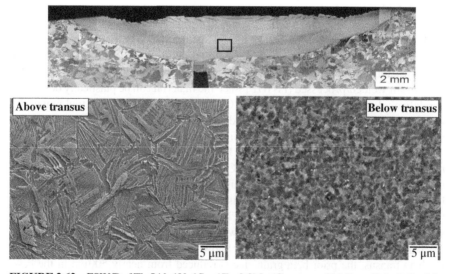

FIGURE 2.62   FSW/P of Ti–5Al–1V–1Sn–1Zr–0.8Mo (Courtesy of Melissa (Rubal) Gould).

(a)

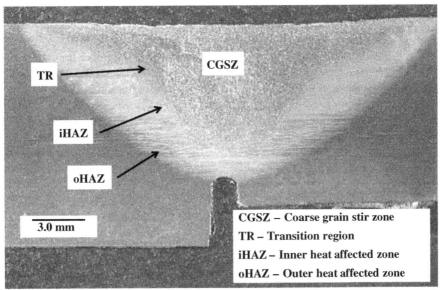

(b)

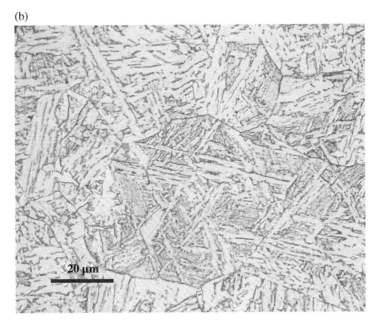

FIGURE 2.63 FSW/P of HSLA-65: (a) regions of the weld and (b) stir zone microstructure (Courtesy of Matt Sinfield).

alpha grain structure develops with an average grain size on the order of 1 μm. This represents a 500× grain refinement relative to the original base metal.

It is also possible to FSW/P steels using appropriate high-temperature tool materials, including tungsten-based tools and polycrystalline cubic boron nitride (PCBN). FSW/P of steels is conducted at temperatures in the range from 1000 to 1200°C where the material is austenitic (fcc). The extreme grain refinement that is achieved in the stir zone of other materials is usually not achieved in steels because of the austenite grain growth that occurs during cooling from the processing temperature. An example of the microstructure associated with a friction stir weld in an HSLA steel (HSLA-65) is shown in Figure 2.63. Note that distinct regions are developed surrounding the stir zone with a HAZ similar to that formed in a fusion weld. In this steel, the stir zone microstructure exhibits a relatively coarse prior austenite grain size and a bainitic microstructure that forms during cooling.

2.7.2 Diffusion Welding

Diffusion welding is accomplished by heating two components to elevated temperature while in intimate contact. This process requires long times at elevated temperature and is usually conducted in vacuum or protective atmosphere to prevent oxidation at the interface. Pressure is generally applied that is of sufficient magnitude to promote some local deformation at the interface.

Although the interfaces in intimate contact must be macroscopically flat, at the microscopic level, they exhibit surface roughness in the form of "asperities" that limit complete interfacial contact, as shown in Figure 2.64. These asperities undergo high local stress when the two surfaces are subjected to a moderate load. This is because the small area of contact distributed over the asperity peaks must support the entire applied load (F), where stress is determined by the applied force divided by the area of contact.

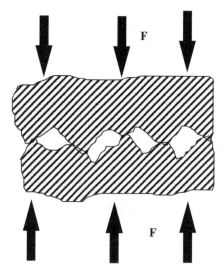

FIGURE 2.64 Microscopic features of a diffusion weld interface prior to bonding.

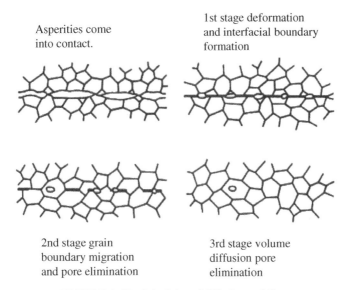

Asperities come into contact.

1st stage deformation and interfacial boundary formation

2nd stage grain boundary migration and pore elimination

3rd stage volume diffusion pore elimination

FIGURE 2.65 Principles of diffusion welding.

As a result of this high stress, the asperities sustain elastic and (at higher load levels) plastic deformation. Elastic deformation is temporary by nature and is removed when the load is removed, forcing the surfaces apart. Plastic deformation is permanent. Once the material is plastically deformed, it cannot revert to its original shape.

Two necessary conditions must be met before a satisfactory diffusion weld can be made: (i) mechanical intimacy of the faying surfaces and (ii) disruption and dispersion of surface contaminants (oxides). This is illustrated in Figure 2.65. Stage 1 involves deformation of asperities. This deformation may be temperature and time dependent, similar to creep. Stage 2 includes boundary migration, recrystallization, and pore size reduction. Stage 3 involves bulk diffusion phenomena including oxide and contaminant dissolution and further pore size reduction.

If the diffusion weld is properly made, there may be no distinguishing features at the bond line. In some cases, there may be oxides present that indicate the location of the original bond interface. In situations where dissimilar metals are diffusion welded, there will be a composition gradient present (interdiffusion zone). Depending on the material system, this may lead to the formation of intermetallic compounds. In some cases, interlayer materials may be introduced that prevent interdiffusion and suppress intermetallic formation.

2.7.3 Explosion Welding

Explosion welding is a form of impact (or collision) welding that can be used to join metallurgically incompatible materials and/or to apply cladding to the surface of a material. The explosion weld forms almost instantaneously, thereby suppressing most metallurgical reactions, such as the formation of embrittling intermetallic phases. A schematic showing the typical setup for explosion welding is shown in Figure 2.66.

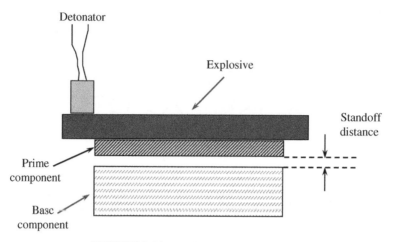

FIGURE 2.66   Explosion welding setup.

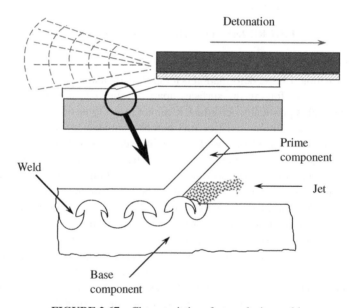

FIGURE 2.67   Characteristics of an explosion weld.

The collision between the components generates kinetic energy that produces local melting, vaporization, and possibly plasma formation. Most of the liquid and vapor is expelled from the joint by a strong jet action at the interaction point. This "jetting" action removes oxides and other contaminants from the surface and produces metallurgically clean interfaces that are easily bonded. The resulting weld interface often has a wavy appearance, as illustrated in Figure 2.67. In general, the bond line of an explosion weld can be revealed using metallographic techniques. Some welds will exhibit the classic wavy appearance, but this is not always the case. Occasionally, local

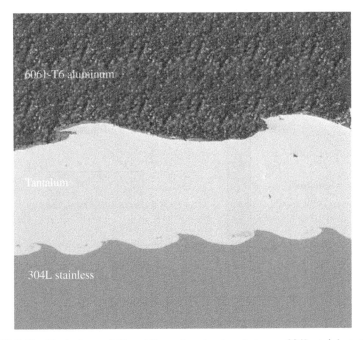

FIGURE 2.68 Explosion weld bond line microstructure between 304L stainless steel and 6061 aluminum using a tantalum interlayer (Courtesy of High Energy Metals, Inc.).

melting can be observed, usually at the tip of the wave. The HAZ of explosion welds is extremely narrow and often undetectable.

An example of an explosion weld interface is shown in Figure 2.68. In this case, aluminum Alloy 6061-T6 is bonded to Type 304L stainless steel. Because of the metallurgical incompatibility of aluminum and steel, a tantalum interlayer is used as a buffer. Note the wavy appearance of both bond lines, but with a different periodicity in the wave pattern.

2.7.4 Ultrasonic Welding

Ultrasonic welding can also be used to join difficult-to-weld materials and is widely used to join both metals and plastics. Ultrasonic waves are transmitted to the interface of the two materials. This results in local heating of the interface. Force is then applied to break down surface asperities and allow metallic bonding to occur. A schematic of the process is shown in Figure 2.69.

The bond line is usually detectable in ultrasonic welds using appropriate metallographic techniques. Because the heat input is very low, there is usually no apparent HAZ. Often, the bond line may contain oxides that are not removed prior to welding or not disrupted by the welding process.

An example of an ultrasonic weld in aluminum is shown in Figure 2.70. The ridges on the surface are the result of the grid pattern on the sonotrode tip and anvil, which

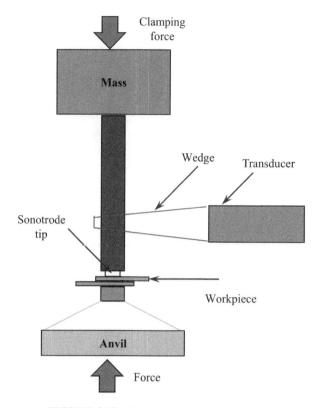

FIGURE 2.69   Schematic of ultrasonic welding.

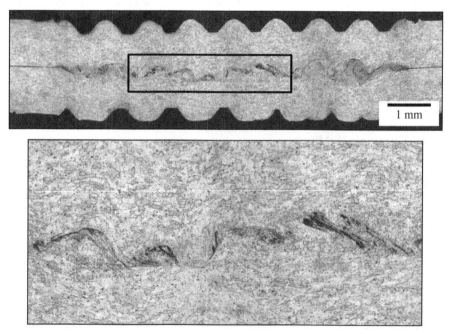

FIGURE 2.70   Ultrasonic weld bond line microstructure in aluminum (Courtesy of Sonobond Ultrasonics, Inc.).

are needed to engage the pieces to be welded. This process also develops a wave pattern at the interface that increases the bond area and improves bond strength. The darker etching features at the bond line are entrapped oxides. Note that there is no evidence of grain growth due to the localized heating at the interface.

REFERENCES

[1] Kou S. *Welding Metallurgy*. 2nd ed. Hoboken NJ: John Wiley and Sons, Inc.; 2003. 176.

[2] Easterling K. *Introduction to the Physical Metallurgy of Welding*. London: Butterworths and Co.; 1983. 73.

[3] Nippes EF. The weld heat-affected zone. Weld J 1959;38 (1):1s–17s.

[4] Savage WF, Nippes EF, Szekeres ES. A study of weld interface phenomena in a low alloy steel. Weld J 1976;55 (9):260s–268s.

[5] Savage WF. Solidification, segregation, and weld imperfections (Houdremont Lecture). Weld World 1980;18 (5/6):89–114.

[6] Savage WF, Nippes EF, Miller TW. Microsegregation in 70Cu–30Ni weld metal. Weld J 1976;55 (6):165s–173s.

[7] Chalmers B. *Principles of Solidification*. New York: Wiley; 1964.

[8] Mullins WW, Sekerka RF. Morphological stability of a particle growing by diffusion or heat flow. J Appl Phys 1963;34:323.

[9] Davies GJ, Garland JG. Solidification structures and properties of fusion welds. Int Met Rev 1975;20:83–105.

[10] David SA, Vitek JM. Correlation between solidification parameters and weld microstructures. Int Met Rev 1989;34:213–245.

[11] Katayama S. Solidification phenomena of weld metal (1st report) – Characteristic solidification morphologies, microstructures and solidification theory. Weld Int 2000;14 (12):25–37.

[12] Katayama S. Solidification phenomena of weld metal (2nd report) – Solidification theory, solute redistribution and microsegregation behavior. Weld Int 2000;14 (12):13–24.

[13] Hallum D, Baeslack WA. Nature of grain refinement in titanium alloy welds by micro-cooler inoculation. Weld J 1990;69 (9):326s–336s.

[14] Simpson RP. Controlled weld pool solidification structure and resultant properties with yttrium inoculation of Ti–6Al–6V–2Sn welds. Weld J 1977;56 (3):67s–87s.

[15] Gutierrez A, Lippold JC. A proposed mechanism for equiaxed grain formation along the fusion boundary in Al–Cu–Li alloys. Weld J 1998;77 (3):123s–132s.

[16] Nelson TW, Lippold JC, Mills MJ. Nature and evolution of the fusion boundary in ferritic-austenitic dissimilar metal welds, Part 1: Nucleation and growth. Weld J 1999;78 (10):329s–337s.

[17] Nelson TW, Lippold JC, Mills MJ. Nature and evolution of the fusion boundary in ferritic-austenitic dissimilar metal welds, Part 2: On-cooling transformations. Weld J 2000;79 (10):267s–277s.

[18] Heiple CR, Roper JR. Mechanism for minor element effect on GTA fusion zone geometry. Weld J 1982;61 (4):97s–102s.

[19] Smith VG, Tiller WA, Rutter JW. A mathematical analysis of solute redistribution during solidification. Can J Phys 1955;33:723–745.

[20] Flemings MC. *Solidification Processing*. New York: McGraw-Hill; 1974. p 36–38.

[21] Brooks JA, Baskes MI. Weld microsegregation and modeling. In: David SA, editor. *Advances in Welding Science and Technology*. Metals Park, OH: ASM International; 1987. p 93–99.

[22] Perricone MJ, Dupont JN. Effect of composition on the solidification behavior of several Ni–Cr–Mo and Fe–Ni–Cr–Mo alloys. Metals Mater Trans A 2005;37A:1267–1289.

[23] Alexandrov BT, Lippold JC, Sowards JW, Hope AT, Saltzmann DR. Fusion boundary microstructure evolution associated with embrittlement of Ni-base alloy overlays applied to carbon steel. Weld World 2013;57 (1):39–54.

[24] Baeslack III WA, Lippold JC, Savage WF. Unmixed zone formation in austenitic stainless steel weldments. Weld J 1979;58 (6):168s–176s.

[25] Gooch TG. Corrosion behavior of welded stainless steel. Weld J 1996;75 (5):135s–154s.

[26] Kostrivas A, Lippold JC. Weldability of Li-bearing aluminum alloys. Int Mat Rev 1999;44 (6):217–237.

[27] Flasche LH, Ahluwalia HS. Localized corrosion of the unmixed zone in Ni-base alloy weldments. 12th International Corrosion Congress, NACE, Vol. 4; Houston, TX; 1993. p 2895–2906.

[28] Pepe JJ, Savage WF. Effects of constitutional liquation in 18Ni maraging steel weldments. Weld J 1967;46 (9):411s–422s.

[29] Pepe JJ, Savage WF. The weld heat-affected zone of the 18Ni maraging steels. Weld J 1970;49(12):545s.

[30] Qian M, Lippold JC. Effect of multiple postweld heat treatment cycles on the weldability of Waspaloy. Weld J 2002;81 (11):233s–238s.

[31] Priester L. *Grain Boundaries: From Theory to Engineering*. Springer, Berlin, Heidelberg; 2013.

[32] Blum BS, Witt RH. Heat-affected zone cracking in A-286 weldments. Weld J 1963;42 (8):365s–370s.

[33] Thompson EG. Hot cracking studies of Alloy 718 weld heat-affected zones. Weld J 1969;48 (2):70s–79s.

[34] Tamura H, Kato N, Watanabe T. Hot cracking in synthetic weld heat-affected zone on Ni–Cr–Mo type high strength steel. Trans JWS 1974;5 (2):180–191.

[35] Cahn J, Balluffi R. On diffusional mass transport in polycrystals containing stationary and migrating grain boundaries. Scripta Metall 1979;13:499–502.

[36] Lippold JC. An investigation of heat-affected zone hot cracking in alloy 800. Weld J 1983;62 (1):1s–11s.

[37] Gleiter H. Theory of grain boundary migration rate. Acta Metall 1969;17:853–862.

[38] Duvall DS, Owczarski WA. Technical note: behavior of solute at mobile heat-affected zone grain boundaries. Weld J 1966;45 (8):356s.

[39] Balluffi RW. Grain boundary diffusion mechanisms in metals. Metal Trans B 1982;13B: 527–553.

[40] Turnbull D, Hoffman RE. The effect of relative crystal and boundary orientations on grain boundary diffusion rates. Acta Metall 1954;2:419–426.

[41] Shewmon P. *Diffusion in Solids*. 2nd ed. Warrendale, PA: TMS Publications; 1989. p 196.

[42] Baeslack III WA, Ernst SC, Lippold JC. Weldability of high-strength, low-expansion superalloys. Weld J 1989;68 (10):418s–430s.

[43] Petch NJ. The cleavage strength of polycrystals. J Iron Steel Inst 1953;174:25–28.

[44] Hall EO. The deformation and ageing of mild steel. Proc Phys Soc B 1951;64:747–753.

[45] Duckworth EW, Baird JD. Mild steels. J Iron Steel Inst 1969;207:854–871.

[46] Callister WD. *Materials Science and Engineering: An Introduction*. 3rd ed. New York: John Wiley and Sons, Inc.; 1994.

[47] Linnert GE. *Welding Metallurgy, Volume 1: Fundamentals*. 4th ed. American Welding Society: Miami, FL; 1994.

[48] Stout RD. *Weldability of Steels*. New York: Welding Research Council; 1987.

3

HOT CRACKING

3.1 INTRODUCTION

Weld "hot" cracking refers to cracking phenomena that occur during fabrication due to the presence of liquid in the microstructure. Hot cracking is most often associated with liquid films that are present along grain boundaries in the fusion zone and the partially melted zone (PMZ) region of the heat-affected zone (HAZ). These liquid films may persist to temperatures well below the equilibrium solidus temperature of the bulk alloy, thus extending the solidification range of the alloy to the "effective" solidus temperature. In many cases, the effective solidus may be well below the equilibrium solidus due to the effect of solute and impurity segregation.

Although widely used in the literature to describe a number of elevated temperature cracking phenomena, the use of the term "hot cracking" here refers to three distinct types of cracking. *Weld solidification cracking* occurs in the fusion zone at the end of solidification. This form of cracking is usually associated with liquid films along solidification grain boundaries (SGBs). *HAZ liquation cracking* occurs in the PMZ region of the HAZ. It is also intergranular and results from continuous liquid films that form at elevated temperature. *Weld metal liquation cracking* is unique to reheated weld metal and is most often observed in multipass welds or after repair welding. These cracks may form at SGBs or migrated grain boundaries (MGBs) and result from the remelting of these boundaries during reheating to elevated temperatures.

As noted above, the term "hot crack" is often used to describe any crack that forms at elevated temperature and thus is not very descriptive. The AWS definition of

Welding Metallurgy and Weldability, First Edition. John C. Lippold.
© 2015 John Wiley & Sons, Inc. Published 2015 by John Wiley & Sons, Inc.

"hot crack" is *a crack formed at temperatures near the completion of solidification.* Often, other forms of cracking that occur at elevated temperature but in the solid state are referred to as "hot cracks," such as ductility-dip cracks. Thus, the definition of the term "hot cracking" is somewhat vague and arbitrarily interpreted. In this text, only those forms of cracking associated with grain boundary liquid films will be considered as hot cracking.

3.2 WELD SOLIDIFICATION CRACKING

Weld solidification cracking occurs preferentially along SGBs and, occasionally, subgrain boundaries in the fusion zone. Weld solidification cracking is a phenomenon that has been the subject of considerable study, in a wide variety of engineering alloys, over the past 40 years. Despite the large body of information currently available concerning weld solidification cracking, the precise micromechanisms responsible for cracking are still not completely understood. Fundamentally, two conditions must be satisfied for weld solidification cracking to occur: (i) thermally and/or mechanically imposed restraint (strain) and (ii) a crack-susceptible microstructure.

The *restraint* factor may be both intrinsic and extrinsic. Intrinsic restraint results from natural shrinkage during solidification since all metals undergo a negative volume change during solidification. Other factors that influence intrinsic restraint include material properties (particularly strength), size and thickness of the workpiece, joint design, and size and shape of the weld bead. Extrinsic restraint is provided primarily by mechanical fixturing.

The *crack-susceptible microstructure* results from the persistence of liquid films along solidification boundaries in the weld metal, as discussed in Chapter 2. Weld solidification cracking can often be minimized or eliminated by reducing the level of mechanical restraint. For example, joint geometries or weld parameter changes that alter the weld bead size and shape are often effective in alleviating cracking. A permanent solution to persistent solidification cracking problems is usually not achieved, however, until the metallurgical basis of the problem is understood and alleviated, often by compositional modification of the weld metal (in the case where filler materials are added) or control of the solidification process.

3.2.1 Theories of Weld Solidification Cracking

Several theories of weld solidification cracking, or "hot tearing" in reference to castings, have been proposed since the 1940s. These include:

1. *The shrinkage-brittleness theory.* This theory was supported by a number of early investigators including Bochvar and Sviderskaya [1], Pumphrey and Jennings [2], Medovar [3], and Toropov [4].
2. *The strain theory.* This theory was originally developed by Pellini [5, 6] also to describe hot tearing in castings but was eventually extended to describe weld solidification cracking.

3. *The generalized theory of supersolidus cracking.* This theory was advanced by Borland [7] in the early 1960s and incorporates aspects of both the shrinkage-brittleness and strain theories.

4. *The modified generalized theory.* A modification to Borland's theory was proposed by Matsuda and coworkers [8, 9] at the Joining and Welding Research Institute (JWRI) at Osaka University, based on direct observation of weld solidification cracking using special high-speed cinematography.

5. *The technological strength theory.* This theory was proposed by Prokhorov [10] in the early 1960s and addresses solidification cracking from a strictly mechanical standpoint.

Central to all these proposed mechanisms is the requirement for some form of liquid film along solidification boundaries during the final stages of solidification. It is the inability of this liquid film to support the shrinkage-induced and extrinsic mechanical strain during solidification and subsequent cooling of the weld that results in the separation of these boundaries, thus forming a solidification crack. As noted previously, the ultimate solution to reducing or eliminating hot cracking lies in eliminating or effectively controlling the liquid films along these boundaries.

3.2.1.1 *Shrinkage-Brittleness Theory*

This theory of solidification cracking was first proposed by the Russian researcher Bochvar in 1947 [1] and was based on his work with aluminum castings. In 1952, this approach was adopted by Medovar [3] of the Paton Welding Institute to describe solidification cracking in welds. At about the same time, Pumphrey and Jennings [2] in England were using a similar argument to explain cracking during casting and welding of aluminum alloys. The theory that evolved from this work came to be known as the shrinkage-brittleness theory.

Central to this theory is the existence of an "effective interval" of solidification that occurs below the "coherency" temperature, as shown in Figure 3.1. At high temperatures within the solidification range, the ratio of solid to liquid is low and the solidification subgrains (cells or dendrites) are completely surrounded by liquid. Below some critical temperature within this solidification range, the solid begins to interact and form a rigid network. The temperature at which this interaction begins is termed the "coherency" temperature, and the remaining solidification temperature range is referred to as the "effective interval."

As defined by the shrinkage-brittleness theory, weld solidification cracking always occurs within this effective interval. That is, cracking can only occur after some solid–solid bridging has occurred, which allows strain to accumulate in the structure. Above this interval, sufficient liquid is available to "heal" cracks that may form. In alloys, cracking susceptibility will be greatest at the composition where the "effective interval" is the largest, as shown in Figure 3.1. The theory also predicts a decrease in cracking susceptibility in alloys that undergo a eutectic reaction at compositions that allow "healing" due to the presence of sufficient quantities of liquid of eutectic composition.

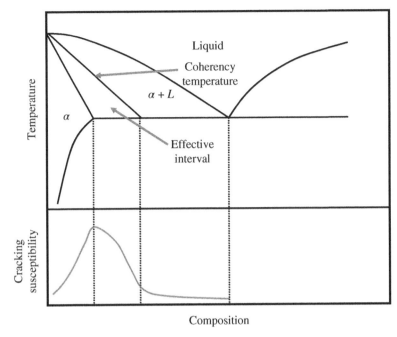

FIGURE 3.1   Schematic illustration of the shrinkage-brittleness theory based on a eutectic phase diagram.

3.2.1.2   Strain Theory   The strain theory of hot tearing in castings was first advanced by W.S. Pellini at the U.S. Naval Research Laboratory in 1952 [5]. In 1954, he used this same approach to describe solidification cracking in welds based on work with stainless steels and nickel-base alloys [6]. The strain theory differs from the shrinkage-brittleness theory in that coherency, or solid–solid bridging is not considered until the very final stages of solidification.

The strain theory considers two stages of solidification, namely, a mushy stage and a liquid film stage, as shown in Figure 3.2. The mushy stage of solidification occurs at temperatures above the film stage where considerable liquid is still present. The film stage is further divided into an early and late period with the late film stage occurring just prior to complete solidification. Weld solidification cracking, based on the strain theory, is not possible during the mushy stage due to the uniform distribution of strain in the solid–liquid mixture. During the film stage, however, strains can be extremely high locally and result in separation along boundaries with continuous liquid films separating solid grains. Cracking is therefore a solid–liquid separation rather than a solid–solid separation as proposed by the shrinkage-brittleness theory. This theory does not consider the concept of crack healing within the solidification temperature range. This theory also predicts the fracture surface will not exhibit any features associated with solid–solid contact, that is, the fracture surface should exhibit smooth dendrites.

Referring to Figure 3.2, the *normal film stage* refers to equilibrium solidification, whereby only liquid films remain in the structure at the end of solidification.

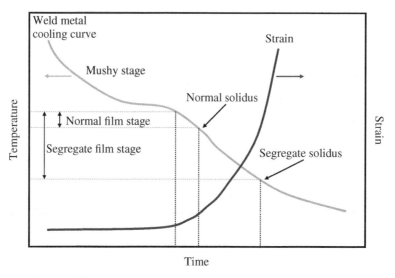

FIGURE 3.2   Strain theory according to Pellini (Adapted from Ref. [6]).

Segregation during nonequilibrium weld solidification extends this range over a wider temperature range, as indicated by the *segregate film stage*. Strain begins to build up in the system during the film stage and will tend to concentrate at boundaries separated by thin liquid films. When this strain exceeds the ductility of the boundary, cracking will occur.

3.2.1.3   Generalized Theory   In 1960, J.C. Borland of the British Welding Research Association (now The Welding Institute (TWI)) proposed a modification of the shrinkage-brittleness theory that included some aspects of the strain theory [7]. This modification, called the "generalized theory," also adopted the idea of a coherency (or coherent) temperature at which solid–solid interaction begins but included an additional stage of solidification below this temperature. Borland termed this as the "critical solidification range," which corresponded to the lower part of the "effective interval" from the shrinkage-brittleness theory, as illustrated in Figure 3.3.

Above the "critical" range but below the coherent temperature, failure of the solid–solid bridges could be accommodated by the presence of sufficient liquid and crack "healing" is possible. Within the "critical" range, healing of cracks was not possible due to the reduced amount of liquid and a more complex and well-developed solid network. Along with this theory, Borland also suggested the importance of liquid wetting characteristics during weld solidification. The importance of grain boundary wetting by liquid films will be discussed later in this chapter.

Referring to Figure 3.3, *Stage 1* defines the mushy stage of solidification. *Stage 2* defines a coherency range where cracks that form may be healed since there is sufficient surrounding liquid. *Stage 3* defines the "critical" range where cracks that form are separated from liquid that can heal them, or there is insufficient liquid present in this range to promote significant healing. Finally, *Stage 4* defines the region where

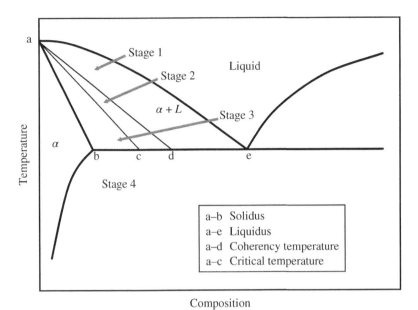

FIGURE 3.3   The generalized theory according to Borland (Adapted from Ref. [7]).

solidification is complete and solidification cracking does not occur due to the absence
of liquid in the system.

3.2.1.4   Modified Generalized Theory   Matsuda and coworkers from the JWRI at
Osaka University further modified the shrinkage-brittleness and generalized theories
based on experiments permitting direct observation of weld solidification cracking
[8, 9]. They observed that Stage 1 occurs over a much narrower temperature range than
originally proposed and that significant solid networks form very rapidly upon cooling
below the liquidus temperature. This is illustrated by the schematic in Figure 3.4. Stage 3,
as defined by Borland, is further divided into a "film stage" (3H) and a "droplet stage"
(3L). Initiation of weld solidification cracking occurs in the liquid film stage and
propagation proceeds in either the film stage or the droplet stage. Initiation is not possible in the droplet stage of weld solidification due to extensive solid–solid contact.

This theory also takes into account the fractographic features of weld solidification
cracking and suggests that the dendritic character of solidification cracks represents
crack initiation and propagation during Stage 3H. The flat fracture features often
associated with weld solidification cracks are representative of crack propagation
during Stage 3L. Later, the possibility of the flat features forming from the result of
solid-state, ductility-dip cracking will be discussed.

Based on direct observation of solidification cracking and metallographic and
fractographic examination of solidification cracks, Matsuda proposed four different
types of fracture behavior, as shown schematically in Figure 3.5:

- Type D: dendritic fracture. This fracture behavior results from separation of
  liquid films at the SGBs and suggests a continuous network of liquid film along

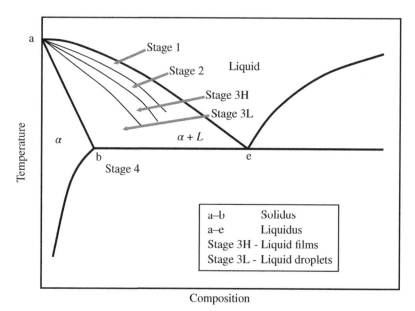

FIGURE 3.4 The modified generalized theory as proposed by Matsuda (Adapted from Ref. [8, 9]).

these boundaries. The fracture surface exhibits the classic "eggcrate" dendritic morphology.

- Type F: flat fracture. This fracture surface shows no evidence of dendritic fracture but (according to Matsuda) was the result of liquid film separation.
- Type D+F: a mixture of the two. This essentially represents a transition region from dendritic to flat fracture.
- Type R: intergranular, solid-state fracture. This is representative of ductility-dip cracking observed in austenitic stainless steels and Ni-base alloys.

3.2.1.5 Technological Strength Theory In the early 1960s, the Russian researcher Prokhorov studied the mechanical behavior of a fusion weld during solidification and proposed the technological strength theory [10]. This theory considers a ductility curve for a material during solidification, as shown in Figure 3.6. The material exhibits a loss in ductility within the brittle temperature range (BTR). Outside the BTR, the ductility of the material is assumed to be sufficient to accommodate the thermally and/or mechanically induced strain during welding. In addition to the ductility curve, the welding-induced strain is also shown in the same diagram by assuming that the strain is linearly proportional to the decrease in temperature upon weld cooling (i.e., thermal contraction strain).

The possibility of cracking is a consequence of the competition between the accumulation of strain and the recovery of ductility during weld cooling. Solidification cracking will occur if the ductility of the material is exhausted by the thermally and/or mechanically induced strain. This theory does not directly take into consideration any

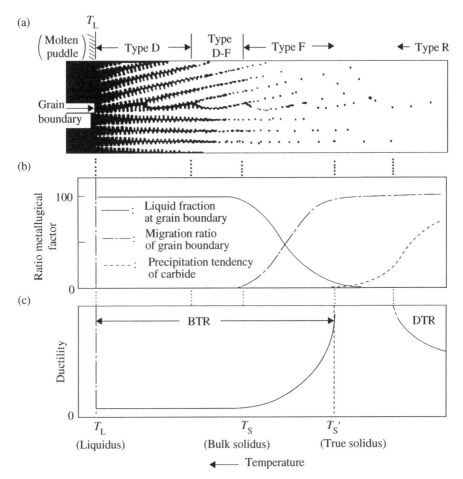

FIGURE 3.5 Fracture behavior and morphology according the modified generalized theory (From Ref. [8, 9]. © ASM).

aspects of weld microstructure. The width and depth of the BTR are influenced by the solidification temperature range and the nature of grain boundary liquid films, but no direct correlations are possible.

Referring to Figure 3.6, *Line A–B* represents the thermal contraction of the material in the BTR. *Line A–C* represents the summation of contraction deformation and other deformations (mechanical) in the system. *Line A–D* represents the critical amount of deformation to cause cracking in the system with the given ductility curve. When the deformation line exceeds (crosses) the ductility curve, the ductility of the system is exhausted and cracking will occur. This would be represented in this diagram by a line with a steeper slope than Line A–D that intersects the ductility curve.

3.2.1.6 Commentary on Solidification Cracking Theories While the generalized theory proposed by Borland [7] has often been referenced to explain the

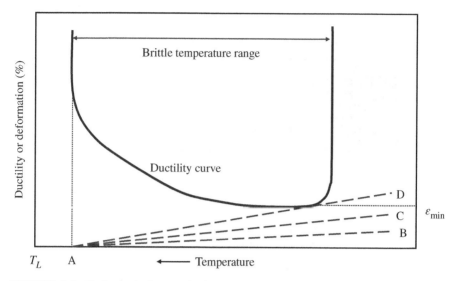

FIGURE 3.6 Technological strength theory as proposed by Prokhorov (Adapted from Ref. [10]).

nature of solidification cracking in metals, there are aspects of this theory that are not consistent with actual observations of solidification cracking in most materials. The concepts of a "coherency" temperature where a solid–solid network begins to form and a "critical solidification range" over which solid–solid bridges are broken without the possibility of liquid healing are inconsistent with fractographic observations of solidification cracking. In fact, it is very unusual to observe any evidence of solid–solid fracture on the surface of a solidification crack when examined in a scanning electron microscope (SEM). Typical SEM micrographs from an aluminum alloy (6061-T6) and a Ni-base filler metal (ERNiCrFe-13, FM 52MSS) are shown in Figure 3.7. The fracture surface of the Ni-base alloy is entirely dendritic, while the aluminum alloy shows only a few locations where solid–solid fracture may have occurred. Fracture behavior will be discussed in more detail later in this chapter and in Chapter 8. The point to be made here is that, in general, solidification crack surfaces show an almost entirely dendritic appearance indicating that the theory of coherency (solid–solid bridging) as critical to solidification cracking is not supported. In Borland's defense, his theory was developed prior to the widespread use of the SEM to study fracture behavior so he did not have the opportunity to couple theory with actual observation of the fracture surface.

Thus, based on fracture appearance, it would appear that the strain theory proposed by Pellini [6] is more appropriate since this theory predicts that liquid films persist until the final stages of solidification. His premise that continuous thin films are present at the time that solidification cracks initiate and propagate is fully supported by metallographic and fractographic evidence, as will be described later in this chapter.

The modified generalized theory of Matsuda [8] also recognized the importance of a "film stage" at the end of solidification in which solidification cracks initiate and propagate, giving rise to dendritic fracture surfaces, as shown schematically in

(a)

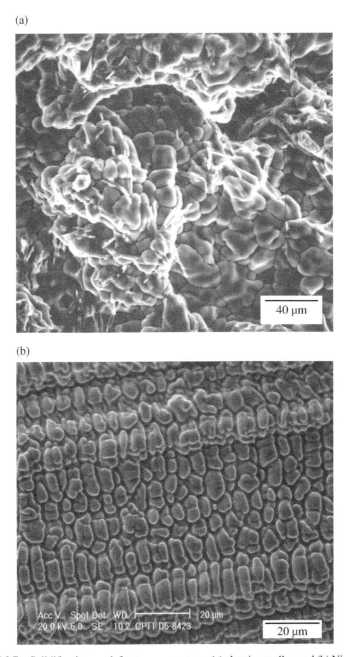

(b)

FIGURE 3.7 Solidification crack fracture appearance: (a) aluminum alloy and (b) Ni-base alloy.

Figure 3.5. The transition from dendritic to flat fracture that is proposed may be an artifact of the testing procedure (Varestraint test) since flat fracture associated with liquid films in the fusion zone is not generally observed in practice. The transition to intergranular, solid-state fracture is unique to systems that are susceptible to ductility-dip cracking, as will be described in Chapter 4.

The technological strength theory, while not addressing the metallurgical or fracture aspects of solidification cracking, is very useful in explaining the ductility/strain contribution. The basic premise of this theory is that the weld metal exhibits a characteristic ductility "signature" that defines a BTR. The ductility within this BTR determines whether cracking will occur in the presence of applied strain during solidification. Thus, referring to Figure 3.6, if the strain during solidification exceeds the value ε_{min} on the ductility curve, the ductility of the system will be exhausted and cracking will occur. If ε_{min} can be increased as a function of composition (e.g., reducing impurities) or the strain decreased such that the ductility is not exhausted, cracking will not occur.

This suggests that if the weld metal ductility can be determined experimentally and the strain estimated based on contraction and mechanical components, the solidification cracking susceptibility of the weld metal can be quantified. This is exactly the approach that has been used by both the JWRI at Osaka University and the Welding and Joining Metallurgy Group at Ohio State University (OSU). These approaches will be described in detail later in this chapter.

3.2.2 Predictions of Elemental Effects

Weld metal composition has a dominant effect on solidification cracking behavior, and the use of simple composition analysis and control is a key element in controlling susceptibility to this form of cracking. The relative potency factor (RPF) was developed by J.C. Borland in conjunction with the generalized theory as a method to determine the *relative* effect of alloying and impurity elements on weld solidification cracking susceptibility, based on the equilibrium solidification behavior of binary alloys [11]. Since RPF is based on equilibrium solidification conditions, it represents only an approximation of nonequilibrium solidification behavior of fusion welds. Because most engineering alloys are multicomponent systems, the RPF may often under- or overestimate the effect of a particular element due to the synergistic influence of other elements. For example, sulfur is particularly effective in promoting solidification cracking in a simple Fe–S or Ni–S alloys but becomes less effective when even a small amount of manganese is added to the system. The RPF is not capable of predicting these interactive effects. The RPF, as defined by Borland, is simply a measure of the solidification temperature range (ΔT) normalized by the nominal alloy composition, C_0. In order to calculate RPF, it is only necessary to determine the liquidus slope, m_L, and the partition coefficient, k, from the phase diagram, as given by the following:

$$\text{RPF} = \frac{\Delta T}{C_0} = m_L \left(1 - \frac{1}{k}\right) \tag{3.1}$$

A maximum potency occurs at the point of maximum solid solubility, C_{Smax}, and represents (for equilibrium solidification) the widest solidification temperature range. The maximum potency factor (MPF) is defined by

$$\text{MPF} = m_L C_{Smax} \left(\frac{1-1}{k} \right) \tag{3.2}$$

The RPF and MPF for some iron binary alloys are shown in Table 3.1. This table includes values originally published by Borland [11] and those calculated using currently available binary phase diagrams. The original values from Borland only considered iron alloy solidification as delta ferrite. Note that the relative potencies calculated using modern phase diagrams are in generally good agreement, although a few of the values are significantly different. For example, boron has a much higher RPF than predicted by Borland, while manganese is considerably lower. RPF and MPF values have also been added for austenite solidification for the Fe–C and Fe–Mn systems.

In general, it should be noted that as the RPF increases, the solidification temperature range increases. The specific potency increases as C_0 approaches C_{Smax} and then decreases for C_0 greater than C_{Smax} for *equilibrium solidification*. The cracking susceptibility peak may be skewed to a composition below C_{Smax} when considering nonequilibrium weld solidification.

The RPF and MPF values provide some indication of elements that promote solidification cracking in steels. It is well known that S, P, and B are deleterious with regard to cracking and these elements are carefully controlled. As noted earlier, these values are only approximations of potency for weld solidification cracking since weld solidification is clearly nonequilibrium in nature and diffusion that can occur in the solid during solidification is not taken into account.

Using modern computational techniques, it is possible to calculate the average partition coefficients (k) during nonequilibrium solidification of multicomponent alloys and to determine the expansion of the solidification range that occurs due to the partitioning. An example of this is shown in Figure 3.8, where the average k-values for a number of elements are determined using the Scheil solidification approximation

TABLE 3.1   Partition coefficient, k, and RPF and MPF for some Fe binary alloys

	B	S	P	C_F	C_A	Mn_F	Mn_A	Nb	Ti	Si	Ni	Zn
k-Value	0.005	0.019	0.25	0.17	0.49	0.74	0.85	0.29	0.64	0.85	0.74	0.5
RPF[b]	917	925	121	322	—	22	—	29	14	2	—	—
RPF[c]	19,741	1084	145	396	96	2	1	23	10	3	1	8
MPF[b]	137	166	338	32	—	26	—	130	97	32	—	—
MPF[c]	395	163	369	36	202	17	49	120	89	40	5	375

[a]Based on solidification as delta ferrite unless indicated by subscript A (austenite solidification).
[b]From Ref. [11].
[c]Calculated from modern phase diagrams.

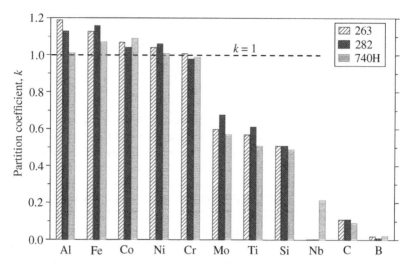

FIGURE 3.8 Partition coefficient, k, values for Ni-base Alloys 263, 282, and 740. Note that values less than 1.0 result in partitioning to the liquid phase (From Ref. [13]).

within the computational software ThermoCalc® [12]. These values are calculated for three Ni-base superalloys (Alloys 263, 282, and 740H) that are being considered for high-temperature power generation applications where weldability is a high priority [13].

Note that for alloying elements with k-values very close to 1 (Fe, Al, Ni, Cr), there will be virtually no partitioning and these elements will have little or no effect on expansion of the nonequilibrium solidification temperature range. Elements with $k < 1$ will partition to the liquid and have the effect of expanding the solidification temperature range. This is especially true for Nb, C, and B in these alloys. Note that this software can account for back diffusion during solidification (see Eq. 2.5) and, thus, the k-value for carbon has been adjusted to higher than equilibrium value to account for this. The same software can also calculate the solidification profile under Scheil solidification conditions (see Eq. 2.3), as shown in Figure 3.9. This allows the nonequilibrium solidification temperature range to be determined and, in eutectic systems, the fraction eutectic calculated.

Similar solidification simulations can be conducted for Fe-base alloys and, to a limited extent, for other alloy systems (such as Ni-base and Al-base alloys) where there is sufficient thermodynamic data to calculate phase diagrams.

There are also a number of composition-based relationships that have been developed to predict solidification cracking susceptibility. Those listed in Table 3.2 have been proposed for cracking in steels [14–17]. A comprehensive list can be found in a review paper by Matsuda [18]. Note that all these relationships show positive factors for carbon, phosphorus, and sulfur. The addition of manganese tends to reduce susceptibility to solidification cracking, presumably due to its interaction with sulfur and mitigation of sulfur-rich liquid films by formation of a high-temperature MnS compound. The strong effect of S and P is due to their low partition coefficient (k) in

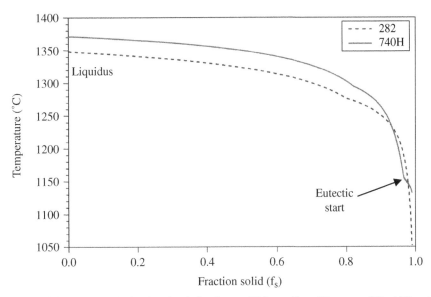

FIGURE 3.9 Scheil solidification simulation for two Ni-base alloys (Courtesy of David Tung).

TABLE 3.2 Cracking susceptibility factor (CSF) relationships for steels

Factor	Steel/weld type	References
$CSF = [P \times (C + 0.142Ni + 0.282Mn + 0.2Cr - 0.14Mo - 0.224V) + 0.195S + 0.0216Cu] \times 10^4$	Low-alloy steels	[14]
$CSF = 42[C + 20S + 6P - 0.25Mo - 720] + 19$	Low-alloy steels	[15]
$CSF = 184C + 970S - 188P - 18.1Mn - (4760S \times C) - (12,400S \times P) + (501P \times Mn) + (32,600C \times S \times P) + 12.9$	Low-carbon SAW weld metal	[16]
$CS_{TWI} = 230C + 190S + 75P + 45Nb - 12.3Si - 5.4Mn - 1$	Not specified	[17]

iron-based alloys, as predicted by the RPF in Table 3.1. The increase in cracking results from a severe depression of the solidus temperature (wider solidification temperature range) and the tendency for these elements to promote grain boundary wetting by liquid films.

3.2.3 The BTR and Solidification Cracking Temperature Range

Considerable experimental work has been conducted to determine susceptibility to weld solidification cracking and develop testing methods that relate cracking susceptibility to composition. Matsuda and coworkers in Japan [8, 19] developed a technique using the transverse Varestraint (Transvarestraint) test that allowed direct measurement of the BTR. This involved the use of *in situ* observation with a

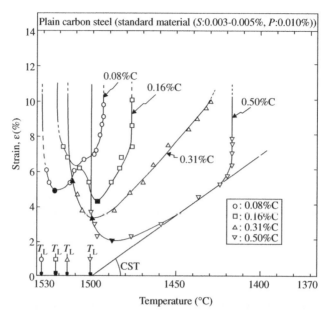

FIGURE 3.10 BTR curves for carbon steel using the slow-bending Transvarestraint technique (From Ref. [18]).

high-speed video camera under slow-bending conditions during testing. This allowed for the observation of crack initiation and propagation in real time. Using temperature data gathered by plunging thermocouples into the molten weld metal, they were able to construct the BTR curves for carbon steels shown in Figure 3.10, where carbon was varied in the range from 0.08 to 0.50 wt%. Note the significant increase in BTR with the highest carbon contents. This is due to the tendency for carbon to promote austenite (fcc) solidification at these high carbon levels and the more pronounced effect of P and S segregation that occurs during austenite solidification. At lower carbon contents, the steel weld metal solidified as delta ferrite (bcc) over a much narrower temperature range. In addition to a wider BTR, the higher carbon content steels also exhibit a lower minimum strain (ε_{min}). Matsuda *et al.* used the concept of critical strain tangent (CST) to describe this behavior. As the CST decreases (lower angle), susceptibility to solidification cracking increases.

Similar curves were developed for a number of austenitic stainless steels, as shown in Figure 3.11. Note that these alloys solidify as either ferrite (F) or austenite (A) and that austenite solidification again leads to higher values of BTR and correspondingly higher susceptibility to solidification cracking. The Transvarestraint technique can also induce solid-state, ductility-dip cracking, and curves that show the ductility-dip temperature range (DTR) are also shown in Figure 3.11 for the alloys that solidify as fcc austenite. This form of cracking will be described in more detail in Chapter 4.

While the slow-bending Transvarestraint test coupled with *in situ* observation provided very good data on the BTR of many alloys, the technique itself requires very specialized (and somewhat expensive) equipment. Data analysis is also somewhat

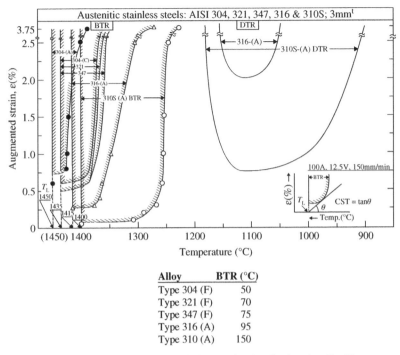

FIGURE 3.11 BTR curves for austenitic stainless steels using the slow-bending Transvarestraint technique (From Ref. [18]. © AWS).

Alloy	BTR (°C)
Type 304 (F)	50
Type 321 (F)	70
Type 347 (F)	75
Type 316 (A)	95
Type 310 (A)	150

cumbersome, requiring some level of expertise and experience with the technique. For these reasons, the JWRI technique has not been widely adopted by other laboratories.

In the 1990s, Lin and Lippold sought to develop a more "user-friendly" approach for measuring BTR [20]. They adopted the same Transvarestraint technique of Matsuda but used the crack length data from the as-tested samples to determine the solidification cracking temperature range (SCTR). A schematic of the SCTR approach for quantifying weld solidification cracking is shown in Figure 3.12. The SCTR represents a subset of the BTR in that it does not consider the liquidus temperature as the starting point for solidification cracking. To determine SCTR, Transvarestraint tests are conducted over a range of strains, and the maximum crack distance (MCD) is measured on the sample surface. The MCD represents the distance from the S–L interface to the end of the crack along a perpendicular orientation to the S–L interface. Details of this technique are provided in Chapter 9.

Unlike the slow-bending Transvarestraint technique, the samples are bent very rapidly in order to promote solidification cracking within the solid–liquid region at the trailing edge of the weld pool. Solidification cracks will only initiate and propagate where liquid films are present. If the temperature gradient during solidification is known or can be measured, the length of the longest crack is indicative of the temperature range over which solidification cracking occurs, hence the term

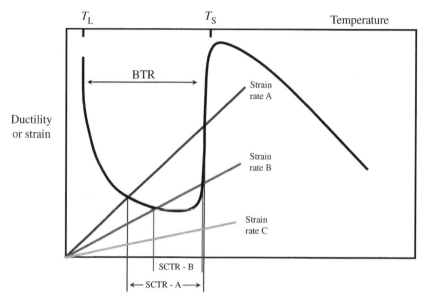

FIGURE 3.12   Schematic representation of the Transvarestraint technique for determining the solidification cracking temperature range (SCTR) (From Ref. [20]. © Springer).

SCTR. As described in Chapter 9, it is important that the MCD be measured at the saturated strain level to insure that the crack has propagated its maximum distance within the susceptible region.

Using this technique, SCTR values have been developed for a number of materials, in particular stainless steels and Ni-base alloys. SCTR data compiled from other sources is provided in Table 3.3 [21, 22].

From Table 3.3, it is again clear that stainless steels solidifying with ferrite (bcc) as the primary phase are much more resistant to solidification cracking than stainless steels and Ni-base alloys that solidify as austenite (fcc). The detailed reasons for this are described in companion texts [21, 22].

As described earlier in Section 3.2.2, it is now possible to use computational methods to calculate the solidification temperature range under nonequilibrium (Scheil–Gulliver) conditions. Table 3.4 compares the calculated solidification range with the measured range using the single sensor differential thermal analysis (SS DTA) technique [23, 24] and the SCTR determined with the OSU Varestraint technique. Note that for three of the Ni-base alloys (617, 625, and Haynes 230), the calculated and measured values are in good agreement and the measured SCTR is equivalent or less than this value. Only the data for Hastelloy X is not in reasonable agreement. More data of this type must be generated in order to develop computational techniques that are more effective in predicting weld solidification cracking susceptibility.

Other evidence showing a relationship between solidification temperature range and cracking susceptibility is provided by DuPont *et al.* [25] for Nb-bearing

TABLE 3.3   SCTR data for stainless steels and Ni-base alloys[a]

Alloy	Solidification mode[b]	Ferrite number[c]	SCTR (°C)
Stainless steels			
Duplex, 2205	F	85	26
Duplex, 2507	F	75	45
Austenitic, 304L	FA	6	31
Austenitic, 316L	FA	4	49
Superaustenitic, AL6XN	A	0	115
Austenitic, 310	A	0	139
Austenitic PH, A-286	A	0	418
Ni-base alloys			
Hastelloy C-22	A	NA	50
Alloy 617	A	NA	85
Haynes 230W	A	NA	95
Hastelloy W	A	NA	145
Hastelloy X	A	NA	190
Alloy 625	A	NA	205

[a]From Refs. [21, 22].
[b]F, ferrite (bcc); FA, ferrite–austenite (primary bcc with fcc at end of solidification); A, austenite (fcc).
[c]Ferrite number measured using magnetic technique or determined with WRC-1992 diagram.

TABLE 3.4   Comparison of nonequilibrium solidification temperature range determined by Scheil–Gulliver and single sensor differential thermal analysis (SS DTA) with the SCTR measured by the Varestraint test[a]

Weld metal	Solidification range (°C)		Varestraint SCTR (°C)
	Scheil–Gulliver	SS DTA	
Alloy 617	160	93	85
Alloy 625	243	97 (306)[b]	205
Hastelloy W	325	162	145
Hastelloy X	160	108	190
Haynes 230W	125	139	95

[a]From Ref. [22].
[b]Value in parentheses includes end of eutectic solidification.

Ni-base alloys. Figure 3.13 illustrates the relationship between solidification cracking susceptibility and solidification temperature range for a number of experimental and commercial alloys. Solidification cracking susceptibility in Figure 3.13 is determined by the maximum crack length measured during the Varestraint test. Not surprisingly, cracking susceptibility increases with solidification range in these alloys.

It should be noted that solidification cracking data collected using the Varestraint test does not represent the contribution of "backfilling" or crack healing in eutectic systems, since the augmented strains applied can overwhelm any backfilling effect. Backfilling of cracks is an important consideration in controlling solidification cracking in some alloy systems and will be discussed later in this chapter.

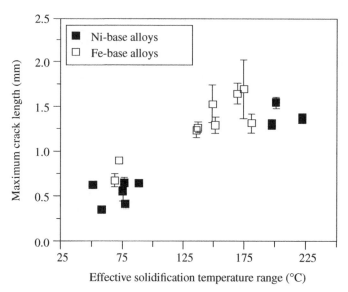

FIGURE 3.13 Relationship between solidification temperature range and cracking susceptibility in Ni-base alloys (From Ref. [25]. © Wiley).

3.2.4 Factors that Influence Weld Solidification Cracking

The factors that influence susceptibility to weld solidification cracking can be divided
into those that affect the solidification conditions and those that influence the restraint
applied during solidification. As noted in the previous section, composition has a
dominant effect on weld solidification behavior and the resultant susceptibility and is
the factor most often manipulated to reduce or eliminate cracking. Restraint is
inherent during solidification due to the negative volume change that occurs during
the liquid-to-solid phase transformation. The thermal contraction that occurs as the
surrounding HAZ cools also adds to this restraint. The level of restraint can be influenced to some degree by controlling the weld geometry and the strength level of the
base metal. Other factors such as weld heat input and weld bead size and shape also
affect the restraint.

3.2.4.1 Composition Control The effect of an individual alloying or impurity
element is usually influenced synergistically by the presence of other elements in the
alloy. As discussed previously, predicting cracking susceptibility based on binary
phase relationships is somewhat risky. As a result, it is usually necessary to consider
all the elements present when attempting to predict weld solidification cracking susceptibility. For many alloy systems, empirical relationships have been developed to
predict cracking susceptibility based on chemical composition. Several relationships
for carbon steels were listed in Table 3.2. The simplest of these was developed by
Bailey and Jones of TWI for C–Mn steels [17]:

$$CS_{TWI}(wt\%) = 230C + 190S + 75P + 45Nb - 12.3Si - 5.4Mn - 1$$

where if $CS_{TWI} < 20$, the material is resistant to cracking. This formula demonstrates the deleterious effect of carbon (over 0.08 wt%) and sulfur and the beneficial effect of silicon and manganese in these steels. Manganese is known to form a stable MnS compound prior to solidification, and, thus, the addition on Mn reduces the "effective" level of S in the weld pool [26, 27]. The effect of silicon is not so straightforward since it is added as a deoxidizer and also improves the flow and wetting characteristics of the weld pool. The RPF of Borland (Table 3.1) indicates a low value for Si in support of the effect predicted in the CS_{TWI} relationship.

The diagram in Figure 3.14 was developed by Karjalainen *et al.* [28] from published weldability data for structural steels. It clearly shows the effect of solidification behavior on cracking susceptibility in these steels. As the Ni-equivalent (Ni_{eq}) value increases, steels with various Cr-equivalent (Cr_{eq}) values transition from resistant to susceptible. This transition reflects the change in solidification behavior from primary ferrite (bcc) at low Ni_{eq} to primary austenite (fcc) at higher levels. This diagram can serve as a guideline when assessing potential susceptibility.

Note that simple C–Mn steels with less than 0.10 wt% C are generally resistant to solidification cracking, as supported by practical experience with these steels and their consumables. For example, a simple steel with 0.10C and 0.5Mn has a $Ni_{eq} = 2.35$ and is in the resistant region of the diagram. Also note that this diagram was developed from weldability test data where restraint levels are generally high (augmented stress or strain tests). Thus, it will tend to be conservative for predicting solidification cracking, that is, higher values of Ni_{eq} can probably be tolerated in low-restraint situations.

Figure 3.15 shows the relationship between the Mn/S ratio and carbon content [29]. Since Mn is effective at tying up S, low Mn/S ratios promote cracking, particularly at higher C contents. The net effect is that higher carbon levels promote solidification as austenite and, in the presence of higher sulfur levels (lower Mn/S), cracking susceptibility increases.

For the austenitic stainless steels (300 series), control of the weld metal ferrite content can insure resistance to weld solidification cracking. In general, maintaining the ferrite number (FN) above FN 3 is usually sufficient to prevent cracking—even in highly restrained welds. This is a direct result of the control of the primary solidification mode. When ferrite is the primary phase of solidification, susceptibility to solidification cracking is quite low. This is explained in part by the narrower SCTR associated with primary ferrite solidification, as shown in Table 3.3.

The WRC-1992 diagram, Figure 3.16, can be used to determine potential susceptibility to weld solidification cracking. The diagram delineates regions where solidification as austenite (A, AF) and ferrite (FA, F) occurs. If the composition of the weld metal can be maintained in the FA or F region, susceptibility to solidification cracking is low. For this reason, austenitic stainless steels whose composition lies in the FA range and duplex stainless steels that generally solidify in the F mode are extremely resistant to weld solidification cracking, even in the presence of high impurity (P + S) levels. For austenitic stainless steels that solidify as austenite, susceptibility to solidification cracking is much higher, particularly if impurity levels are high.

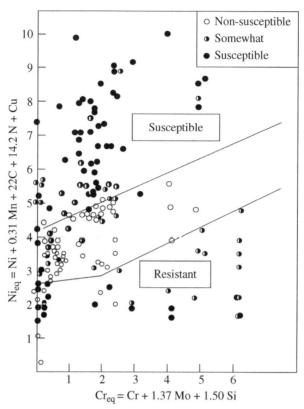

FIGURE 3.14 Cracking susceptibility of steels based on composition (From Ref. [28]. © ASM).

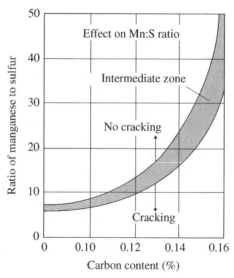

FIGURE 3.15 Effect of Mn/S ratio and carbon content on solidification cracking of steels (From Ref. [29]. © Elsevier).

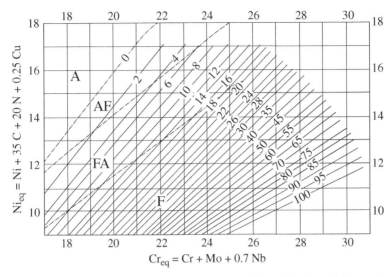

FIGURE 3.16   The WRC-1992 diagram for austenitic and duplex stainless steels.

The Suutala diagram, Figure 3.17, is also useful for predicting the susceptibility of austenitic stainless steels to solidification cracking [30]. This diagram was developed by evaluating a wide range of published austenitic stainless steel weld metal cracking studies. The equivalents used are those developed by Hammar and Svensson [31]. As the Cr_{eq}/Ni_{eq} increases above a critical level, the resistance to cracking shows an almost step change increase, irrespective of the impurity level. This increase results from the shift in solidification behavior from primary austenite to primary ferrite at a ratio of approximately 1.48.

At extremely low $S+P$ contents, cracking resistance is high across the range of compositions. Achieving these low levels of impurities is not generally possible using conventional melting practices. Argon–oxygen decarburization (AOD) melting practice can effectively reduce sulfur content, but has little or no effect on phosphorus levels. As a result, solidification cracking in austenitic stainless steels is best eliminated by controlling solidification behavior.

Composition effects on the solidification cracking susceptibility of austenitic stainless steels are covered in detail in a companion text [21].

In material systems that exhibit a eutectic reaction, solidification cracking generally shows an increase to a peak in susceptibility and then a decrease, as shown schematically in Figure 3.18. This behavior is related to both the solidification temperature range and the fraction eutectic generated in the system at the end of solidification, as described by the shrinkage-brittleness (Fig. 3.1) and Borland's generalized theory (Fig. 3.3). In *Region 1*, both the temperature range and fraction eutectic are increasing. Susceptibility is initially low because insufficient liquid is present to coat the SGBs and the solidification temperature range is small. In *Region 2*, eutectic films coat the boundaries and the solidification temperature range is at its maximum. In *Region 3*, sufficient liquid of eutectic composition is generated to promote crack healing. This

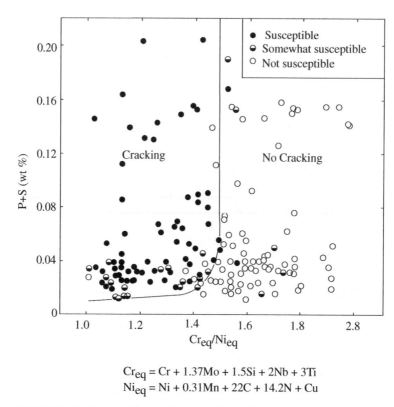

$$Cr_{eq} = Cr + 1.37Mo + 1.5Si + 2Nb + 3Ti$$
$$Ni_{eq} = Ni + 0.31Mn + 22C + 14.2N + Cu$$

FIGURE 3.17   Suutala diagram for austenitic stainless steels (From Ref. [30]).

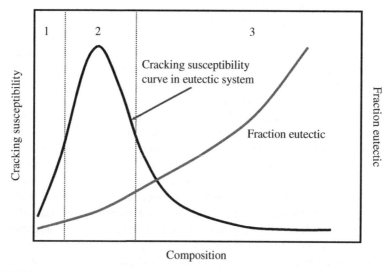

FIGURE 3.18   Solidification cracking susceptibility curve in an alloy system containing a eutectic reaction.

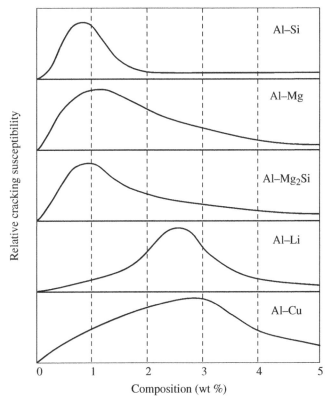

FIGURE 3.19 Solidification cracking susceptibility of aluminum binary alloy systems (From Refs. [32–36]).

is known as "eutectic healing" and can be used to reduce cracking susceptibility, even when the solidification temperature range is large. This approach is particularly important in the aluminum alloy systems.

Considerable solidification cracking data was developed in the 1940s and 1950s in conjunction with the development of the shrinkage-brittleness theory, as summarized in Figure 3.19 [32–36]. All of these aluminum alloy systems show essentially the same behavior illustrated in Figure 3.18, with an initial increase in cracking followed by a decrease as the fraction eutectic liquid present at the end of solidification increases. This effectively demonstrates the importance of crack backfilling in eutectic systems as a potential method for reducing cracking susceptibility.

Using the relationship for determining eutectic fraction described in Section 2.3.4.2 and Equation 2.4, it is possible to calculate (for a binary system) the fraction eutectic that is present at the peak cracking susceptibility for each alloy system in Figure 3.19. As shown in Table 3.5, the fraction of eutectic liquid that would be present at the end of solidification is in the range from 2.4 to 5.5%. This suggests a critical amount of liquid along the SGB to promote cracking, presumably by allowing the formation of a continuous liquid film. At lower eutectic fractions, the liquid at the grain boundary

TABLE 3.5   Fraction eutectic present in aluminum binary alloy systems at the maximum of cracking shown in Figure 3.19

System	k-Value	C_{Eut} (wt%)	Cracking maximum (wt%)	Fraction eutectic (%)
Al–Cu	0.17	33.0	3	5.5
Al–Mg	0.42	35.5	4	2.4
Al–Si	0.13	12.6	0.75	3.9

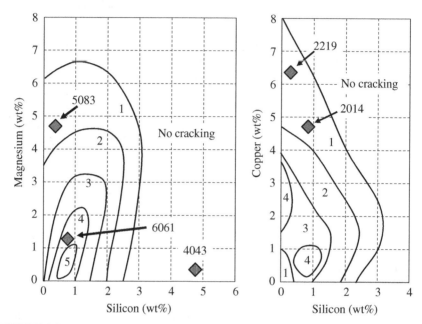

FIGURE 3.20   Solidification crack contour plots for Al–Mg–Si and Al–Cu–Si systems. The location of specific alloys is indicated (From Ref. [35]).

will not be continuous, and at higher fractions, crack healing is possible. Calculations of fraction eutectic necessary for crack healing based on these plots suggest that levels on the order of 10% or higher are needed.

In addition to the plots shown in Figure 3.19, contour cracking maps were also developed in the 1940s and 1950s, as shown in Figures 3.20 [35] and 3.21 [33]. The numbers represent maximum crack length in the type of weldability test where cracking developed by shrinkage (self-restraint tests). The locations of a few commercial 2000-, 5000-, and 6000-series alloys are superimposed on these diagrams to indicate the inherently poor weldability of some of the aluminum base metals. In particular, aluminum Alloy 6061, one of the most popular of the aluminum structural alloys, is effectively unweldable by autogenous processes or when using matching filler metals. Alloy 6061 is usually welded using an Alloy 4043 filler metal that contains high levels of silicon to promote eutectic healing of solidification

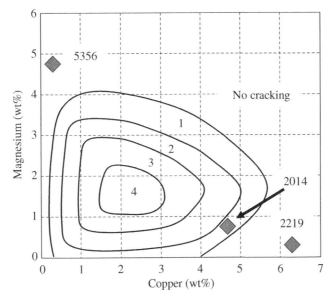

FIGURE 3.21 Solidification crack contour plot for the Al–Mg–Cu system. The location of specific Al–Mg and Al–Cu alloys is indicated (From Ref. [33]).

cracks. Aluminum Alloy 2219 was designed specifically for weldability and is used in a number of aerospace and military applications. The high Cu content of this alloy promotes eutectic healing at the end of solidification.

Other empirical relationships based on chemical composition can be found in the literature for a wide variety of general classes of engineering materials. The reader is referred to a companion text for information on the cracking susceptibility of Ni-base alloys [22]. Again, these relationships are meant to be a guideline only. Other factors, particularly the level of weld restraint, may significantly influence their predictive capabilities.

3.2.4.2 Grain Boundary Liquid Films

Liquid grain boundary *films* are required for weld solidification cracking. As the amount of liquid that solidifies isothermally during the final stage of weld freezing increases, the susceptibility to cracking generally decreases. This phenomenon is the basis for the design of many brazing and weld filler metal compositions. For example, aluminum alloy Filler Metals 4043 and 4047 are near-eutectic Al–Si alloys frequently used for joining a variety of aluminum alloys. In this and other alloys that form a large amount of eutectic liquid along the SGBs, strain does not build up along a localized boundary film and cracking can usually be avoided.

This phenomenon also explains, in part, the decrease in cracking susceptibility beyond the maximum solid solubility, C_{Smax}, in Borland's RPF argument and may also explain why the peak in cracking susceptibility may be shifted to compositions below C_{Smax} in many alloy systems. The *critical level* of grain boundary liquid appears to be that amount that barely, but completely, wets the boundary at a temperature sufficiently below the bulk solidification temperature. For aluminum alloys, the critical level appears to be in the range from 2.4 to 5.5 vol%, as described in Table 3.5. For other alloy systems, the amount will depend on the wetting characteristics of the liquid, as shown in Figure 2.45.

In practice, it is usually impossible to know the values of the relative interfacial energies in complex alloy systems, and thus, the boundary wetting characteristics are very difficult to quantify. Controlling boundary wetting as a means of reducing weld solidification cracking is usually not practical, although reducing the level of sulfur or increasing oxygen is thought to reduce the wetting of grain boundaries in many ferrous alloys. Impurities such as phosphorus and boron are also thought to improve the wetting characteristics, as is silicon that is added to many weld filler metals to improve fluidity.

The presence of a thin liquid film along a solidification boundary severely reduces the ductility of the boundary, but the boundary will still have some level of strength. This is illustrated in Figure 3.22, which shows both strength and ductility in the solidification temperature range. Within the nonequilibrium BTR, there is an increase in strength as the boundary films become very thin. This is analogous to a thin film of water between two panes of glass. The film can support some load perpendicular to the film thickness, but not in shear. This behavior again reinforces the importance of restraint in the system and how it is applied on liquated weld metal boundaries.

3.2.4.3 Effect of Restraint   Weld restraint arises from both intrinsic (internal) and extrinsic (external) contributions. The primary intrinsic restraint results from the

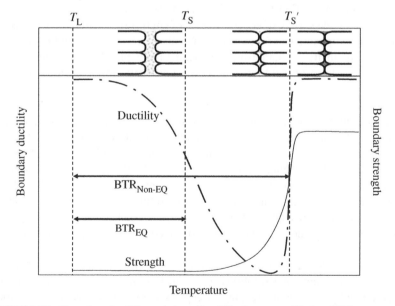

FIGURE 3.22   Boundary ductility and strength within the nonequilibrium BTR.

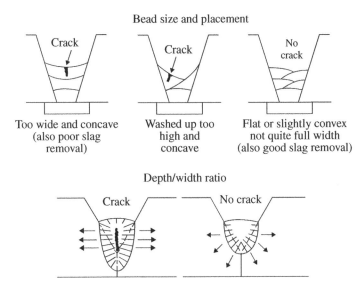

FIGURE 3.23   Effect of weld bead geometry on solidification cracking susceptibility (From Ref. [37]).

volumetric reduction (shrinkage) associated with solidification—this is in the range from 3 to 8% for metallic systems. The weld bead shape and the properties of the surrounding HAZ and base material also contribute to intrinsic restraint.

The macroscopic puddle shape can have a significant influence on susceptibility to solidification cracking. Elongated, "teardrop"-shaped weld pools are generally susceptible to centerline cracking due to both enhanced segregation of alloying and impurity elements (macroscopic solidification) and geometric considerations (see Fig. 2.16). The centerline is normal to the maximum restraint arising both intrinsically and extrinsically. This situation can usually be eliminated by reducing the weld travel speed, pulsing the welding current, or weaving the bead from side to side.

An important factor that influences the susceptibility of a material to weld solidification cracking may simply be the profile of the weld bead. The profile often affects the level of restraint that builds up in the weld during the final stages of solidification. The schematic in Figure 3.23 demonstrates this effect in a simple V-groove weld joint [37]. Concave welds that bridge the entire joint produce more restraint than smaller convex beads. Maintaining a flat or slightly convex weld profile may avoid solidification cracking in some situations. Large concave welds are common in actual fabrication since their use optimizes process efficiency by increasing deposition rates and reducing welding times. When severe cracking problems occur, an intermediate approach is often required, which reduces the level of restraint in the solidifying weld. This is usually best accomplished by lowering the heat input and using more passes to fill the joint. Adjusting the depth/width ratio can also reduce restraint within the weld during solidification. With some processes, such as electron beam welding, this may be difficult due to the inherent nature of the process.

In alloy systems that are strengthened by precipitation, such as Ni-base superalloys, weld solidification cracking susceptibility can generally be reduced by reducing the base metal strength via solution annealing. A postweld aging treatment is then used to restore base metal and weldment strength.

The level of weld restraint that is present during weld solidification and the manner in which that strain is accommodated along boundaries that are wet with a liquid film are difficult to quantify and, more importantly, control. In general, weld restraint can only be controlled qualitatively by proper joint design, fixturing, welding process/ parameters, and material selection and preparation. As noted previously, simple changes in weld geometry may be helpful in preventing cracking in materials that have marginal susceptibility to weld solidification cracking. Finite element models can be used to predict the level of weld stress based on geometry and material properties. Such an approach may be useful in reducing weld stresses to a level where cracking is avoided.

3.2.5   Identifying Weld Solidification Cracking

Identifying weld solidification cracking in most materials is relatively straightforward. When observed metallographically, cracking almost always occurs along SGBs. In alloys where cracking is severe, cracks may also occur along subgrain boundaries (cells and dendrites), but this is unusual. In structural steels and titanium alloys, where diffusion and elevated temperature transformations have removed any remnants of the solidification structure, this metallographic evidence may not be present and examination of the fracture surface (fractography) may be required. This is usually conducted using an SEM at magnifications of 500× or higher.

The fracture surface of solidification cracks is usually quite distinct. They tend to show a cellular or dendritic morphology, often described as an "eggcrate" appearance. This results from the separation of opposing cellular or dendritic fronts along a thin liquid film at the end of solidification. As described earlier, it is unusual to observe any evidence of solid–solid bridging on the fracture surface since this would manifest itself by a ductile rupture fracture mode. The dendritic to flat fracture described by Matsuda (Fig. 3.5) is also not usually observed in actual weld failures resulting from solidification cracking. In general, a change from dendritic to flat fracture is probably indicative of a transition to solid-state, ductility-dip cracking, particularly in austenitic (fcc) weld metal such as austenitic stainless steels and Ni-base alloys. Again, an SEM is usually required to resolve these features since they are on the scale of microns.

A few examples of metallographic and fractographic characterization of weld solidification cracking are provided. These are meant to be representative cases. Additional fractographic analysis of solidification cracking is provided in Chapter 8.

Materials that solidify as austenite (fcc) are generally more susceptible to solidification cracking than those that solidify as ferrite (bcc). Among these materials are many of the austenitic stainless steels, some of the structural steels, and all of the Ni-base alloys. The example shown in Figure 3.24 is a Ni-base superalloy, Alloy 718 (Ni–20Cr–15Fe–5Nb–3Mo), that is precipitation strengthened by additions of Nb. Nb segregates strongly in most Ni-base alloys resulting in a eutectic reaction with a final solidification structure consisting of NbC, Nb-rich Laves phase, or a mixture of

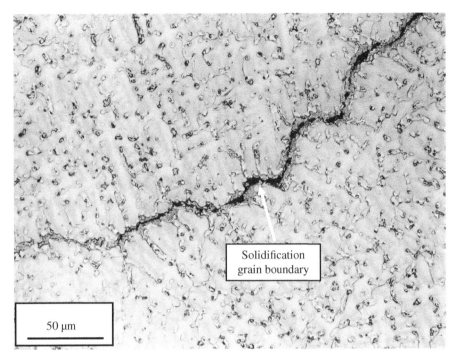

FIGURE 3.24 Solidification crack in Ni-base Alloy 718.

these phases. The net effect of Nb segregation is the expansion of the solidification temperature range and an increase in susceptibility to cracking. The remnants of a continuous liquid film along an SGB are apparent in Fig. 3.24 with NbC and possibly Laves phase present.

Aluminum alloys are also susceptible to solidification cracking. All the commercial alloys solidify as an fcc phase (α-phase). Most aluminum alloys have a large solidification temperature range and a high thermal expansion (and contraction) coefficient. The combination of these promotes cracking in aluminum alloys.

Most of the aluminum alloys are based on eutectic systems and form some fraction eutectic at the end of solidification. In small amounts, this eutectic can wet the grain boundaries and promote cracking. In some alloys, the eutectic content can be controlled such that crack healing occurs. This is the case for Alloy 2219, nominally an Al–6.5Cu alloy. A solidification crack in Alloy 2219 is shown in Figure 3.25. The arrows in the micrograph show a region along an SGB where eutectic liquid has healed, or backfilled, a portion of the cracked boundary. This alloy generates approximately 10% eutectic liquid at the end of solidification that promotes crack healing.

The metallography/fractography pairs shown in Figure 3.26 show the effect of eutectic liquid and backfilling on the fracture surface appearance. These are both Ni–30Cr filler metals with different levels of Nb that influence the amount of eutectic liquid that forms at the end of solidification. The arrows in the optical metallography photos indicate the SGBs at the tip of the solidification cracks.

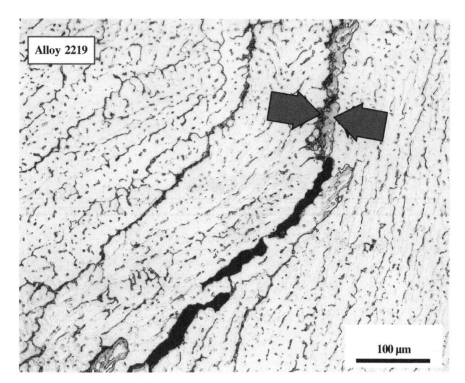

FIGURE 3.25 Solidification crack in aluminum Alloy 2219. Arrows indicate region of crack healing.

In the left pair in Figure 3.26, the weld metal has a relatively low fraction (<5%) of eutectic liquid present at the end of solidification. The resultant fracture surface shows a very distinct dendritic nature. The pair on the right is from weld metal that forms about 10% eutectic liquid at the end of solidification. This high fraction of liquid results in a coating of the fracture surface by the liquid due to backfilling, somewhat obscuring the dendritic nature of the fracture surface.

The final example, Figure 3.27, shows a metallographic/fractographic pair of weld solidification cracking in Ni-base Filler Metal 625 with a nominal composition of Ni–22Cr–9Mo–3.5Nb. As noted previously, Nb partitions strongly during solidification of Ni-base alloys and promotes the formation of a γ/NbC eutectic and/or Laves phase at the end of solidification. There is enough eutectic present at the end of solidification to promote some backfilling along the SGBs, as indicated by the arrows. The fracture morphology reflects the presence of this eutectic liquid on the surface— note that the dendritic features are not so well defined. This is because the original dendritic surface has been masked by a coating of eutectic liquid.

As noted in Table 3.3, the SCTR of Filler Metal 625 is on the order of 205°C. This would indicate a relatively high susceptibility to solidification cracking. However, Filler Metal 625 is often selected for its resistance to solidification cracking. This appears to be in contradiction to its high SCTR value. The discrepancy has to do with

(a)   (b)

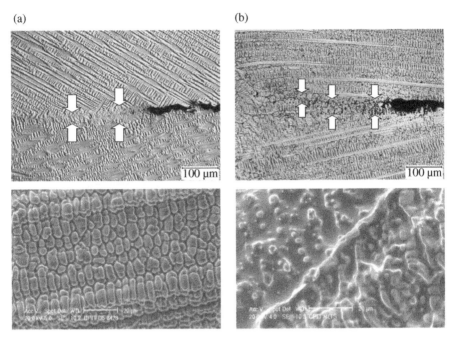

FIGURE 3.26   Solidification cracks in Ni–30Cr weld metals: (a) low fraction eutectic liquid and (b) high fraction eutectic liquid.

the manner in which SCTR is determined—at relatively high strains using the Transvarestraint test. These high strains force the solidification crack to propagate across the entire solidification range. At lower strains, the presence of eutectic liquid allows crack healing to occur, and Filler Metal 625 is resistant to cracking. In actual practice, this filler metal is found to be susceptible to solidification cracking when weld restraint levels are high or under adverse bead shape conditions (high D/W ratios and/or concave bead shapes).

Weld solidification cracking is also possible in materials that solidify as a bcc crystal structure. This includes most carbon and low-alloy steels; ferritic, martensitic, and duplex stainless steels; and titanium alloys. In these alloy systems, rapid diffusion in the solid at elevated temperatures eliminates the partitioning that distinguishes the cellular and dendritic microstructures of austenite (fcc) solidification. An example of a solidification crack in a martensitic stainless steel (Alloy HT9) is shown in Figure 3.28. Note that there is evidence of liquid films along the SGBs in this weld metal. Electron probe microanalysis shows strong segregation of alloying elements Cr and Mo to these boundaries, along with the impurity element phosphorus.

Solidification cracking can also occur in duplex stainless steels under high-restraint conditions. An example of a solidification crack in Alloy 255 is shown in Figure 3.29. Cracking occurs at the end of solidification when the structure is fully ferritic, but some austenite forms at the SGB during cooling and completely envelops the solidification crack.

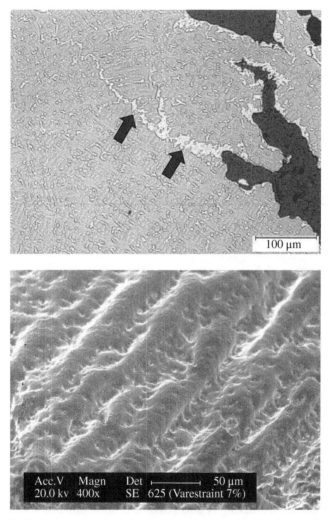

FIGURE 3.27 Metallographic/fractographic pair showing solidification cracking and crack healing (arrows) in Ni-base Alloy 625 weld metal.

Solidification cracking in titanium alloys is similar to the duplex stainless steels. Cracking occurs along beta (bcc) SGBs at the end of solidification, but these boundaries then transform to alpha (hcp) during cooling. A solidification crack along a prior beta grain boundary in the α–β titanium alloy Ti–6Al–6V–2Sn is shown in Figure 3.30 [38].

3.2.6 Preventing Weld Solidification Cracking

Based on the previous section, methods for eliminating or reducing susceptibility to solidification cracking are somewhat apparent. Control of solidification behavior, when possible, is always desirable. With steels, solidification as ferrite (bcc crystal

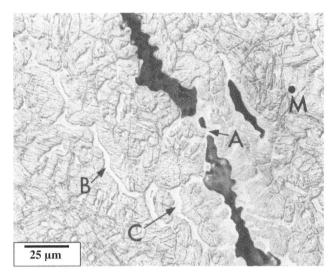

Element	A	B	C	M
Cr	18.2	17.0	17.2	12.0
Mo	2.9	2.5	2.4	0.9
Ni	0.5	0.5	0.4	0.4
Mn	0.8	0.7	0.8	0.6
V	0.7	0.6	0.6	0.3
W	0.7	0.7	0.7	0.4
S	0.04	0.08	0.05	0.04
P	0.24	0.27	0.19	0.02

FIGURE 3.28 Solidification cracking in a martensitic stainless steel (Alloy HT9). Composition (in wt%) along solidification grain boundaries (A, B, C) relative to the bulk weld metal (M).

structure) will result in an improvement in cracking resistance relative to solidification as austenite (fcc). When that option is not available, such as with some austenitic stainless steels, reduction of impurity elements (S, P, and possibly B) is helpful. Other materials that solidify as a bcc crystal structure, such as titanium, also have good resistance to solidification cracking.

In weld metal that undergoes a eutectic reaction at the end of solidification, this reaction will result in the formation of some fraction of eutectic liquid. Control of the volume fraction and distribution of liquid films at the end of solidification is advised, although the methods to do this are not always straightforward based on the composition of the base and filler metals. As shown in Fig. 3.18, cracking resistance in eutectic systems can be controlled by either minimizing the eutectic liquid films or increasing fraction eutectic to the level that crack backfilling can occur. In general, reducing susceptibility to weld solidification cracking is best accomplished by minimizing the solidification temperature range, but this is not always an option.

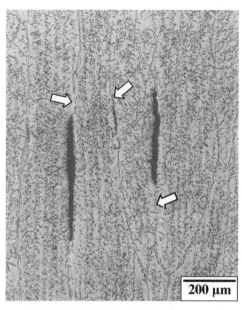

FIGURE 3.29 Solidification cracking in Alloy 255 duplex stainless steel. Arrows indicate cracking along prior ferrite solidification grain boundary.

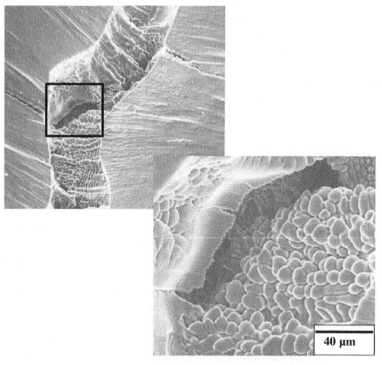

FIGURE 3.30 Solidification cracking in a titanium alloy (From Ref. [38]).

Restraint can be controlled to some extent through joint design and weld procedure. In general, lower heat input and smaller welds with convex bead shape reduces susceptibility. The intrinsic restraint arises primarily from solidification shrinkage, but control of factors such as base metal strength can prove beneficial. Many precipitation-strengthened alloys, such as the Ni-base superalloys, are welded in the solution-annealed condition to reduce intrinsic restraint. Locating welds in low-restraint areas or using fixturing to apply compressive forces (or reduce tensile restraint) may be helpful in certain applications.

In most cases, the selection of base and filler metals that are resistant to cracking by nature of their composition is the best solution to a solidification cracking problem. When this is not possible, the use of bead shape control or restraint-control techniques is warranted. For many materials, including the Ni-base alloys, the use of very small weld beads is usually successful in preventing cracking. This results in a large economic penalty since overall welding time will increase dramatically—particularly for heavy section joints.

3.3 LIQUATION CRACKING

Liquation cracking along grain boundaries in the PMZ region of the HAZ is termed *HAZ liquation cracking*. HAZ liquation cracks are typically small, often extending only 2–3 grain diameters into the HAZ beyond the fusion boundary. These cracks are, by definition, intergranular and normally form during cooling from peak temperatures above the effective solidus temperature of the material. HAZ liquation cracking is the direct result of localized melting along HAZ/PMZ grain boundaries. As discussed in Chapter 2, there are two general mechanisms that have been proposed to explain the onset of liquation along these boundaries, namely, the *penetration mechanism* and the *segregation mechanism*.

Liquation cracking that forms in reheated weld metal is termed weld metal *liquation cracking*. Weld metal liquation cracks form due to the remelting of either SGBs or MGBs in the weld metal very close to the fusion boundary between the two weld passes. This will be discussed in a subsequent section.

3.3.1 HAZ Liquation Cracking

HAZ liquation cracking via the *penetration mechanism* occurs by a sequence of three events. First, local liquation must occur in the microstructure. This will occur above some critical temperature dependent on the nature of the base material and the HAZ thermal history. Second, thermally driven grain growth must occur that allows interaction between the liquation event and the grain boundary (assuming the liquation event does not occur at the grain boundary). In the HAZ adjacent to the fusion boundary, some grain growth almost always occurs. Third, the liquid must be able to wet, or penetrate, the grain boundary in order to make it susceptible to cracking. A schematic illustration of the penetration mechanism was shown in Figure 2.38.

The penetration mechanism for grain boundary liquation requires both a liquation phenomenon and grain boundary motion. As shown in Figure 2.38, liquation (in this case, constitutional liquation of a particle) and boundary motion must be simultaneous. When the boundary encounters the liquated particle, it will be "pinned" and further motion inhibited. Depending on the wetting characteristics of the liquid/ boundary combination, the liquid may then penetrate along the boundary. This gives rise to grain boundary liquid films. The degree of penetration depends on the temperature field, the wetting characteristics, and the amount of liquid. In some alloys, continuous grain boundary liquid films may form in the PMZ.

Localized melting in the PMZ may occur via a number of different mechanisms. At temperatures very close to the solidus, all polycrystalline materials will undergo incipient melting at the grain boundary. This type of melting occurs at grain boundaries, since these are high-energy sites, that is, the boundary energy combined with the thermal energy contributes to allow melting. In most materials, incipient melting occurs only a few degrees below the bulk solidus temperature, although the presence of impurity elements may expand this range. In general, incipient melting is not very important in the formation of the PMZ.

If the base material is in the as-cast condition, local variations in the melting temperature will exist due to segregation, or partitioning, during solidification. As a result, the interdendritic regions of a casting will melt at a lower temperature than the bulk, leading to the formation of a PMZ. Materials that are thermomechanically processed following casting often exhibit some local compositional banding. This may be particularly true in plate materials that experience a series of directional rolling operations. Because of these local fluctuations in composition, the melting temperature will also fluctuate. In the PMZ, it is possible that some regions may melt at a lower temperature than an adjacent region due to this phenomenon. Full solution annealing of the base metal is often required to eliminate this banding.

A number of materials undergo a phenomenon described as "constitutional liquation." As described in Section 2.5.1, this form of liquation is manifested by localized melting at a particle/matrix interface. It is somewhat unique to welding since it occurs under transient thermal conditions, normally requiring relatively rapid heating rates.

An example of HAZ liquation cracking arising from constitutional liquation is shown in Figure 3.31. The material is Alloy 718 (55Ni–18Cr–18.5Fe–5Nb–3Mo–1Ti–0.4Al) and HAZ grain boundary melting and some cracking is apparent just adjacent to the fusion boundary (arrows). This liquation occurs via a penetration mechanism, following the constitutional liquation of NbC. Because of this extensive PMZ, the HAZ liquation cracking susceptibility of Alloy 718 is quite high.

Another example of HAZ liquation cracking resulting from constitutional liquation is shown in Figure 3.32. The material in this case is Alloy 800 (Fe–20Cr–35Ni–0.5Al–0.5Ti, 0.1C), a high temperature, heat-resisting stainless steel. The extensive grain boundary liquid films in this alloy result from the constitutional liquation of TiC.

HAZ liquation cracking is also observed in materials that contain no particles that undergo constitutional liquation, and, thus, an alternative mechanism must exist, which results in HAZ grain boundary liquation. Basically, the mechanism must

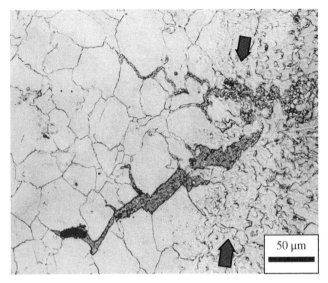

FIGURE 3.31   HAZ liquation cracking in Alloy 718. Arrows indicate the location of the fusion boundary.

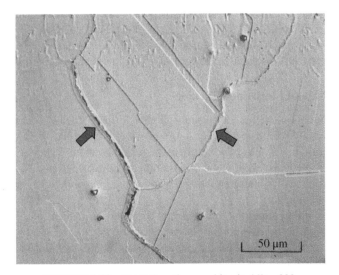

FIGURE 3.32   HAZ liquation cracking in Alloy 800.

explain the enrichment of the grain boundaries with solute and/or impurity elements in sufficient concentration to significantly reduce the melting temperature of the boundaries relative to the surrounding matrix. These boundaries will then melt when the local HAZ temperature exceeds some critical liquation temperature. This is known as the *segregation mechanism*, as described in Section 2.5.2, and shown schematically in Figure 2.42.

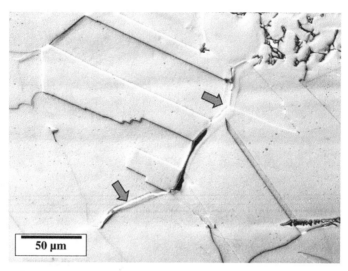

FIGURE 3.33   HAZ liquation cracking in Type 304L stainless steel (From Ref. [39]).

An example of HAZ liquation cracking in a Type 304L stainless steel is shown in Figure 3.33 [39]. Grain boundary melting is evident by the widening of the boundary, as indicated by the arrows. Unlike the grain boundary liquid films associated with constitutional liquation (Figs. 3.31 and 3.32), the segregation mechanism typically results in melting over a much shorter distance from the fusion boundary. HAZ liquation cracks formed by this mechanism are generally very small and often go undetected.

3.3.2   Weld Metal Liquation Cracking

Weld metal liquation cracking represents another form of liquation cracking that is specific to reheated weld metal (i.e., multipass welds). Similar to HAZ liquation cracks, they form in very close proximity to the fusion boundary. They are most often observed in single-phase weld metal of austenitic stainless steels and Ni-base alloys. These cracks are always intergranular and may occur along *both* SGBs and MGBs in the reheated weld metal.

Localized melting along the SGBs results from the enrichment of alloy/impurity elements arising from partitioning during macroscopic weld solidification (see Fig. 2.26). The HAZ in multipass welds in effect contains a "built-in" segregated microstructure and does not require transport of impurity/solute elements to locally reduce the melting temperature. This is very similar to localized interdendritic melting that occurs when welding on castings. Impurity segregation along MGBs is thought to promote liquation and potential cracking at these boundaries. Boundary "sweeping" or pipeline diffusion may be responsible for this segregation, as discussed in Section 2.5.2.

An example of a weld metal liquation crack in fully austenitic stainless steel weld metal is shown in Figure 3.34. This crack is present along an MGB. Note that the crack is quite short in cross section, on the order of 0.1 mm in length, which is typical for these types of cracks.

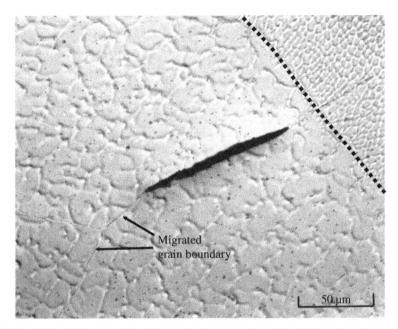

FIGURE 3.34 Weld metal liquation crack along a migrated grain boundary in austenitic stainless steel weld metal.

It should be noted that there is some confusion in the literature regarding terminology for weld metal liquation cracks. Much of the older literature refers to these as "microfissures," indicating that they form at elevated temperature and are usually quite short in length when observed in metallographic cross section. It is likely that many of what were identified as "microfissures" were actually solid-state, ductility-dip cracks. Since both weld metal liquation cracks and ductility-dip cracks are prevalent in single-phase austenitic (fcc) weld metal, it can be difficult to distinguish between the two unless careful metallography and/or fractography is conducted. There is more discussion on ductility-dip cracking in Chapter 4.

3.3.3 Variables that Influence Susceptibility to Liquation Cracking

There are a number of variables that can influence susceptibility to liquation cracking, including composition, grain size, base metal heat treatment condition, weld heat input, and filler metal selection.

3.3.3.1 Composition Composition has the strongest influence on liquation cracking susceptibility, but often, changes in composition are not a possible solution. Many base metals are inherently susceptible to HAZ liquation cracking due to elements that are intentionally added. This is particularly true of materials such as Ni-base Alloy 718 and stainless steel Type 347 that have intentional additions of Nb and Ni-base alloy Waspaloy and stainless steel A-286 that contain Ti. In the case of

both Nb and Ti, these elements form MC-type carbides that are susceptible to constitutional liquation in the HAZ. For materials such as these, methods other than composition control must be used to prevent cracking.

Liquation cracking due to a segregation mechanism is often due to impurity segregation to grain boundaries. The most important impurity elements in steel and
Ni-base alloys are P and S and in some cases B. Often, liquation cracking can be
reduced or eliminated by simply reducing the level of impurity elements in the base
metal or weld metal. It should be noted that B is added intentionally to some base
metals to improve creep resistance, so restriction of B (particularly in Ni-base superalloys) may have consequences with regard to service properties.

3.3.3.2 *Grain Size*   In general, as the grain size increases, HAZ liquation cracking
susceptibility increases. The increased grain boundary area resulting from a small
grain size reduces the likelihood of complete boundary wetting and produces a
stronger microstructure that is better able to support the weld restraint that accumulates during cooling. Even if liquid films are present, increased boundary area will
prevent strain localization and potential cracking.

As grain size increases, the grain boundary area decreases. If liquid is present
along the grain boundaries, the coarse-grained material will be more susceptible to
cracking since the strain will be localized on fewer boundaries and the unit strain per
boundary will be high. As shown schematically in Figure 3.35, HAZ liquation
cracking due to segregation effects is generally influenced by both grain size and
impurity content. Similar behavior is observed in systems that undergo constitutional

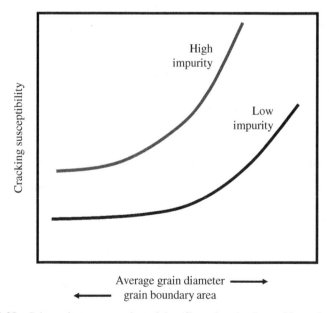

FIGURE 3.35   Schematic representation of the effect of grain size and impurity content on
HAZ cracking susceptibility.

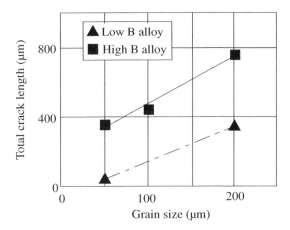

FIGURE 3.36 Effect of grain size and boron content on HAZ liquation cracking in Alloy 718 (From Ref. [40]. © Maney).

liquation, as demonstrated by the work of Guo *et al.* [40] with Ni-base Alloy 718, as shown in Figure 3.36.

Grain size also has an important influence on weld metal liquation cracking susceptibility. The SGB grain size is controlled to a large extent by the original solidification conditions, but MGBs may migrate during multipass welding. In general, grain size in the weld metal of single-phase alloys is controlled by weld heat input and is usually much larger than the HAZ/PMZ grain size.

3.3.3.3 Base Metal Heat Treatment The strength level of the base metal can affect the restraint in the HAZ during solidification. This is particularly the case when welding precipitation-strengthened materials, such as Ni-base superalloys. Solution heat treatment dissolves "constituent" particles that may subsequently liquate and also minimizes intrinsic restraint by reducing the strength level of parent material. (Welding in the solution-annealed condition is also used to reduce weld solidification cracking.) One possible consequence of such a heat treatment may be an increase in grain size that will increase susceptibility. Some balance between these competing effects must be considered.

3.3.3.4 Weld Heat Input and Filler Metal Selection Since HAZ and weld metal liquation cracking occur over a fixed temperature range, weld heat input will influence the temperature gradient in the HAZ and subsequently control the extent to which liquation occurs. As shown in Figure 2.37, a steep temperature gradient in the HAZ (low weld heat input) can reduce the extent of grain boundary melting in the HAZ.

Another factor that has a more limited effect on HAZ liquation cracking susceptibility is the strength of the filler metal. Since much of the restraint in the HAZ is associated with solidification shrinkage in the weld metal, lower-strength filler metals can generate less shrinkage stress that is translated to the HAZ or underlying weld metal.

This approach is usually not an option in most situations since matching filler metal composition or other factors associated with filler metal selection tend to dominate.

3.3.4 Identifying HAZ and Weld Metal Liquation Cracks

HAZ liquation cracks are always located along HAZ/PMZ grain boundaries, usually in close proximity to the fusion boundary or continuous across the fusion boundary into the fusion zone. They normally form on cooling from the peak HAZ temperature since this is when strain accumulates in the system due to thermal contraction.

(a)

(b)

FIGURE 3.37 Fractography of HAZ liquation cracking: (a) duplex stainless steel and (b) HSLA steel (Courtesy of Jeremy Caron).

Metallographically, these cracks may be evident on the surface and in a weld cross section. They are usually very small and may escape detection by nondestructive testing techniques. The fracture surface appearance is intergranular with occasional "decoration" of grain faces by partially melted particles. The degree of decoration is dependent on the nature of grain boundary liquation. In alloys where the PMZ liquid films are very thin, the intergranular fracture surface will exhibit very clean (minimal liquid) grain faces.

Two examples of HAZ liquation crack surfaces from SEM fractography are shown in Figure 3.37. In the first case, a duplex stainless steel, the intergranular fracture is very "clean," showing little evidence of liquid films. These steels are very resistant to HAZ liquation cracking, since grain boundary melting occurs over a narrow temperature range. The fractograph in Figure 3.37a is actually from a hot ductility test conducted at the nil ductility temperature (see Chapter 9).

In the second case, evidence of resolidified liquid films is observed on the fracture surface (Fig. 3.37b). This fracture surface is from an HSLA steel that has a much wider liquation temperature range than the duplex stainless steel.

By definition, weld metal liquation cracking always occurs in reheated weld metal, such as in a multipass weld or multilayer cladding operations. Cracking can be observed along *both* SGBs and MGBs and may be observed along both types of boundary in the same weldment. Cracking may occur either on heating or on cooling when heated above the liquation temperature, depending on how strain accumulates in the system upon reheating. Weld metal liquation cracks may exhibit either an intergranular appearance if cracking occurs along MGBs or dendritic appearance if the cracking occurs along weld metal SGBs. In some situations, both fracture modes may be observed on the same fracture surface indicating a transition from SGB to MGB fracture.

3.3.5 Preventing Liquation Cracking

Many of the same guidelines for preventing weld solidification cracking are applicable for HAZ and weld metal liquation cracking. Control of impurity levels in base and filler metals is always recommended. Base metals that contain constituent particles such as TiC and NbC in an austenite matrix can lead to constitutional liquation. Often, Ti and Nb are intentional alloy additions and, therefore, difficult to avoid in some alloys.

Fully austenitic microstructures are the most susceptible. The presence of some ferrite in the austenitic stainless steel HAZs and welds will minimize both HAZ and weld metal liquation due to its influence on wetting characteristics. Minimizing grain size is beneficial, although this generally requires the use of low heat input or HED processes and may not always be practical. There is evidence that grain boundary orientation (or misorientation) may influence liquation. Low-energy, "special" grain boundaries may be more resistant to the formation of liquid films. Research is ongoing to evaluate how to take advantage of this effect by manipulating orientation.

Reducing restraint is an obvious, but difficult to control, factor influencing cracking. Many high-strength, precipitation-hardened alloys (Ni-base superalloys) are commonly welded in the solution heat-treated condition. This reduces the intrinsic restraint in the HAZ. Weld metal liquation cracking can often be eliminated by using multiple stringer beads rather than large welds. This has both a metallurgical and restraint benefit.

REFERENCES

[1] Bochvar AA, Sviderskaya ZA. Izv Akad Nauk 1947;3:349–355.

[2] Pumphrey WI, Jennings PH. J Inst Metals 1948;75:235–256.

[3] Medovar BI. On the nature of weld hot cracking. Avtom Svarka 1954;7 (4):12–28.

[4] Toropov VA. On the mechanism of hot cracking of welds. Metallo Obra Metallov 1957;6:54–58.

[5] Pellini WS. Foundry 1952;80:124–133.

[6] Apblett WR, Pellini WS. Factors which influence weld hot cracking. Weld J 1954;33 (2):83s–90s.

[7] Borland JC. Generalized theory of super-solidus cracking in welds and castings. Brit Weld J 1960;7 (8):508–512.

[8] Matsuda F, Nakagawa H, Sorada K. Dynamic observation of solidification and solidification cracking during welding with optical microscope. Trans JWRI 1982;11 (2):67–77.

[9] Matsuda F. Solidification crack susceptibility of weld metal. In: David SA, Vitek JM, editors. *Recent Trends in Welding Science and Technology*. Metals Park, OH: ASM International; 1990. p 127–136.

[10] Prokhorov NN, Prokhorov NN. Fundamentals of the theory for technological strength of metals while crystallising during welding. Trans JWRI 1971;2 (2):205–213.

[11] Borland JC. Suggested explanation of hot cracking in mild and low alloy steel welds. Brit Weld J 1961;8:526–540.

[12] ThermoCalc® is a registered trademark of ThermoCalc Software, Inc.

[13] Tung D, Lippold JC. Weld solidification behavior of Ni-base superalloys for use in advanced supercritical coal-fired power plants. In: Huron ES *et al.*, editors. *Superalloys 2012*. Hoboken, NJ: John Wiley & Sons; 2012. p 563–567.

[14] Cottrell CLM. Factors affecting the fracture of high-strength steels. JISI 1965;203:598.

[15] Morgen-Warren EJ. Metal Tech 1974;1 (6):271–278.

[16] Garland JG, Bailey N. Weld Res Int. 1975;5 (3):1–33.

[17] Bailey N, Jones SB. Solidification cracking of ferritic steel during submerged arc welding. Weld J 1978;57 (8):217s–231s.

[18] Matsuda F. Hot crack susceptibility of weld metal. Proceedings of the 1st US–Japan Symposium on Advances in Welding Metallurgy; San Francisco, CA: American Welding Society; 1990. p 19–36.

[19] Senda T, Matsuda F, Takano G, Watanabe K, Kobayashi T, Matsuzaka T. Fundamental investigations on solidification crack susceptibility for weld metals with trans-varestraint test. Trans JWS 1971;2 (2):135–151.

[20] Lippold JC. Recent developments in weldability testing (keynote). In: T. Boellinghaus and H. Herold, editors. *Hot Cracking Phenomena in Welds*. Berlin, New York: Springer; 2005. p 271–290.

[21] Lippold JC, Kotecki DJ. *Welding Metallurgy and Weldability of Stainless Steels*. Hoboken, NJ: John Wiley & Sons, Inc.; 2005.

[22] DuPont JN, Lippold JC, Kiser SD. *Welding Metallurgy and Weldability of Nickel Base Alloys*. Hoboken, NJ: John Wiley & Sons, Inc.; 2009.

[23] Alexandrov BT, Tatman JK, Murray G, Lippold JC. Non-equilibrium phase transformation diagrams in engineering alloys. In: *Trends in Welding Research VIII*. ASM International; 2009. p 467–476. Proceedings of the 8th International Conference; ASM International, Metals Park, OH April 2008, Callaway Gardens, GA.

[24] Lippold JC, Sowards JW, Alexandrov BT, Murray G, Ramirez AJ. Weld solidification cracking in Ni-base alloys. In: *Hot Cracking Phenomena in Welds II*. Berlin: Springer; 2008. p 147–170.

[25] DuPont JN, Lippold JC, Kiser SD. *Welding Metallurgy and Weldability of Nickel Base Alloys*. Hoboken, NJ: John Wiley & Sons Inc.; 2009. p 218.

[26] Nakagawa H, Matsuda F, Senda T. Effect of sulfur on solidification cracking in weld metal of steel (Report 2). Trans JWRI 1974;5 (2):39–44.

[27] Nakagawa H, Matsuda F, Senda T, Matsuzaka T, Watanabe K. Effect of sulfur on solidification cracking in weld metal of steel (Report 2). Trans JWRI 1974;5 (2):45–60.

[28] Karjalainen LP, Kujanpää VP, Suutala N. Hot cracking in iron base alloys: effect of solidification mode. In: S.A. David, editor. *Advances in Welding Science and Technology*. Metals Park, OH: ASM International; 1986. p 145–149.

[29] Lancaster JF. *The Metallurgy of Welding*. 6th ed. Cambridge, UK: Abington Publishing; 1999.

[30] Kujanpää V, Suutala N, Takalo T, Moisio T. Correlation between solidification cracking and microstructure in austenitic and austenitic-ferritic stainless steel welds. Weld Res Int 1979;9 (2):55.

[31] Hammar O, Svensson U. *Solidification and Casting of Metals*. London: The Metal Society; 1979. p 401–410.

[32] Singer ARE, Jennings PH. Hot shortness of aluminum-silicon alloys of commercial purity. J Inst Metals 1947;73:197–212.

[33] Pumphrey WI, Lyons JV. Cracking during casting and welding of more common binary aluminum alloys. J Inst Metals 1948;74:439.

[34] Dowd JD. Weld cracking in aluminum alloys. Weld J 1952;31 (10):448s–456s.

[35] Jennings PH, Singer ARE, Pumphrey WI. Hot shortness of some high purity alloys in the systems Al-Cu-Si and Al-Mg-Si. J Inst Metals 1948;74:227.

[36] Dudas JH, Collins FR. Preventing weld cracks in high-strength aluminum alloys. Weld J 1966;45 (6):241s–249s.

[37] Blodgett OW. Why do welds crack? How weld cracks can be prevented. Weld Innov 1985;2 (3):4.

[38] Baeslack WA III. Observations of solidification cracking in Ti alloy weldments. Metallography 1980;13:277–281.

[39] Lippold JC, Varol I, Baeslack III WA. An investigation of heat-affected zone liquation cracking in austenitic and duplex stainless steels. Weld J 1992;71 (1):1s–14s.

[40] Guo H, Chaturvedi M, Richards NL. Effect of nature of grain boundaries on intergranular liquation during weld thermal cycling of a Ni-base alloy. Sci Technol Weld Join 1999;3:257–259.

4

SOLID-STATE CRACKING

4.1 INTRODUCTION

There are a number of cracking mechanisms that are associated with the "true" HAZ or reheated weld metal. These include ductility-dip cracking (DDC), reheat and postweld heat treatment (PWHT) cracking, strain-age cracking (SAC) (Ni-base alloys), lamellar (or delamination) cracking, and Cu-contamination cracking (CCC). These cracking mechanisms span the entire range of materials systems, including steels, stainless steels, Ni-base alloys, Cu-base alloys, and aluminum alloys.

4.2 DUCTILITY-DIP CRACKING

This form of solid-state cracking occurs in a number of engineering materials. The presence of a ductility dip in austenitic (fcc) alloys was reported as early as 1912 by Bengough [1]. In 1961, Rhines and Wray [2] reported that a ductility dip occurs in copper alloys, nickel alloys, austenitic stainless steels, titanium, and aluminum. DDC was originally identified as a problem associated with the hot working of materials—such as the hot deformation of cast ingots to form wrought bar or plate products [2, 3]. DDC was identified as a problem during welding in the 1970s and 1980s. Early papers documented DDC in austenitic stainless steels, although this form of cracking was deemed "anomalous" [4]. In fact, DDC in welds has been around for quite some time and was often misinterpreted as some form of hot cracking, often

Welding Metallurgy and Weldability, First Edition. John C. Lippold.
© 2015 John Wiley & Sons, Inc. Published 2015 by John Wiley & Sons, Inc.

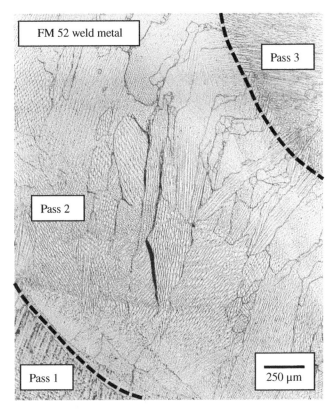

FIGURE 4.1 Example of ductility-dip cracking in multipass Ni–30Cr weld metal (ERNiCrFe-7).

termed "microfissuring" [5, 6]. Since the late 1990s, severe DDC has been encountered in Ni-base weld metals used in the nuclear power industry [7–9]. This has resulted in considerable study of the DDC phenomenon and a desire to develop DDC-resistant filler metals.

DDC may occur in both the weld metal and HAZ but is generally considered a weld metal problem. In weld metals, DDC is always intergranular (IG) and generally occurs in reheated weld metal during multipass welding, as shown in Figure 4.1. The term ductility dip refers to the sharp reduction in ductility that susceptible materials exhibit in a temperature range between the alloy solidus and approximately half the solidus temperature. The mechanism responsible for DDC is still the subject of considerable debate. Oddly enough, this type of cracking is frequently observed in austenitic (fcc) materials, which have very low levels of impurities (sulfur, phosphorus, and boron). Because grain boundary liquation does not play a role in DDC, the importance of impurity segregation to grain boundaries is reduced.

A schematic representation of the "ductility signature" of a material that exhibits a ductility dip is shown in Figure 4.2. In many materials, ductility will be very high at temperatures just below the solidus temperature (T_s) and then gradually decrease

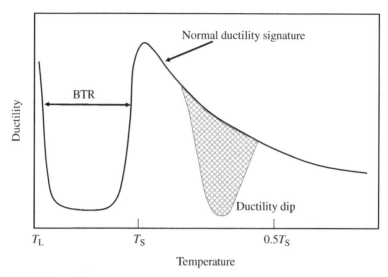

FIGURE 4.2 Ductility–temperature relationship for a material exhibiting a solid-state ductility dip.

upon cooling to room temperature. In materials susceptible to DDC, a sharp drop in ductility occurs between approximately $0.9T_S$ and $0.6T_S$ (for stainless steels and Ni-base alloys, the range is between 800 and 1150°C). Both the width and depth of this "dip" define the DDC susceptibility of the material.

As with other cracking mechanisms, there must be a component of restraint that leads to crack initiation and propagation. Many weld filler metals that are designed to be resistant to weld solidification cracking can be susceptible to DDC. As will be described later in this section, this is particularly the case with many Ni-base filler metals. This situation can be depicted using the ductility temperature signature shown in Figure 4.3. Note that the brittle temperature range (BTR) for this material is quite narrow and the minimum strain for solidification cracking (ε_{min}) is quite high, indicating good resistance to solidification cracking. By superimposing two restraint situations in Figure 4.3, it can be seen how the combination of strain accumulation and inherent ductility affects cracking susceptibility.

Figure 4.3 demonstrates how a combination of metallurgical and restraint control can be used to prevent cracking based on exhaustion of the material ductility at elevated temperature. As will be discussed later, the metallurgical factors used to reduce or eliminate weld solidification cracking may in fact increase susceptibility to DDC.

Since failure due to DDC usually occurs at elevated temperatures where rapid grain growth occurs, most theories suggest that cracking is associated with large grain size and the behavior of the grain boundaries (such as grain boundary precipitation and boundary sliding). The absence of impurities and second-phase particles removes any obstacles for grain boundary motion, and the resultant grain size in susceptible weld metals can be extremely large. As a result, the structure contains very little grain boundary area, and it is hypothesized that strain is concentrated at these boundaries [10].

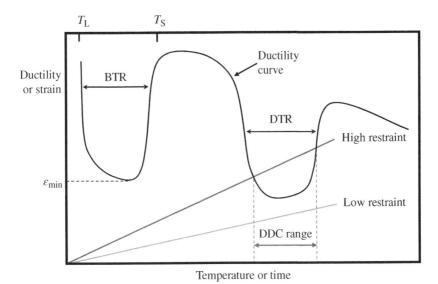

FIGURE 4.3 Temperature–ductility curve for a material resistant to solidification cracking but susceptible to ductility-dip cracking under high-restraint conditions.

In support of this hypothesis, fracture due to the ductility dip is IG but sometimes with ductile features (usually termed ductile IG).

From a welding standpoint, the two material systems that show the highest susceptibility to DDC are the austenitic stainless steels and Ni-base alloys. DDC has also been reported in copper and titanium alloys, but this is unusual in practice.

The micrograph in Figure 4.4 shows a solidification grain boundary (SGB) and migrated grain boundary (MGB) pair in Ni-base alloy weld metal. The MGB represents the crystallographic component of the SGB that has migrated during cooling from the solidification temperature or during reheating, as described in Section 2.3.3.3. Note that the MGB is very straight. In the weld metal, DDC <u>always</u> occurs along the MGB.

The photomicrograph in Figure 4.5 is a higher-magnification region from Figure 4.1 that shows the nature of cracking in an actual multipass, thick-section weldment. As indicated previously, DDC in weld metal occurs preferentially along MGBs. Note that the grain boundary exhibits a local fine grain microstructure that results from strain localization at the boundary. This indicates that the local boundary strain levels are sufficiently high to promote recrystallization at elevated temperature.

4.2.1 Proposed Mechanisms

A number of theories have been proposed to describe the mechanism of DDC, as summarized in Table 4.1. The first theory of DDC was proposed by Rhines and Wray [2] based on studies of the hot working behavior of base metals. They believed that the loss of intermediate-temperature ductility was caused by grain boundary shearing,

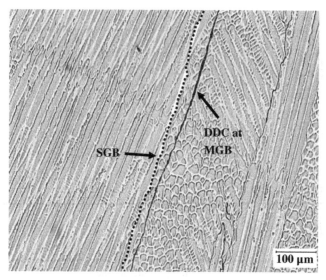

FIGURE 4.4   Ductility-dip cracking along a migrated grain boundary (MGB) in a Ni-base alloy weld metal. The dotted line represents the solidification grain boundary (SGB).

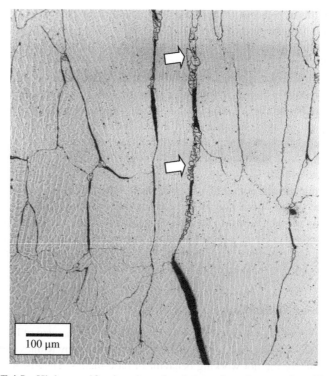

FIGURE 4.5   High-magnification view of grain boundary character from Figure 4.1.

TABLE 4.1 Ductility-dip cracking theories

Name	Description	Year
Rhines and Wray [2]	Grain boundary shearing up to recrystallization temperature	1961
Yamaguchi *et al.* [11]	Sulfur segregation and embrittlement	1979
Zhang *et al.* [12, 13]	Combination of effects up to recrystallization temperature	1985
Ramirez and Lippold [14, 15]	Grain boundary sliding, microvoid formation, boundary tortuosity	2004
Nishimoto *et al.* [16–18]	Impurity segregation	2006
Noecker II and DuPont [19, 20]	Grain boundary sliding, carbide distribution and morphology	2007
Young *et al.* [21]	Precipitation-induced cracking	2008

similar to creep rupture failures that occur at lower temperatures over longer exposure times. According to their theory, at temperatures below the recrystallization temperature, grain boundary voids have time to join by grain boundary shearing and cause a fracture. Above the recrystallization temperature where recrystallization occurs because of mechanical working, the formation of new grain boundaries makes the joining of the voids difficult, presumably due to the creation of more grain boundary area. This mechanism is in general agreement with that recently proposed by Ramirez and Lippold [14] and Noecker and Dupont [19, 20] for weld metals, as discussed in more detail later in this section.

Yamaguchi *et al.* [11] proposed that higher sulfur levels increased the tendency for a ductility dip in Ni-base superalloys at temperatures between 950 and 1150°C (1740 and 2100°F), where sulfur segregated to and embrittled the grain boundaries and which cracked under an applied stress. Similar sulfur segregation mechanisms have also been proposed by Matsuda [23] and more recently by Nishimoto *et al.* for Ni-base Alloy 690 weld metals [16–18]. Recent work by Collins *et al.* [24] with Ni-base FM 82 (ERNiCr-3) also showed that sulfur additions increase susceptibility to DDC. While sulfur and other impurities may contribute to DDC, the work by Ramirez and Lippold concluded that differences in susceptibility could not be simply explained by impurity (S and P) content since many materials with very low impurity content were also susceptible [22].

Zhang *et al.* [12] reported that the combined effects of grain boundary precipitation, grain boundary sliding, grain boundary migration, and grain boundary serration affect the DDC performance of the low-expansion alloy Invar (Fe–36Ni). They also suggested that recrystallization and decreased flow stress were factors in the recovery of ductility at elevated temperatures [13]. Young *et al.* [21] have suggested that DDC in high-Cr, Ni-base filler metals results from grain boundary carbide precipitation and associated "precipitation-induced cracking." While this mechanism may have some relevance to high-Cr, Ni-base alloys that form grain boundary carbides, it cannot explain DDC that occurs in materials where grain boundary precipitation does not occur.

Work at Ohio State University (OSU) on a variety of austenitic stainless steels and Ni-base alloys has shown that precipitation behavior and grain boundary "tortuosity"

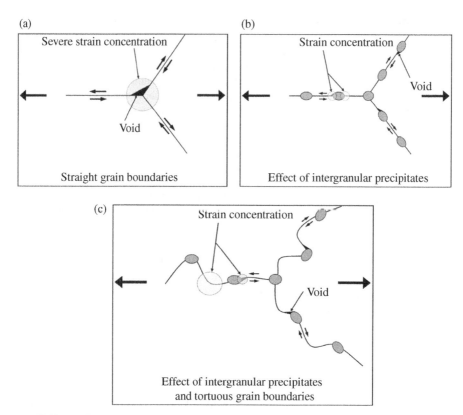

FIGURE 4.6   Schematic of grain boundary character as a function of precipitation.

have a major influence on DDC susceptibility [14, 15, 22, 24–26, 27–30]. It was concluded that DDC is essentially an elevated temperature, grain boundary sliding phenomenon. They also found that impurity (P, S, O, and H) segregation, grain boundary precipitation, and boundary tortuosity affect DDC susceptibility and that controlling the nature of grain boundaries is the key to avoiding DDC in Ni-base weld metals.

The DDC theory proposed by Ramirez and Lippold is shown schematically in Figure 4.6. When there is no grain boundary precipitation, the grains are able to grow and the grain boundaries tend to be very straight. When strain is applied to this microstructure within the ductility-dip temperature range (DTR), grain boundary sliding occurs and stresses are high at the grain boundary triple points, leading to void formation at these locations. Eventually, cracks propagate along the grain boundaries.

When precipitates form along the grain boundaries in the solid state, stresses will now concentrate at the triple points and also at the precipitate–grain boundary interface. This results in void formation along the grain boundary where precipitates are present. In the situation where precipitates form at the end of solidification, the grain boundaries are pinned by these precipitates resulting in a "tortuous" grain boundary. Because of this tortuosity, the boundary resists sliding due to the mechanical locking effect of the tortuous boundary.

(a)

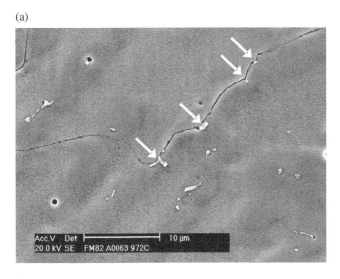

(b)

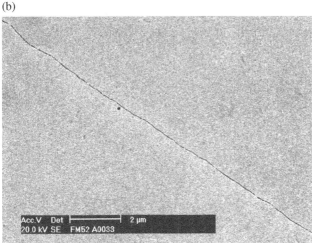

FIGURE 4.7 Migrated grain boundaries in Ni-base weld metal. (a) Filler Metal 82 and (b) Filler Metal 52. Arrows in (a) show carbides pinning the boundary.

It has been shown that weld metals that exhibit "clean" and straight grain boundaries (as shown in the upper left of Figure 4.6) are most susceptible to DDC, while those that contain tortuous grain boundaries are the most resistant. Note that the grain boundaries in Figure 4.5 are extremely straight. Obviously, designing the microstructure in order to control grain boundary character is a key factor in preventing DDC.

The weld metal grain boundary characteristics of two Ni-base FM are shown in Figure 4.7. FM 82 (ERNiCr-3) is a nominal Ni–20Cr–3Mn–2.5Nb composition that is widely used in the power generation industry. Because it contains Nb, it forms NbC at the end of solidification via a eutectic reaction. These NbC precipitates pin the weld

(a)

(b)

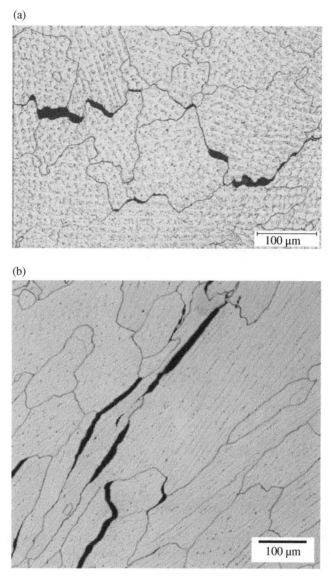

FIGURE 4.8   Ductility-dip cracks in strain-to-fracture samples tested at 950°C and 5% strain. (a) Filler Metal 82 and (b) Filler Metal 52.

metal MGBs (arrows) and produce a tortuous (nonstraight) grain boundary. FM 52 (ERNiCrFe-7) is a nominal Ni–30Cr–9Fe composition with low carbon (~0.02 wt%) that forms no precipitates at the end of solidification. As a result, the MGBs easily move away from the SGBs and become very straight.

The photomicrographs shown in Figure 4.8 are from strain-to-fracture (STF) test samples tested at approximately 5% total strain within the DTR. This test will be described later in this section and in detail in Chapter 9. Note that the FM 82 grain boundaries are somewhat tortuous and the cracks are quite short. In contrast, the FM 52 grain boundaries are very straight, allowing more grain boundary sliding and promoting larger cracks. The strain threshold to produce DDC is much lower for FM 52 than for FM 82.

A technique for evaluating grain boundary orientation is called electron back-scatter diffraction (EBSD). It is also commonly referred to as orientation imaging microscopy (OIM). This technique uses diffraction patterns generated in the scanning electron microscope (SEM) to identify crystal orientation differences in a crystalline material. Thus, the high-angle grain boundaries in the weld metal or HAZ can be easily delineated. The EBSD patterns shown in Figure 4.9 are from two Ni-base weld metals that exhibit a large difference in DDC susceptibility.

Note that the DDC-resistant weld metal in Figure 4.9a exhibits very tortuous MGBs. This boundary type was developed by the formation of precipitates (NbC) at the end of solidification that promote boundary pinning. In contrast, the weld metal in Figure 4.9b forms no precipitates at the end of solidification, and pinning of the MGBs does not occur. This results in large grains with very straight boundaries.

The data generated from EBSD analysis can also be used to estimate the strain distribution in the sample, as shown in Figure 4.10. The difference in the grain boundary appearance in these two figures results from the accumulation of strain at the weld metal MGBs, where strain accumulation results in a widening of the boundary in the EBSD analysis. The weld metal sample in Figure 4.10a is in the as-welded condition, while that in 4.10b has been strained to 5.5% at 950°C (1740°F) in the STF test.

One of the interesting features of austenitic (fcc) materials in the DDC temperature range is that the strain tends to concentrate at the grain boundary, rather than being distributed uniformly throughout the microstructure. This strain localization is what gives rise to grain boundary sliding and preferential void formation and cracking at the grain boundaries.

4.2.2 Summary of Factors That Influence DDC

The factors that influence DDC in stainless steels and Ni-base alloys include the nature of the grain boundary, strain localization, temperature, precipitation behavior, impurity segregation, and restraint. The effect of these factors is discussed in the following paragraphs.

Grain Boundary Character. DDC always occurs along high-angle grain boundaries. In single-phase austenitic (fcc) weld metal, DDC occurs at the MGBs. In the case where there is no mechanism to pin the crystallographic boundary following solidification, these boundaries will straighten and be distinct from the SGB, as shown in Figure 4.4. If the boundary is pinned at the end of solidification, the crystallographic portion of the SGB cannot pull away from the SGB and will not straighten. This gives rise to the "tortuous" grain boundaries that are shown in Figures 4.7a and 4.8a. This is the case in austenitic stainless steel weld metals that contain ferrite and Ni-base weld metals where a precipitation reaction (carbide or Laves-phase formation) occurs at the end of solidification.

(a)

(b)

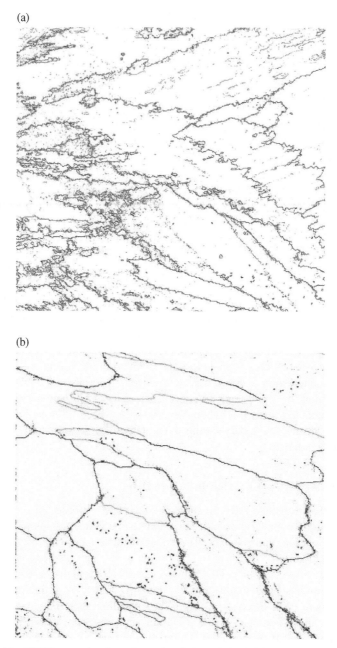

FIGURE 4.9   EBSD maps showing migrated grain boundaries in Ni-base weld metal. (a) Filler Metal 52MSS and (b) Filler Metal 52M.

(a)

(b)

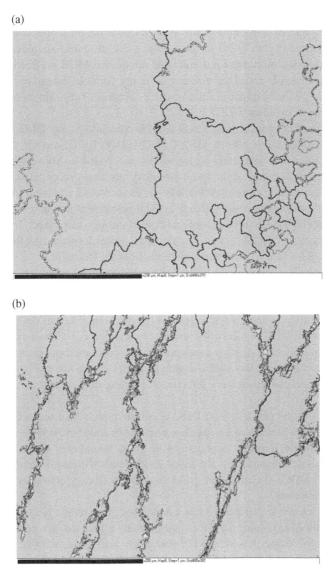

FIGURE 4.10   EBSD strain patterns for a Ni-base weld metal. (a) As-welded and (b) 5.5% strain at 950°C.

Since DDC is a grain boundary sliding mechanism, grain boundary tortuosity provides a mechanical locking effect along the entire length of the boundary (Fig. 4.6c). When the crystallographic boundary pulls away to form a MGB (Figs. 4.6a and 4.4), grain boundary sliding is facilitated, and voids will open at strain concentration points, such as triple points. In Ni-base weld metals, grain boundary tortuosity and sliding can be controlled by precipitation, either by pinning the boundaries or by providing local grain boundary locking or both. The use of alloying additions that promote precipitation at the appropriate temperature is the most effective method for reducing susceptibility to DDC in these alloys. In austenitic stainless steel weld metals that contain a mixture of austenite and ferrite, the MGB is effectively pinned at the ferrite–austenite interface creating a highly tortuous boundary. In systems where pinning of the boundary is not possible, control of restraint becomes much more important in avoiding DDC.

Temperature. For the stainless steels and Ni-base alloys, the DDC temperature range extends from about 800 to 1150°C (1470–2100°F). Below 800°C, the grain boundaries do not slide and strain localization is not possible. Above approximately 1150°C, strain localization at the grain boundary promotes recrystallization. When recrystallization occurs, more grain boundary area is created, and local strain is dissipated. During welding, it is impossible to avoid this temperature range since the weld metal and parts of HAZ will be heated and cooled through this range. There is some debate over whether DDC that occurs in multipass welds forms during the heating or cooling cycle. This topic is discussed under the heading Restraint.

Composition. The composition of the weld metal and base metal has an important influence on susceptibility to DDC. In the weld metal, the composition will influence the solidification behavior and as-welded microstructure. As noted previously, weld metals that solidify as fcc austenite are the most susceptible to DDC. There have been no reports of DDC in steels that solidify as ferrite (bcc). Composition can also influence the partitioning of alloy elements during solidification, potentially leading to the formation of secondary solidification products. As shown in Figures 4.7a and 4.8a, these second phases are very effective at pinning the boundary and preventing migration.

The effect of impurity levels on DDC is controversial. As discussed previously, some investigators have shown a negative effect of S and P in Ni-base weld metals. [11, 16–18]. While impurity segregation to grain boundaries may contribute to DDC, it is not a dominant factor, and many weld metals with low impurity levels are susceptible to DDC. The effect of grain size and boundary character tends to be much more important.

Oxygen is known to diffuse to grain boundaries in austenitic (fcc) materials and may potentially have a harmful effect, although its effect has not been quantified. Limited work by Nissley based on STF weldability tests suggested that oxygen segregation to grain boundaries can be detrimental [31].

Another interesting composition effect is the influence of hydrogen. Experience with Ni-base filler metals of the ERNiCrFe-7 type has shown that the addition of hydrogen to the shielding gas (such as 98Ar–2H$_2$ mixtures) increases susceptibility to DDC [24]. For many Ni-base filler metals, the addition of H$_2$ improves the wetting characteristics of the molten weld pool, thereby increasing productivity. It is not clear how hydrogen actually affects the DDC mechanism, since it is quite mobile in the DDC temperature range.

Restraint. Restraint is an important factor in all weld cracking phenomena but is particularly important with respect to DDC because it must be present at elevated temperature over a relatively narrow temperature range. As shown schematically in Figure 4.3, restraint must result in a critical level of strain that exhausts the available ductility within the DTR. In fact, while the schematic in Figure 4.3 suggests that DDC occurs during cooling from the solidification temperature range, it is unusual for DDC to occur in single-pass welds. Rather, DDC is almost always encountered in multipass welds due to the reheating of previously deposited weld metal. A good example of this is shown in Figure 4.1.

Presumably, residual stress builds up in the weld deposit during cooling to room temperature, and this stress is then relieved during reheating by the subsequent pass. The relaxation of stress leads to strain accumulation at the grain boundaries, and if it exceeds some critical threshold, the grain boundaries will fail. The EBSD strain maps in Figure 4.10 clearly show the localization of strain along MGBs in the weld metal. It has also been demonstrated that DDC can be suppressed in highly restrained, thick-section weldments by minimizing weld bead size, which in turn reduces the level of restraint in the weld metal and the degree of stress relaxation during reheating. Unfortunately, this is often not a viable solution to mitigate DDC since it increases welding time and production costs.

4.2.3 Quantifying Ductility-Dip Cracking

A number of tests have been used to determine susceptibility to DDC. Most prevalent are the hot ductility test, the spot Varestraint test, the double-spot Varestraint test, and the STF test. The hot ductility, spot Varestraint, and STF tests are described in some detail in Chapter 9. The slow-bending Varestraint test used by Matsuda [32] to develop BTR curves for solidification cracking also yielded ductility-dip temperature curves for DDC, as shown in Figure 3.11. The spot Varestraint test is normally used to evaluate HAZ liquation cracking susceptibility in base metals but can also be used to evaluate weld metal liquation and DDC. The double-spot Varestraint test was originally developed by Lippold and Lin at Edison Welding Institute [33] in order to isolate DDC from solidification cracking during standard transverse Varestraint testing. With this test, an initial gas tungsten arc spot weld was applied to a sample and then a second spot weld applied within this initial weld upon performing a standard spot Varestraint test (see Chapter 9). Weld metal DDC could then be generated in the initial spot weld upon the application of sufficient strain. While this was an improvement over the standard Varestraint test, there were still three major complicating issues: (i) weld metal liquation cracking and DDC could not be separated, (ii) it was difficult to determine the temperature range over which cracking occurred, and (iii) some materials only exhibited DDC at the highest achievable strain (~10%) so determining a strain threshold was difficult. Despite these issues, the double-spot Varestraint test did provide some useful DDC susceptibility data for a number of Ni-base filler metals. For example, Kikel and Parker [34] were able to compare the DDC susceptibility of FM 52 and Alloy 690 with that of FM 82 and 625.

The Gleeble™ hot ductility test was used by both Noecker and Dupont [19, 20] and Young et al. [21] to quantify DDC in high-Cr, Ni-base weld metals. Since this is essentially a hot tensile test where the sample is pulled to failure, no information on the critical threshold strain to initiate DDC is obtained. In the case of Young et al. [21], susceptibility to DDC was determined by the drop in on-cooling ductility after heating to an arbitrary peak temperature.

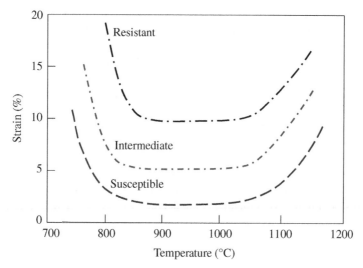

FIGURE 4.11 Schematic of strain-to-fracture behavior for predicting ductility-dip cracking susceptibility of austenitic (fcc) materials.

The STF test was developed by Nissley and Lippold [35] to avoid the shortcomings of the Varestraint and hot ductility tests for determining susceptibility to DDC. The details of this test are provided in Chapter 9. It is a Gleeble-based test that allows weld metals or base metals to be evaluated using microstructure, strain, and temperature as the principal variables. Using this test, strain–temperature envelopes can be developed that show the regime over which DDC occurs for a given material. A schematic representation of strain–temperature DDC behavior for susceptible, intermediate, and resistant materials is shown in Figure 4.11. Actual data for STF testing of stainless steels and Ni-base alloys can be found elsewhere [24, 27–30].

Relative susceptibility to DDC can be determined by the minimum threshold strain for cracking across the entire temperature range and the severity of cracking above the threshold strain. For example, in Figure 4.11, the minimum threshold strain for susceptible materials is on the order of 2%, while that of intermediate materials is approximately 5%. Although this difference in minimum threshold strain may seem small, actual practice has shown that weld metals with threshold strains of approximately 5% are moderately susceptible to DDC, while those that approach or exceed 10% tend to be quite resistant.

The minimum strain for cracking in the STF test tends to be in the range from 900 to 1000°C. A more efficient method for comparing DDC susceptibility among materials has been to simply test over a range of strain at 950°C to determine the threshold strain for cracking and the increase in cracking as a function of strain above the threshold. This yields data as shown schematically in Figure 4.12 for susceptible, intermediate, and resistant materials, where the number of cracks on the sample surface is plotted as a function of applied strain.

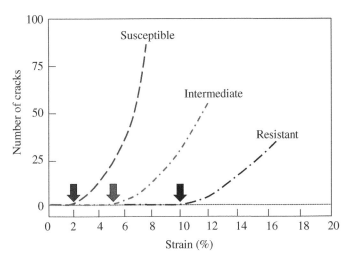

FIGURE 4.12   Schematic of ductility-dip cracking as a function of applied strain at 950°C for austenitic (fcc) materials.

4.2.4   Identifying Ductility-Dip Cracks

Ductility-dip cracks can occur in both the weld metal and HAZ, although they are most common in the weld metal. In the weld metal, they are <u>always</u> along MGBs. In cases where the restraint is very high, there may be some recrystallization observed along these boundaries. A typical DDC in fully austenitic (fcc) weld metal is shown in Figure 4.13. The MGBs are clearly evident and distinct from the SGBs. In cases where the MGB is still in very close proximity to the SGB, it may be difficult to distinguish DDC from solidification cracking. In general, ductility-dip cracks tend to be very straight since they follow the MGB. An example of DDC and solidification cracking that exist in the same sample in a high-Cr, Ni-base weld metal (FM 52 M) is shown in Figure 4.14. This is from a Transvarestraint weldability test sample tested at 5% strain. In this weld metal, there is very little separation between the SGBs and MGBs (little boundary migration occurs), so the different crack types appear to occur at the same microstructural location. The solidification crack tends to have a more wavy appearance since it follows the SGB, while the DDC is very straight because it is restricted to the MGB. Also, there is considerable liquid film evident along the solidification crack path.

Because DDC is most often associated with reheated weld metal, it may also be difficult to differentiate ductility-dip cracks from weld metal liquation cracks that can also form at MGBs. In general, weld metal liquation cracks tend to be very short and are located in close proximity to the interpass boundary, as illustrated in Figure 3.34. In cases where interpretation by metallographic evidence alone may be difficult, it is usually necessary to examine the fracture surface characteristics using an SEM.

The fracture surface, when observed in the SEM, can usually distinguish DDC from hot cracks (solidification and liquation), since the DDC will not show any

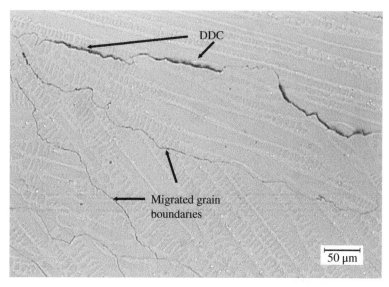

FIGURE 4.13 Typical ductility-dip cracking along migrated grain boundaries in Ni-base weld metal.

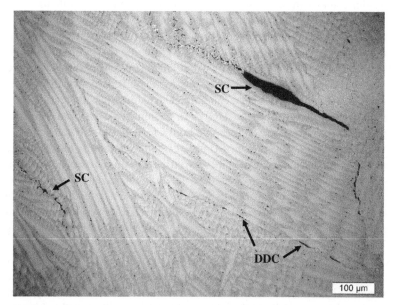

FIGURE 4.14 Cracking in high-Cr, Ni-base weld metal (Filler Metal 52M) tested at 5% strain in the Transvarestraint test. Both solidification cracking (SC) and ductility-dip cracking (DDC) are present in this sample (Courtesy of Adam Hope).

evidence of liquid films. The fracture appearance of DDC varies as a function of temperature and material type. The IG nature of the fracture is always apparent macroscopically. At high magnifications, there may be evidence of microductility (ductile IG), "wavy" features, and occasionally slip lines that are apparent on the fracture surface. In some cases, DDC may actually propagate from solidification cracks in the weld metal since the BTR and DTR in some alloys may almost overlap. Some examples of fracture surfaces associated with DDC in Type 310 stainless steel weld metal are shown in Figure 4.15. Additional examples of DDC fractography are provided in Chapter 8. For more detail on DDC fractography, the reader is referred to the work of Collins *et al.* [25, 26], Ramirez [14], and Matsuda [36].

4.2.5 Preventing DDC

From a welding standpoint, DDC is most frequently found in the fusion zone, so the discussion here focuses primarily on weld metals. The most effective method for avoiding weld metal DDC is to select filler metals that are resistant to this form of cracking. For austenitic stainless steels, these include filler metals such as Types 308 and 309 that contain sufficient ferrite in the deposit to pin MGBs and considerably increase boundary tortuosity. Weld metals of this type will usually exhibit threshold strain for cracking levels above 10% (see Fig. 4.11). As described in Chapter 3, these weld metals are also very resistant to solidification cracking. For austenitic stainless steel weld deposits that are fully austenitic, control of DDC is more difficult, and the general approach used is to reduce the restraint (residual stress) in the deposited weld metal, as discussed in the following.

Ni-base weld metals can also exhibit a range of DDC behavior, depending on composition and solidification behavior. Alloys that form no second phase at the end of solidification to control the character of the MGBs are very susceptible to DDC. This includes FM 52, which exhibits large grains and very straight MGBs, as shown in Figures 4.1 and 4.5. The addition of elements that promote the formation of a carbide at the end of solidification via a eutectic reaction will greatly improve resistance to DDC. For example, FM 82 and 625, which contain additions of Nb and form NbC at the end of solidification, are quite resistant to DDC. The NbC that is present at the end of solidification is effective in pinning the MGBs and making these boundaries very tortuous and resistant to sliding in the DDC temperature range. Recently, FM 52MSS (ERNiCrFe-13) was introduced as a "DDC-resistant" filler metals for nuclear power applications. It also contains Nb and approximately 4 wt% Mo and has shown extremely good resistance to DDC with threshold strains in the STF test exceeding 10%.

In general, weld metals that form a second phase at the end of solidification are effective in pinning the MGBs and show good resistance to DDC. This is a form of "grain boundary engineering" that uses the mechanical locking effect of a tortuous grain boundary to oppose boundary sliding. Similar advantages may be gained by reducing grain size, but this is often difficult in fusion welding.

Impurities (particularly S) have been shown to have an adverse effect on DDC, but reducing impurities alone will not prevent DDC under high-restraint conditions. For example, the FM 52 weld metal shown in Figure 4.1 had very low levels of $S+P$ but

(a)

(b)

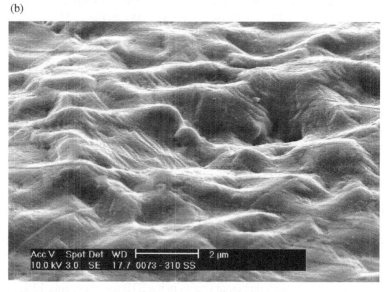

FIGURE 4.15 Fractographic features of ductility-dip cracks in Type 310 stainless steel STF samples. (a) Microscopic wavy appearance, 950°C, and (b) microscopic wavy appearance with emergent slip steps, 1100°C (Courtesy of Nathan Nissley).

was obviously very susceptible to DDC under high-restraint conditions. In general, impurities have a secondary additive effect on susceptibility to DDC. If possible, control of the grain boundary character (tortuosity) is more effective in avoiding DDC. There is also some evidence that oxygen segregation to grain boundaries may have a negative effect, but this has not been well documented [31]. A better understanding of oxygen effects is recommended since oxygen may in fact be added to shielding gases, particularly for gas metal arc welding. There is also the possibility of oxygen pickup during the welding process. Finally, the use of shielding gases that contain H, such as 98Ar–2H$_2$, can lead to increased susceptibility to DDC, as shown by Collins *et al.* [24]. Such shielding gases are sometimes recommended since they improve the wetting characteristics of the molten pool and help avoid lack-of-fusion defects. H$_2$-bearing shielding gases should be avoided in situations where DDC is a potential problem.

In situations where DDC cannot be avoided by filler metal selection and grain boundary engineering, susceptibility must be reduced by reducing restraint. If the residual stress in the solidified weld metal can be reduced, relaxation of these stresses that occurs by reheating during multipass welding does not lead to DDC. Reduction of weld heat input and the use of small weld beads for thick-section, highly restrained joints can be effective, although at a penalty to productivity. It has been shown that even highly susceptible filler metals such as FM 52 can be free of DDC if restraint is controlled through optimization of heat input and bead size and placement. This is very similar to the approach that can be used to avoid weld solidification cracking.

4.3 REHEAT CRACKING

Reheat cracking and stress relief cracking, as the terms imply, are associated with PWHT or stress relief heat treatments following welding. These heat treatments are designed to temper martensitic microstructures, reduce residual stresses, or both. Reheat cracking also describes solid-state cracking in multipass welds where subsequent weld passes provide the "reheating." This form of cracking also includes "underbead cracking" that is associated with cladding of some pressure vessel steels. In general, reheat cracking is usually associated with steels that must be subjected to stress relief following welding due to code requirements. Another form of PWHT cracking, called "strain-age" cracking, is associated with precipitation-strengthened Ni-base alloys. This form of cracking will be described in a separate section.

The subjects of reheat, stress relief, and underbead cracking have received considerable attention since the 1950s, and a number of good review papers are available [37–39]. Much of the interest in reheat cracking was driven by PWHT cracking in austenitic stainless steels and Cr–Mo–V pressure vessel steels. This was generally associated with large forgings or thick-section components fabricated for the power generation industries. Pressure vessel steels that were clad with stainless steels for use in steam generators and large pressure vessels in the nuclear power industry were susceptible to "underclad" cracking.

Cracking via the reheat cracking mechanism is most frequently observed in low-alloy steels containing secondary carbide formers (Cr, Mo, and V) and that form untempered martensite in the HAZ. Cracking in these steels normally occurs along prior austenite grain boundaries in the coarse-grained region of the HAZ. Austenitic stainless steels containing carbide formers (Nb, Ti) and/or high carbon content for

TABLE 4.2 Steels susceptible to reheat cracking

Alloy designation	Type	Composition
A508, Class 2	Steel forging	0.6Mo, 0.40Cr, 0.25C
A514, Grade F	Steel plate	0.5Mo, 0.5Cr, 0.05V, 0.15C
A517, Grade F	Steel plate	0.50Mo, 0.5Cr, 0.15C, 0.003B
A533, Grade B	Steel plate	0.5Mo, 0.20C
A710 (HSLA-80, HSLA-100)	Steel plate	2.0Ni, 0.5Cr, 0.5Mo, 1.0Cu, 0.05C
F/P22	Steel forging/plate	2.25Cr–1.0Mo–V
Type 347	Stainless steel	18Cr–11Ni–0.6Nb–0.04C
Type 321	Stainless steel	18Cr–10.5Ni–0.4Ti–0.04C
Alloy 800H	High-alloy SS	21Cr–32Ni–0.4Ti–0.4Al–0.1C

elevated temperature service can also be susceptible to reheat cracking. Another form of cracking, called "relaxation" cracking, occurs in these same steels via a similar mechanism but at lower temperatures and longer times. Relaxation cracking can occur after months or years in susceptible steels used in elevated temperature service.

Other material classes may also be susceptible. Technically, any material that exhibits a strong precipitation reaction in the stress relief temperature range will be susceptible to this form of cracking. Reheat cracking has not been reported in aluminum or titanium alloys. Materials in which reheat cracking, including stress relief, PWHT, underbead, or relaxation types, has been observed are listed in Table 4.2.

4.3.1 Reheat Cracking in Low-Alloy Steels

In low-alloy steels, there are five necessary conditions for reheat cracking to occur.

1. *Elevated temperature thermal excursion.* In the HAZ, regions susceptible to reheat cracking are heated into the austenite phase field (above the A_3 temperature) and into a temperature range where coarsening of the austenite grains occurs.

2. *Carbide dissolution.* The time–temperature relationship in the austenite phase field must be such that carbide dissolution is relatively complete. It is particularly critical that the alloy carbides (Cr, Mo, V) at least partially dissolve.

3. *Residual stress.* As the weldment cools to room temperature, considerable residual stress accumulates in the structure.

4. *Reheating into the critical temperature range.* Reheating into the temperature range between 300 and 675°C (570 and 1250°F) is particularly damaging. This range represents the regime in which carbides reprecipitate.

5. *Creep or stress relaxation during reheating.* This requires that sufficient residual stress is present in the as-welded structure and that the reheat temperature is high enough to promote stress relaxation.

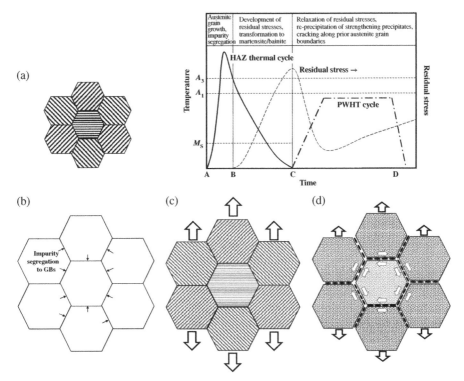

FIGURE 4.16 Schematic of thermal cycle associated with reheat cracking in low-alloy steels, where A–D show the microstructure present at different stages of the HAZ and PWHT thermal cycles.

For austenitic stainless steels, such as Type 347, the HAZ thermal cycle results in some grain growth and carbide dissolution. There is no on-cooling transformation; at room temperature, there is simply an austenitic (fcc) microstructure with residual stress.

A schematic of the weld and PWHT thermal cycle that can give rise to reheat cracking in low-alloy steels is shown in Figure 4.16. During the weld thermal cycle, regions of the HAZ are heated to a temperature where coarsening of the austenite grains occurs and there is sufficient thermal driving force for carbide dissolution. Upon cooling, this region (typically the CGHAZ) transforms to martensite and/or bainite. Since the carbides were completely (or partially) solutionized, the hardness of this region will be in the range from 40 to 50 HRC, requiring a PWHT to temper the martensite. For most low-alloy steels susceptible to reheat cracking, the PWHT temperature will be in the range from 595 to 705°C (1100 to 1300°F). During PWHT, the alloy carbides reprecipitate and residual stresses are relaxed.

It is the combination of stress relaxation and precipitation hardening that leads to cracking along the prior austenite grain boundaries in the CGHAZ. The precise mechanism for reheat cracking is the subject of some debate. Some have suggested that <u>intragranular</u> precipitation of carbides during reheating strengthens the grain interiors relative to the boundaries and promotes localized deformation along the boundaries as simultaneous stress relaxation occurs. In support of this hypothesis, it has been shown that steels that contain molybdenum and vanadium and form intragranular carbides with these elements are particularly susceptible to reheat cracking [40–45]. Impurity segregation to grain boundaries may also contribute. Several studies have shown that the segregation of impurity elements to the austenite grain boundaries during the HAZ thermal cycle results in boundary decohesion during PWHT [40, 46, 47]. These impurities include the typical phosphorus and sulfur but also others that are unique to this form of cracking, including copper (Cu), tin (Sn), antimony (Sb), and arsenic (As).

In support of the impurity segregation effect, Hippsley *et al.* [48, 49] showed that two distinct failure modes can occur as a function of reheat temperature and impurity content. In 2.25Cr–1Mo steel that contained low levels of impurities, reheat cracking was observed to occur in the temperature range from 525 to 575°C (975 to 1070°F) by a ductile IG failure mode. In a phosphorus-doped (540 ppm) companion alloy, fracture behavior was temperature dependent. In the temperature range from 325 to 375°C (620 to 710°F), the fracture surface exhibited smooth IG features. At higher temperature, 425–500°C (800–930°F), failure was by a ductile IG mode. MnS particles were associated with the ductile dimples. Thus, at lower temperatures, reheat cracking is predominantly influenced by impurity segregation and the resultant grain boundary decohesion due to the presence of impurities. While at higher temperatures or when the material is relatively "clean," the failure mode is predominantly influenced by intragranular strengthening and the precipitation of particles (such as MnS) at the prior austenite grain boundaries.

The effect of sulfur segregation and the role of MnS particles at the grain boundaries are controversial. Early theories, such as those of Hippsley *et al.* [48, 49], propose that MnS particles form along prior austenite grain boundaries in the CGHAZ during cooling from elevated temperature and serve as void nucleation sites during reheating. This is supported by the fact that small MnS particles are often associated with the small ductile dimples that are characteristic of some IG fracture surfaces. It was later proposed by Shin and McMahon that these reprecipitated MnS particles dissolved during the reheat thermal cycle and diffused along the grain boundary in advance of the crack tip [50]. Surface analysis conducted using Auger spectroscopy showed that phosphorus segregation was associated with 1.25Cr–0.5Mo and 2.25Cr–1Mo steels that are susceptible to reheat cracking, while sulfur segregation was associated with crack-resistant heats of these steels [51].

It has also been proposed that the mechanism for reheat cracking in low-alloy steels involves the formation of "precipitate free" zones at the prior austenite grain boundaries [52]. Since grain boundaries are normally preferred sites for precipitate formation, it was proposed that the area just adjacent to the grain boundary would be precipitate free and a potential zone of weakness. Combined with the intragranular precipitation hardening, strain would localize in the precipitate free zone resulting in an apparent IG failure. This mechanism would explain the ductile IG features that are often observed on the fracture surface of reheat cracks.

In summary, the mechanism for reheat cracking in low-alloy steels containing Cr, Mo, and V is quite complicated and dependent on a number of composition and restraint-related variables. There is no single accepted mechanism for reheat cracking, and it is clear that the mechanism probably varies based on alloy type and impurity content, weld thermal history, prior austenite grain size, and PWHT conditions. It is agreed that cracking is associated with the prior austenite grain boundaries in the CGHAZ and that intragranular reprecipitation of carbides results in strain localization at these grain boundaries during reheating. Cracking usually occurs in the HAZ of weldments rather than in the weld metal.

The role of grain boundary impurity segregation and the formation of grain boundary precipitates on susceptibility to reheat cracking are still not clear. There are generally two forms of failure: cavitation and low ductility IG fracture. The cavitation mode usually involves a grain boundary precipitate nucleating a void that then grows and coalesces with adjacent voids, leading to a ductile IG fracture mode. The low ductility IG fracture mode is associated with grain boundary sliding and the accumulation of strain at triple points. This is promoted by impurity segregation that reduces the cohesive strength of the boundary with a resultant failure morphology consisting of smooth IG fracture features.

As with many weld cracking phenomena, empirical relationships have been developed for reheat cracking susceptibility that relate susceptibility to the chemical composition of the material. A list of empirical relationships for prediction of reheat cracking susceptibility compiled by Vinckier and Pense [53] is provided in Table 4.3. Note that most of these relationships are specific to certain composition ranges and/ or steel types.

Haure and Bocquet [54] developed the following relationship to describe the effect of carbon and carbide formers in low-alloy steels with $C \leq 0.18\,wt\%$ and $Cr \leq 1.5\,wt\%$:

$$\Delta G \text{ (in wt\%)} = 10C + Cr + 3.3Mo + 8.1V - 2$$

where controlling $\Delta G < 2$ avoids cracking.

Brear and King [55] have developed an empirical relationship that relates reheat cracking susceptibility to the impurity content of low-alloy steels, specifically for A533, Type B:

$$CERL = 0.2Cu + 0.44S + 1.0P + 1.8As + 1.9Sn + 2.7Sb$$

As this value increases, the susceptibility to reheat cracking increases. As shown in Table 4.3, virtually all the relationships dealing with impurity levels include factors for phosphorus (P), arsenic (As), tin (Sn), and antimony (Sb). Cu and S may also be included, but many of these relationships do not contain S presumably because it can be tied up by manganese to form MnS during cooling.

As discussed earlier, the different mechanisms for reheat cracking suggest that empirical relationships based on either carbide formers or impurities are not mutually exclusive. The presence of sufficient carbide-forming elements (Cr, Mo, and V) is

SOLID-STATE CRACKING

TABLE 4.3 Empirical relationships for reheat cracking susceptibility based on composition

Parameter	Relationship	Constraints	Sources
ΔG	$Cr + 3.3Mo + 8.1V - 2$	Susceptible if $\Delta G > 0$ $C \leq 0.18\%$, $Cr \leq 1.5\%$	[1]
ΔG_1	$10C + Cr + 3.3Mo + 8.1V - 2$	Susceptible if $\Delta G > 2$ $C \leq 0.18\%$, $Cr \leq 1.5\%$	[2]
P_{SR}	$Cr + Cu + 2Mo + 10V$ $+ 7Nb + 5Ti - 2$	Susceptible if $P_{SR} > 0$ 0.1–0.25%C, 0–1.5% Cr, 0–0.2%Mo, 0–1.0%Cu, 0–0.15% V, Nb, Ti	[3]
CERL	$0.2Cu + 0.44S + 1.0P + 1.8As$ $+ 1.9Sn + 2.7Sb$	Susceptibility increases with value Valid for A533, Type B steels	[4]
X	$10P + 5Sb + 4Sn + As + Cu$	Susceptibility increases with value	[5]
R	$P + 2.43As + 3.57Sn + 8.16Sb$	Susceptibility increases with value Valid for 0.5Cr–Mo–V steels	[5]

[1]Nakamura H, Naiki T, Okabayashi H. Relation between stress-relief cracking and metallurgical properties of low alloy steels. Trans JWS 1970;1(2):60–71.
[2]Haure J, Bocquet P. Fissuration sous les revêtments inoxyables des pièces pour cuvées sous pression (Cracking below stainless steel cladding under tension). Convention No. 6210-75/3/303, Creusot Loire; September 1975.
[3]Ito Y, Nakanishi M. Study on stress relief cracking in welded low alloy steels, IIW Doc. X-668-72; 1972.
[4]Brear JM, King BL. An assessment of the embrittling effects of certain residual elements in two nuclear pressure vessel steels (A533B, A508). Philos Trans R Soc London A 1980;295:291.
[5]Hrivnak I, Magula A, Zajac J, Smida T. Mathematical evaluation of steel resistance to reheat cracking. IIW Doc. IX-1346-85; 1985.

essential since intragranular reprecipitation of these carbides promotes grain boundary strain localization. Impurity segregation to the prior austenite grain boundaries lowers the boundary cohesive strength and allows reheat cracking to occur at lower levels of stress relaxation. In the absence of intragranular carbide precipitation, impurity segregation alone will not promote reheat cracking. This is why reheat cracking is not observed in plain-carbon steels.

The Cu precipitation-strengthened HSLA-80 and HSLA-100 alloys have also been reported to be susceptible to reheat cracking. Lundin *et al.* [56] concluded that both intragranular eta-Cu precipitation and impurity segregation to prior austenite grain boundaries were responsible for the enhanced reheat cracking susceptibility of these steels. It was found that upon PWHT, the CGHAZ hardness increased significantly due to the formation of eta-Cu precipitates. The fracture surfaces exhibited smooth IG fracture along the prior austenite grain boundaries, along with isolated microductility features. The extent of microductility was shown to decrease with an increase in PWHT temperature, indicating enhanced segregation of embrittling species to grain boundaries at higher temperatures.

4.3.2   Reheat Cracking in Stainless Steels

As indicated earlier, a form of reheat cracking has also been observed in austenitic stainless steels. In particular, the HAZ and fusion zone of Type 347 stainless steel have been reported to be susceptible, particularly in thick sections [57–63]. This is a "stabilized" grade of stainless steel that contains niobium to reduce susceptibility to IG corrosion. Reheat cracking can also occur in other stainless steels, including Ti-stabilized Type 321 and high carbon grades such as 304H and 316H. There have also been reports of reheat cracking in high-alloy stainless steels used for elevated temperature service, in particular Alloy 800[1] and its high-carbon variant Alloy 800H [58, 64, 65]. Cracking normally occurs during the postweld stress relief that is often required for thick-section stainless steel weldments. Many of these alloys have also been shown to be susceptible to relaxation cracking, as discussed in the following. Reheat cracking has also been observed in austenitic stainless steel weld metals, such as Type 308, that are deposited using the flux-cored arc welding (FCAW) process. Cracking in these weld deposits occurs due to the presence of bismuth, which is added to facilitate slag separation from the weld metal [66].

The mechanism for cracking in Type 347 is associated with the precipitation of NbC during the reheating cycle. Since intragranular precipitation occurs in the same temperature range as stress relaxation, locally high strains concentrate at grain boundaries and promote cracking. Grain growth in the HAZ increases cracking susceptibility. It is not clear whether there is any effect of impurity segregation in this alloy. The fracture mode is typically ductile IG, exhibiting extremely fine ductile dimples. It is interesting that this form of cracking can also occur in the weld metal of the austenitic stainless steels. In low-alloy steels, it is almost always in the HAZ. The presence of delta ferrite in the weld deposit does not seem to influence suscepti-bility to this form of cracking in Type 347.

An example of reheat cracking that has occurred in Type 347 weld metal is shown in Figure 4.17. Cracking occurred after postweld stress relief of a large structure constructed from Type 347 stainless and a matching filler metal. The ASME code required that this structure be stress relieved in the temperature range from 850 to 900°C (1560 to 1650°F) before it could be put into service. This resulted in severe reheat cracking in the weld metal. The fracture occurs along MGBs in the weld metal, as indicated by the IG features of the fracture surface.

The cracking susceptibility of Type 347 weld metal exhibits a C-curve cracking response, as shown in Figure 4.18. The two curves shown represent the onset of cracking when a weld metal sample of Type 347 was loaded to either 75% or 100% of its yield strength at a given temperature and held at that temperature until fracture occurs. By plotting the fracture time at a given temperature, the reheat cracking envelope can be determined. For example, at 900°C (1650°F), reheat cracking occurs within 2000 s when yield strength-level stresses are present. The cracking "envelope" described by these two C curves represents the precipitation temperature range of NbC in stainless steel.

---

[1]Note that Alloy 800 is often considered a Ni-base alloy, even though the nominal Fe content is higher than the Ni content.

(a)

(b)

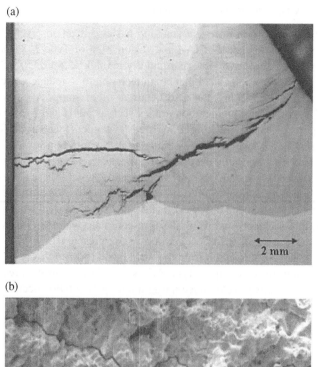

FIGURE 4.17   Example of reheat (stress relief) cracking in Type 347 stainless steel.

Another example of reheat cracking is provided in Figure 4.19. In this case, thick-section welds in Alloy 800H (Fe–20Cr–32Ni–0.5Ti–0.5Al–0.1C) were performed with Weld A (ENiCrFe-2) shielded metal arc electrodes. The nominal composition of Weld A is Ni–15Cr–8Fe–1.5Mo–1.5Nb–0.05C. PWHT following welding is required for stress relief and to avoid relaxation cracking in service. Stress relief was performed at 900°C (1650°F) and led to the cracking shown in Figure 4.19. Surprisingly, the cracking occurred in the weld metal rather than the Alloy 800H HAZ. As shown by the fractography in Figure 4.19b, the cracking occurs along MGBs in the weld metal. It is hypothesized that the presence of Nb in

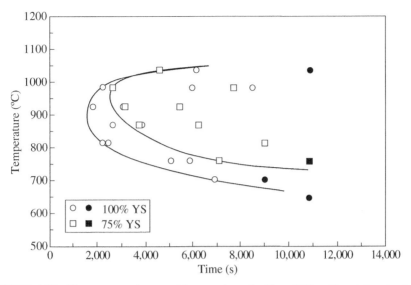

FIGURE 4.18 Temperature–time cracking envelope for Type 347 weld metal under two stress conditions.

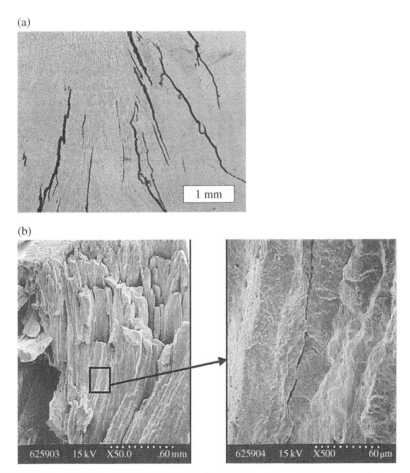

FIGURE 4.19 PWHT cracking in ERNiCrFe-2 (Weld A) weld metal after stress relief at 900°C. (a) Cracking along weld metal migrated grain boundaries and (b) fracture surface morphology.

the filler metal resulted in the precipitation of NbC during PWHT and promoted cracking at the grain boundaries. The precipitation temperature range for NbC in this weld metal is very similar to that for Type 347, and thus, the time–temperature cracking envelope shown in Figure 4.18 is roughly the same for the Weld A filler metal. Additional details regarding the nature of reheat cracking in this dissimilar combination can be found in a companion text [67].

Another manifestation of reheat cracking has been observed in thick-section welds made using Type 308 filler metal applied using the FCAW process [66]. Many flux-cored electrodes contain additions that facilitate the removal of the flux from the weld surface. One of these elements is bismuth (Bi). In the weld metal, Bi apparently segregates to grain and interphase boundaries and reduces the ductility of the weld metal at temperatures above 700°C (1290°F). Reheat cracking occurs along the austenite–delta ferrite interface. At temperatures above 850°C (1560°F), the bismuth segregation promotes local melting, and weld metal liquation cracking is possible.

4.3.3 Underclad Cracking

Underclad cracking is a special form of reheat cracking that occurs during the cladding of low-alloy steels, particularly pressure vessel steels. Cracking via reheating of the coarse-grained microstructure produced during cladding may occur either during welding of the adjacent clad layer or as a result of postweld stress relief. A review of underclad cracking has been conducted by Vinckier and Pense [53] and includes the schematic in Figure 4.20 as an illustration of the crack susceptible region beneath the clad layer.

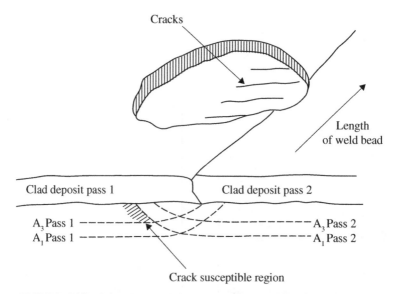

FIGURE 4.20   Schematic of underclad cracking (From Ref. [53]. © WRC).

Underclad cracking in conventional engineering designs can occur during the cladding of pressure vessel steels (in particular A508, Class 2, forgings) with austenitic stainless steel filler metal. The difference in the thermal expansion coefficient between the stainless steel cladding and low-alloy steel base material enhances the residual stress due to welding and increases the level of creep stresses that are generated during reheating relative to welds produced with a matching filler material. In general, the same mechanisms and cautions apply for underclad cracking as previously cited for reheat cracking.

As illustrated in Figure 4.20, reheating of the CGHAZ from the first cladding pass by the second pass leads to stress relaxation and cracking. The most susceptible region tends to be that reheated to just below the A_1 temperature (subcritical HAZ). Regions of the original CGHAZ that are reheated above the A_1 temperature transform to austenite and are immune to reheat cracking. Thus, underclad cracking tends to occur in a narrow band below the clad layer. Tensile residual stresses on the order of 60–65 ksi have been reported for Type 308 SS clad onto A508 forgings [53]. The relaxation of these residual stresses leads to high local plastic strain with a peak in strain in the subcritical reheated region of the original CGHAZ, as illustrated in Figure 4.21.

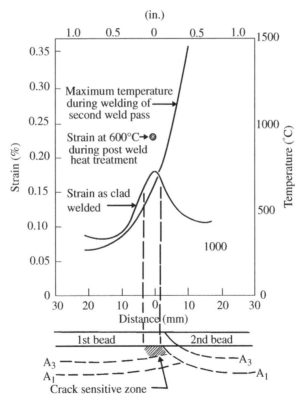

FIGURE 4.21   Residual stress distribution and temperature profile associated with underclad cracking (From Ref. [53]. © WRC).

Most of the research on underclad cracking has been conducted on the A508, Class 2 forgings used for large pressure vessel fabrication. Other forging grades, such as A533, have been shown to be less susceptible due to a lower content of secondary carbide-forming elements (Cr, Mo, V). For example, A533 has the same Mo content as A508 but contains no Cr or V.

4.3.4 Relaxation Cracking

Relaxation cracking is another manifestation of reheat cracking that is related to elevated temperature service exposure. It typically occurs in austenitic stainless steels that are used at service temperatures in the range from 550 to 750°C (1020 to 1380°F). Susceptible materials are the same as indicated for susceptibility to reheat cracking: Types 321, 347, 304H, and 316H and Alloy 800H. The cracking mechanism is the same as for reheat cracking but occurs at much longer times. It is a particularly insidious form of cracking because it normally occurs after hundreds or thousands of hours of service. Work by van Wortel at the TNO Metals Research Institute in the 1990s, supported through a large joint industrial program, evaluated a number of these materials and developed a test methodology to quantify susceptibility to relaxation cracking [64].

Relaxation cracks were found to have slightly different features than reheat cracks, primarily because they form over much longer exposure times. In Alloy 800H, the crack along the HAZ grain boundary exhibited a Ni-rich filament surrounded by a Cr-rich oxide layer. In advance of the crack, small voids (or cavities) were present along the grain boundaries. Large grain boundary carbides were present, surrounded by a precipitate free zone and then a dense distribution of matrix carbides, as shown in Figure 4.22. The formation of matrix carbides results in an increase in hardness (Fig. 4.23) and, combined with stress relaxation, leads to grain boundary failure.

The relaxation cracking test developed at TNO identified specific temperature ranges over which relaxation cracking was most prevalent. For the 300-series austenitic stainless steels, this range was 525–600°C (980–1110°F), and for Alloy 800H, 550–650°C (1020–1200°F). Higher residuals stresses and large grain size were both found

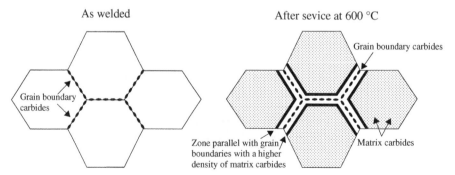

FIGURE 4.22   Effect of service exposure on the precipitation behavior of Alloy 800H after exposure for 6000 h at 600°C (1110°F) (From Ref. [64]. © NACE).

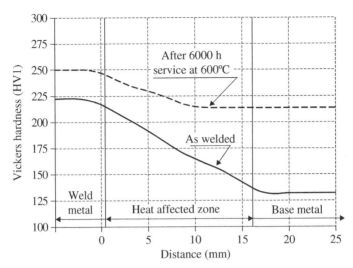

FIGURE 4.23 Hardness variation in Alloy 800H after service exposure at 600°C (1110°F) (From Ref. [64]. © NACE).

to accelerate failure. The use of a stress relief heat treatment prior to service exposure is an obvious method to avoid relaxation cracking, but in heavy-section weldments, the use of such a heat treatment usually results in reheat (stress relief) cracking.

4.3.5 Identifying Reheat Cracking

Reheat cracking has distinct characteristics that usually allow it to be distinguished from other forms of HAZ cracking in low-alloy steels. These include composition effects and both metallographic and fractographic features.

Composition. Virtually all low-alloy steels that are susceptible to reheat cracking contain secondary carbide-forming elements. In particular, the addition of Cr, Mo, and V are most often associated with steels that are sensitive to reheat cracking. Analysis of the fracture surface using analytical techniques such as SEM/XEDS and scanning Auger microscopy often reveals the presence of impurity elements such as S, P, Cu, As, Sn, and Sb. While S and P are also associated with HAZ liquation cracking that may be present in these steels, the presence of Cu, As, Sn, and Sb on the fracture surface is a strong, often irrefutable, indication of reheat cracking.

For austenitic stainless steels, reheat cracking is usually associated with the Nb-stabilized alloy Type 347. This alloy is also susceptible to HAZ liquation cracking, as noted in Chapter 3, so metallographic and fractographic evidence is needed to clarify the mechanism. The presence of bismuth in FCAW consumables for austenitic stainless steels has also been shown to promote reheat cracking in the weld metal.

Metallographic Features. In low-alloy steels, reheat cracking occurs along <u>prior austenite grain boundaries</u> in the CGHAZ. Typical examples of reheat cracking are

(a)   (b)

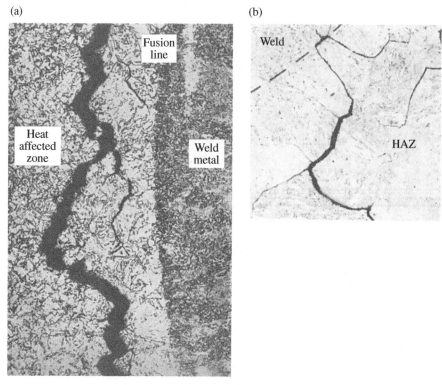

FIGURE 4.24 Examples of reheat cracking. (a) 1Cr–1Mo–0.35 V steel (From Ref. [68]) and (b) A517F (From Ref. [69]) (© TWI).

shown in Figure 4.24 [68, 69]. At room temperature, the HAZ in these steels exhibits transformation products (normally martensite and bainite) that disguise the prior austenite grain boundaries. Special metallographic techniques are often required to reveal these boundaries. The crack path has an IG pattern that may distinguish reheat cracks from hydrogen cracks, which are normally transgranular in these steels. Because cracking susceptibility increases with prior austenite grain size, cracking is normally very close to the fusion boundary. In many cases, the cracks propagate parallel to the fusion boundary only 1 or 2 grain diameters from the boundary. In austenitic stainless steels, reheat cracking may occur in both the weld metal and HAZ. If it is in the HAZ, it is usually very close to the fusion boundary where grain growth has occurred.

Fractographic Features. Two general types of fracture have been associated with reheat cracking. A classical IG fracture (flat and relatively featureless grain faces) is generally associated with susceptible materials that fail at low temperatures and/or that have a relatively high impurity content. At higher temperatures (>500°C) or when the impurity content is low, the failure mode is generally a ductile IG mode. The IG dimples have been associated with grain boundary precipitates, often MnS. As noted earlier, the fracture faces may exhibit high levels of impurities. Examples of ductile IG and low ductility IG fracture are provided in Figure 4.25.

(a)

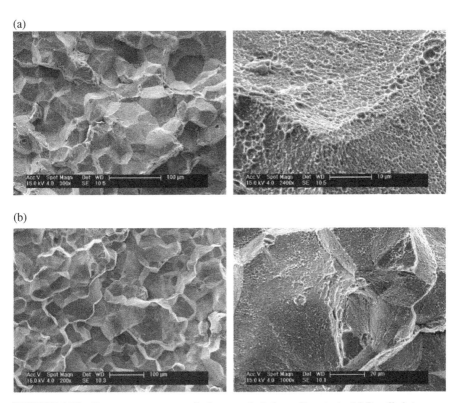

(b)

FIGURE 4.25   Fracture appearance of reheat cracks in low-alloy steels. (a) Ductile intergranular and (b) low ductility intergranular. Note difference in magnification (Courtesy of Katie Strader and Xiuli Feng, OSU).

4.3.6   Quantifying Reheat Cracking Susceptibility

Not surprisingly, there has been considerable effort over the years to develop test techniques that quantify susceptibility to reheat cracking. These tests can be separated into two groups, those that use self-restraint and those that are simulative in nature and use externally applied load (or strain). Among the self-restraint tests, the Lehigh restraint test [70], Y-groove test [66], modified implant test [71], and BWRA test [72] are most often cited. These tests do not typically provide quantitative results; rather, they tend to be of the "go–no go" type. Sectioning is required after testing to determine if reheat cracks are present. These tests can be useful in identifying welding procedures and PWHT conditions for avoiding reheat cracking. For example, Meitzner and Pense [52] used the Lehigh restraint test to develop a C-curve cracking envelope for stress relief cracking in A517, and Nishimoto *et al.* [66] used the Y-groove test to develop a similar curve for Type 308 FCAW deposits containing bismuth, as shown in Figure 4.26.

In order to better quantify susceptibility to reheat cracking, tests have been developed that simulate the HAZ microstructure and then apply a stress to promote cracking. Most of the simulation tests use a thermomechanical simulator, such as the

(a)

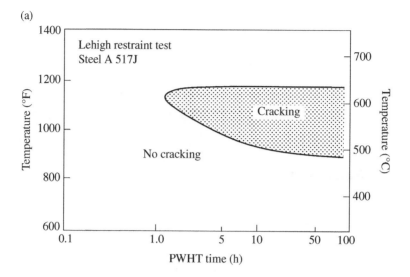

(b)

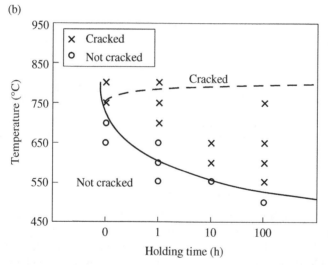

FIGURE 4.26 Effect of stress relief temperature on cracking susceptibility. (a) The Lehigh restraint test for the HAZ of A517 (From [52]. © AWS) and (b) Y-groove test for Type 308 FCAW weld metal (From Ref. [66]. © IIW)

Gleeble [73], to apply stress (and/or strain) to a sample after an appropriate HAZ thermal cycle. Tests developed by Balaguer *et al.* at RPI [74], Nawrocki *et al.* at Lehigh [75], and Norton and Lippold at Ohio State [76] all used the Gleeble to develop a representative HAZ microstructure in a small tensile sample, and then a constant load was applied at various PWHT temperatures. Ideally, this test can develop a "C-curve" cracking response that indicates susceptibility to reheat cracking. Such a response curve developed by Nawrocki *et al.* [75] is shown in Figure 4.27. Typically,

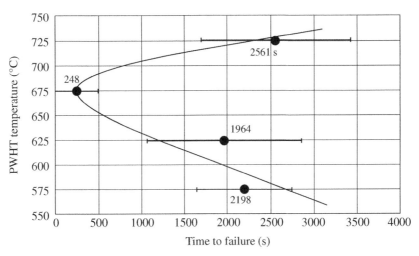

FIGURE 4.27 C-curve cracking response for 2.25Cr–1Mo steel developed using the Gleeble (From Ref. [75]© AWS).

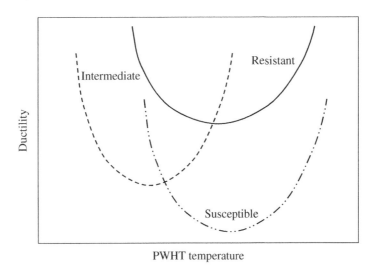

FIGURE 4.28 Illustration of ductility curves for reheat cracking based on the technique of Norton and Lippold [76].

these tests are subject to large scatter in terms of failure time. In order to reduce scatter and better define material ductility as a function of PWHT temperature and time, Norton and Lippold [76] used an approach where the sample was pulled to failure after up to 4 hours of exposure at the PWHT temperature. Although developed for studying strain-age cracking in Ni-base superalloys, the test works equally well for steels. An illustration of the types of ductility curves that can be obtained using this approach is shown in Figure 4.28. This test is described in more detail in Chapter 9.

4.3.7 Preventing Reheat Cracking

Just as the mechanism for reheat cracking is quite complicated and varies as a function of the material composition, methods to avoid reheat cracking can also be complicated. The discussion here refers primarily to reheat and stress relief cracking that occurs during fabrication. Some of the same methods for prevention can apply to relaxation cracking that occurs during service, but this is not always the case. Prevention methods are discussed in terms of (i) composition control, (ii) effect of welding conditions, (iii) control of residual stresses, (iv) control of stress relaxation during reheating, (v) effect of stress concentration, and (vi) "buttering" of the substrate.

Composition Control. The steels listed in Table 4.2 are inherently susceptible to reheat cracking because they contain secondary carbide formers. Reheat cracking can be reduced or eliminated by choosing steels that have reduced susceptibility, for example, by using the relationship developed by Haure and Bocquet [54] or others as listed in Table 4.3. Often, selection of an alternate material is not an option, since the base material has been specified for use in a certain application. In general, as the secondary carbide former content and carbon content increase, the material becomes more susceptible to reheat cracking. Impurities have been shown to contribute to the reheat cracking mechanism by diffusing to grain boundaries and lowering the boundary cohesive strength. Reducing impurity content can improve cracking resistance in most of the susceptible materials, but impurity control alone cannot insure resistance to reheat cracking. Many of the impurity elements that promote reheat cracking in low-alloy steels (in particular As, Sn, and Sb) do not appear on composition analysis reports that are provided by the material supplier, so it may be necessary to conduct (or specify) additional analysis.

Effect of Welding Conditions. The effect of welding conditions on susceptibility to reheat cracking can be profound, and much research has been conducted on a number of steels to understand the effect. There are two basic, and conflicting, approaches. The first approach is to minimize weld heat input in order to reduce HAZ grain size. In both low-alloy and stainless steels, larger grains in the HAZ have been shown to increase susceptibility, since less grain boundary area results in higher-strain localization at the boundaries. Finer grains help to better distribute the strain and minimize void formation and/or boundary sliding. This approach works well with the stainless steels, such as Type 347 and Alloy 800H. The alternative approach is to use preheat and higher heat input in order to affect the transformation behavior in the CGHAZ. This approach does not apply to the stainless steels. For low-alloy steels with intermediate hardenability, the use of preheat and high heat input will slow the cooling rate in the CGHAZ and reduce the hardness by limiting the formation of martensite. As the CGHAZ hardness decreases, relaxation of residual stress during reheating will be more uniformly distributed. It has also been argued that shallower temperature gradients in the HAZ resulting from higher heat input lead to a wider CGHAZ and more grain boundary area over which stress relaxation can be accommodated.

Control of Residual Stresses. Reheat cracking is almost always associated with thick-section weldments where the level of residual stress following welding is quite high—approaching the yield strength of the base or weld metal. Prediction and measurement of residual stresses in welds have been the subject of considerable research, and the reader is referred to other authoritative texts that address these issues. It is well known, however, that the use of preheat can be a simple and effective method to reduce residual stress. Other techniques, including the control of bead size and placement and welding sequence, can be effective at reducing residual stresses. Another effective approach is to select filler metal whose strength undermatches that of the base metal. In this manner, restraint resulting from weld shrinkage during cooling concentrates in the weld metal rather than the CGHAZ. This assumes that the weld metal is of a composition that is not susceptible to reheat cracking. This approach may not be effective with the austenitic stainless steels, such as Type 347, since these steels are usually welded with a matching filler metal. The example shown in Figure 4.17 demonstrates how weld metals can be as susceptible as the base metal to reheat cracking.

Control of Stress Relaxation during Reheating. Even in the presence of high residual stresses, cracking can be avoided if stress relaxation can be managed during the reheating cycle. This is a difficult problem in the low-alloy steels, since the temperature range in which significant stress relaxation occurs overlaps the range where secondary carbides begin to reform in the microstructure. PWHT cannot typically be conducted above the precipitation temperature range because the lower critical temperature (A_1) is an effective upper bound for PWHT, that is, PWHT above the A_1 results in reformation of austenite.

For the austenitic stainless steels, this is not the case and possible methods exist for relaxing residual stresses. Since the nose of the carbide precipitation curve is at relatively high temperatures (850–950°C), it may be possible to hold the weldment at temperatures below this range to allow partial relaxation of residual stresses before heating to the normal stress relief temperature. Even fractional reduction of the residual stress may be successful in avoiding reheat cracking in some situations. The other possibility is to heat the structure rapidly to a temperature above the carbide precipitation temperature range. This will allow rapid relaxation of the stresses while avoiding the formation of intragranular precipitates. This solution may have limited usefulness for large structures, since the ability to heat rapidly enough to avoid the nose of the precipitation curve may be limited. As illustrated in Figure 4.18, heating rates exceeding 30°C/min may be required to avoid cracking in heavy-section Type 347 weldments.

Effect of Stress Concentration. An obvious approach for reducing susceptibility to reheat cracking (and hydrogen-induced cracking) is to eliminate stress concentration in the weldment. This may include elimination of slag intrusions at the weld toe, grinding or blending the weld toe, or use of other material removal techniques to eliminate stress concentrations. Since cracking usually initiates in the HAZ very close to the fusion boundary, attention to this area of the weldment is very important. Welds with partial penetration, lack-of-fusion, or other process-related defects can also greatly increase stress concentration. Defects of this type that are open to the surface are the most damaging. Some attempts to locally alter stress concentration and residual stress have used various "peening" techniques. Peening can generate local compressive stresses on the weld surface and potentially mitigate initiation of reheat cracks. Peening probably has little effect in large, thick-section weldments since peening only affects the structure within a few millimeters from the surface. Many of the austenitic stainless steels that are susceptible to reheat cracking are also susceptible to HAZ liquation cracking, including Types 321 and 347 and Alloy 800. Liquation cracks can also increase stress concentration in the coarse-grained region of the HAZ. Careful inspection prior to PWHT is advised to assure that these types of defects are not present.

"Buttering" of the Substrate. A technique that has been used to avoid reheat cracking and other forms of base metal HAZ cracking (such as lamellar cracking) is often called "buttering." With this approach, a layer of resistant weld metal is applied to the base metal substrate. This layer (or layers) is applied at low heat input and subsequently stress relieved to eliminate residual stresses in the susceptible CGHAZ of the base metal. When welding is conducted after application of the butter layer, the CGHAZ is contained within the butter layer, which is a composition resistant to reheat cracking. While this approach is usually very effective, it adds another fabrication step and increases fabrication costs significantly.

4.4 STRAIN-AGE CRACKING

Strain-age cracking (SAC) is a form of reheat, or PWHT, cracking that is specific to the precipitation-strengthened Ni-base alloys. It is a solid-state cracking phenomenon that is most often observed in the HAZ just adjacent to the fusion boundary, although it is possible for SAC to occur in the weld metal of these alloys. In most cases, it occurs during PWHT but is also possible (although unlikely) during reheating in multipass welds. This form of cracking is most prevalent with the γ' ($Ni_3(Al, Ti)$)-strengthened alloys, and many of these alloys are considered "unweldable" because of this cracking phenomenon. The term "strain-age" refers to the simultaneous effect of relaxation of stresses causing high local strain and the age hardening of the structure by precipitate formation. Although SAC is similar to the reheat cracking mechanism that occurs in steels, there are some important differences that are unique to the Ni-base superalloys.

The rate of γ' precipitation is influenced both by composition (Ti+Al content) and base metal condition. For example, even small amounts of cold work in the base metal will accelerate precipitation. Figure 4.29 from Wilson and Burchfield [77] shows the rate of hardening due to precipitation in three γ'-strengthened alloys (René 42, M-252, and Astroloy). Note that hardening of these alloys occurs extremely rapidly after solutionizing and holding at the aging temperature. In contrast, Alloy 718, which is strengthened by γ'', Ni_3Nb, hardens initially at a much slower rate. It will be seen that the rate of precipitation (hardening) is a key factor in controlling susceptibility to SAC.

A schematic illustration of a representative thermal history for welded and postweld heat-treated Ni-base superalloys is shown in Figure 4.30 [78]. During the weld thermal cycle, strengthening precipitates (and other constituents) that are present in the base metal dissolve in the austenite matrix, and some grain growth will occur,

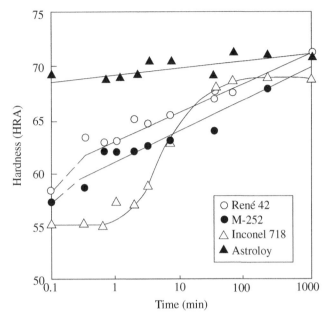

FIGURE 4.29   Hardening rate as a result of precipitation for several Ni-base superalloys (From Ref. [77]. © AWS).

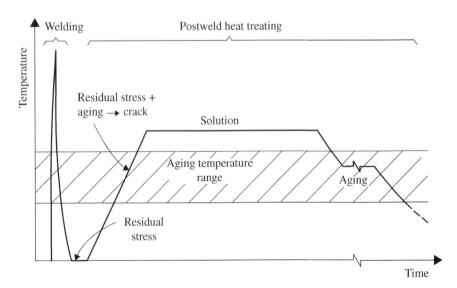

FIGURE 4.30   Schematic of the thermal history during welding and postweld heat treatment of Ni-base superalloys (From Ref. [78]. © Wiley).

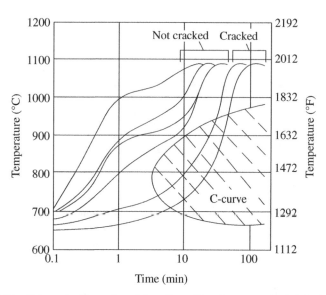

FIGURE 4.31 Schematic illustration of the effect of heating rate and precipitation behavior on susceptibility to strain-age cracking (From Ref. [79]. © AWS).

depending on weld heat input. Since the as-welded fusion zone and HAZ are effectively solutionized, significant softening occurs. A PWHT consisting of solution annealing and aging must be applied to strengthen the weldment and base metal to the original base metal strength level. The solution heat treatment also serves to relieve residual stresses resulting from the welding process. Ideally, the weldment is heated to an appropriate solution-annealing temperature where alloying additions go back in solution (or homogenize in the weld metal due to solidification segregation) and residual stresses relax, and then cooled to an aging temperature where precipitation is controlled such that the required mechanical properties are achieved. A more detailed explanation of the SAC mechanism is provided in the next section.

In practice, it may be difficult (or impossible) to prevent the precipitation of γ' during heating to the solution-annealing temperature. This a function of the alloy composition and the differences in hardening rates, as illustrated in Figure 4.29. The relationship between the heating cycle and precipitation is illustrated in Figure 4.31 [79]. Strengthening precipitates, such as γ' and γ'', exhibit a distinct "C-curve" temperature–time regime where precipitation is possible. If the weldment can be heated rapidly enough to avoid intersecting the precipitation curve, then precipitation will not occur, and solution annealing of the weldment can be achieved. If the weldment cannot be heated rapidly enough (or if the C curve is shifted to the left in Figure 4.31), precipitation will occur, and the alloy will begin to harden. For Ni-base superalloys, precipitation during heating overlaps the temperature range where significant stress relaxation occurs, and this can lead to locally high strains at the grain boundaries. If these strains are sufficiently high, grain boundary failure will occur and a strain-age crack will form.

4.4.1 Mechanism for Strain-age Cracking

As described earlier, the term SAC is derived from the fact that both local strain and aging must occur nearly simultaneously. The term should not be confused with the metallurgical phenomenon of "strain aging" observed in carbon steels. SAC in welds has been studied extensively in various Ni-base superalloys, and the severity of this problem has led to the development of SAC-resistant alloys, such as Alloys 718 and 706.

It is generally agreed that SAC in Ni-base superalloys results from low ductility in the HAZ accompanied by high-strain accumulation in the same region [80–83]. Such reduction of ductility is associated with the development of grain "stiffening" and/or grain boundary weakening during PWHT. Most investigators attribute this to intragranular precipitation hardening combined with precipitate free zones at grain boundaries or IG carbide precipitation. If the decrease in ductility during PWHT occurs before or at a faster rate than stress relief, the "embrittled" region in the HAZ may crack due to its inability to accommodate the redistribution of strain associated with the stress-relieving process.

The following general observations have been made regarding SAC:

- It is always IG.
- It is most prevalent in the HAZ adjacent to the fusion line and in some cases is associated with the partially melted zone (PMZ).
- It occurs during postweld heating to the solution-annealing temperature due to simultaneous precipitation and local strain accumulation at grain boundaries.

The stresses that cause cracking may have three origins: (i) weld residual stress, (ii) thermally induced stresses arising from the difference in the coefficient of thermal expansion between the base material and weld metal, and (iii) stresses from dimensional changes caused by precipitation. In general, the precipitates have a different lattice parameter than the matrix, and their formation will lead to a local grain boundary stress.

Based on the published literature, the metallurgical contributors to SAC are the following:

- The rate of hardening (strengthening). Materials that harden more slowly allow better accommodation of the stresses (i.e., Alloy 718 as shown in Figure 4.29).
- Intragranular precipitation resulting in hardening of the grain interior leading to stress concentration at the grain boundaries. This mechanism was originally proposed by Prager and Shira [84] based on the work of Younger and Baker [85] on austenitic steels.
- "Transient embrittlement" of the HAZ due to the precipitation of IG carbides. According to this theory [86–88], the embrittling reaction is thought to result from dissolution of carbides during the weld thermal cycle and subsequent reprecipitation in continuous "films" of $M_{23}C_6$-type carbides along grain boundaries during heat treatment. These carbide "films" are not capable of resisting the stresses caused by the precipitation of γ', and presumably, failure occurs at the carbide/matrix interface at the grain boundary.

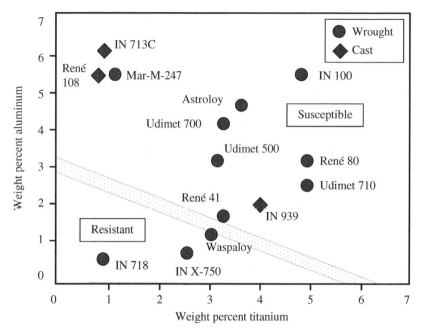

FIGURE 4.32   Effect of Al and Ti content on susceptibility to strain-age cracking in Ni-base superalloys (Modified from Ref. [84]. © WRC).

- Partial melting along grain boundaries adjacent to the fusion line. This may be due to impurity segregation or constitutional liquation. Note that Ti that is added as a strengthening agent can also promote constitutional liquation if Ti-rich, MC carbides are present. Boron that is added to improve creep resistance in these alloys also promotes grain boundary liquation.

The mechanism for SAC in Ni-base alloys is still not precisely defined, although it is clear that both compositional and restraint factors play a role. It is well known, for instance, that certain alloys are more resistant to SAC than others. This resistance is generally attributed to the rate and nature of the precipitation reaction(s) that promote strengthening. The γ'-strengthened alloys are the most susceptible, and the influence of Ti and Al content has been well documented. The relationship of Ti and Al content to SAC was originally proposed by Prager and Shira [84], and a diagram based on their work that includes additional modern alloys is shown in Figure 4.32. The original diagram contained a band running from approximately 6–7 at% Al to 6–7 at% Ti, separating alloys that are resistant (below) from those that are susceptible (above). It is not clear how this diagram was developed, but it probably represented the results of different weldability tests and practical experience. Some diagrams of this type have replaced the band with a line, but the use of a transition band from resistant to susceptible is most appropriate since susceptibility to SAC is a strong function of restraint. For example, in alloys that are marginally susceptible to SAC

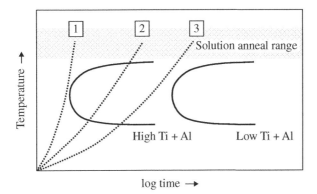

FIGURE 4.33   Schematic illustration of the effect of (Ti+Al) content and heating rate to the solution-annealing temperature.

(such as Waspaloy and René 41), it is known that minimizing the level of residual stress prior to PWHT is very effective in preventing SAC, as discussed in Section 4.4.2.3. Based on this diagram, it can be seen that the higher Al+Ti contents promote a stronger and more rapid precipitation of γ'. This, in effect, shifts the nose of the precipitation curve to much shorter times, making it difficult to suppress precipitation during postweld heating to the solution-annealing temperature range. This is shown schematically in Figure 4.33 as a function of (Ti+Al) content for several heating rates.

Duvall and Owczarski [89] demonstrated the effect of (Ti+Al) content in a study of the PWHT cracking susceptibility of Waspaloy and Alloy 718. They showed that HAZ cracking obeyed a C-curve behavior, as shown in Figure 4.34, and that the C curve for Alloy 718 was displaced to longer times. The C curve for Waspaloy (containing 3 wt% Ti and 1.4 wt% Al) represents the γ' precipitation regime, while the Alloy 718 (containing 0.9 wt% Ti, 0.5 wt% Al, and 5 wt% Nb) C curve represents the precipitation regime for γ''. These results again demonstrate the beneficial effect of the sluggish precipitation reaction of γ'' for avoiding SAC during PWHT of Ni-base superalloys. In this same study, they found no indication of the "transient embrittlement" phenomenon associated with $M_{23}C_6$ precipitation at grain boundaries. Instead, the ductility was kept at a moderately low level during aging in the cracking temperature range by a combination of microstructural interactions produced during welding and heat treatment. Changes in cracking susceptibility between different heats of Waspaloy resulted from changes in ductility, which were caused by γ' precipitation and IG carbide precipitation. Within groups of susceptible and nonsusceptible microstructures, wide differences in the amount and morphology of carbides were observed.

Norton and Lippold [76] used a Gleeble-based test to study the SAC susceptibility of Waspaloy and Alloy 718. The details of this test are described in Chapter 9. In this test, samples are initially subjected to a HAZ thermal cycle and then cooled to room temperature under restraint, so that considerable room temperature residual stress was present. The sample was then immediately heated into the aging temperature range where the stresses were allowed to relax and precipitation of γ' (Waspaloy) or γ''

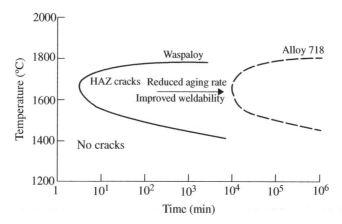

FIGURE 4.34   C-curve strain-age cracking behavior for Waspaloy and Alloy 718 showing much greater tolerance for PWHT cracking for Alloy 718 due to slower aging rate (From Ref. [89]. © AWS).

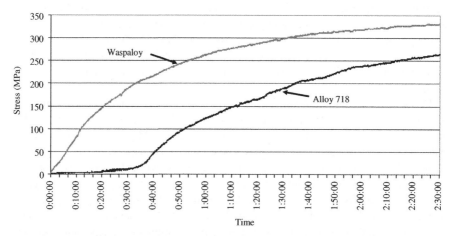

FIGURE 4.35   Effect of postweld aging time on stress accumulation in the simulated HAZ of Waspaloy and Alloy 718 (From Ref. [76]. © ASM).

(Alloy 718) occurred with hold time. Since the sample was fixed, precipitation resulted in increased stress in the sample, as shown in Figure 4.35. Note that the starting residual stress has been subtracted from this data to allow for easier comparison of the aging behavior.

After a predetermined time (up to 4 hours), the samples were then pulled to failure at the test temperature and their ductility was measured. These tests resulted in the development of 3-dimensional C curves based on test temperature, time, and strength/ductility. This data could then be used to generate 2-dimensional ductility C curves for specific time–temperature conditions. An example of this for both alloys after 3 h of PWHT is shown in Figure 4.36. This data again shows the beneficial

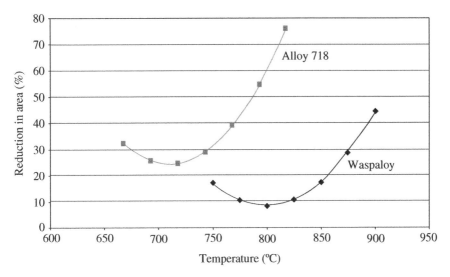

FIGURE 4.36 Postweld heat treatment ductility curves for Waspaloy and Alloy 718 after 3 hours of aging following a simulated HAZ thermal cycle (From Ref. [76]. © ASM).

effect of γ'' versus γ' precipitation with respect to SAC. Although both alloys show a drop in ductility in the PWHT temperature range, the minimum ductility for Waspaloy is much lower. As a result, much lower strains resulting from stress relaxation are required to promote SAC in Waspaloy relative to Alloy 718.

Analysis of these samples also clearly reveals the nature of SAC in Ni-base superalloys. The photomicrograph in Figure 4.37 shows the region near the fracture in a Waspaloy sample. Cracks are IG and tend to initiate at grain boundary triple points. Higher-magnification examination of the grain boundaries in the SEM revealed no evidence of carbides or continuous carbide precipitation, suggesting that the "transient embrittlement" phenomenon is not occurring in Waspaloy. This is in agreement with the conclusion of Duval and Owczarski [90]. Examination of the fracture surfaces of Waspaloy and Alloy 718 samples indicated that the fracture morphology is either smooth or ductile IG. An example of both of these fracture morphologies can be seen on the fracture surface of an Alloy 718 sample in Figure 4.38.

Based on previous research, a mechanism for SAC in Ni-base superalloys is shown schematically in Figure 4.39. It can be described in 4 stages, as shown in schematics A through D in the figure. In *Stage A*, the HAZ is heated to subsolidus temperatures. If strengthening precipitates are present in the base metal, most of these will dissolve on heating to the peak temperature. At the highest temperatures in the HAZ, many of the Ni-base superalloys will undergo some grain boundary liquation just adjacent to the fusion boundary. There will also be some grain growth in the HAZ, the degree of which is determined by the starting base metal microstructure and the HAZ thermal cycle. As the weld cools to room temperature during *Stage B*, the liquid films solidify (there could also be possible HAZ liquation

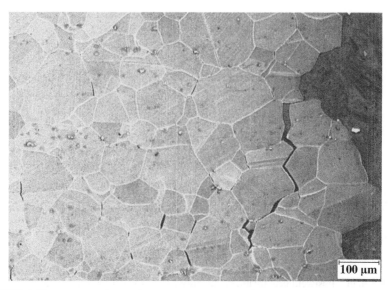

FIGURE 4.37   Intergranular strain-age cracking in the simulated HAZ of Waspaloy (From Ref. [76]. © ASM).

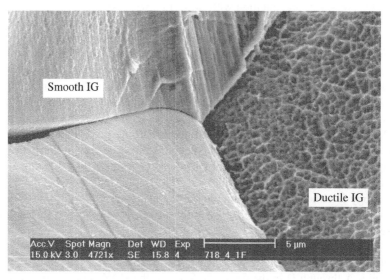

FIGURE 4.38   Fracture morphology of strain-age cracking in Alloy 718 (From Ref. [76]. © ASM).

cracking), and residual stress starts to accumulate in the system. Typically, there is little on-cooling reprecipitation, and the HAZ is essentially in the solution-annealed condition at room temperature. Upon reheating to the solution-annealing tempera-ture, some stress relaxation will occur as the weld is heated above approximately $0.5T_S$, as shown in *Stage C*. Presumably, the residual stress does not completely

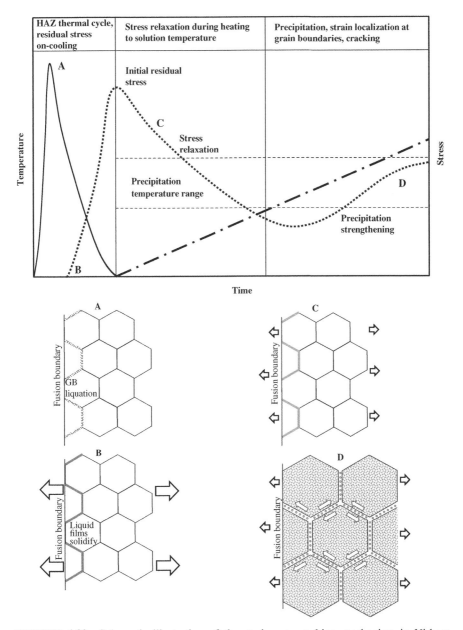

FIGURE 4.39 Schematic illustration of the strain-age cracking mechanism in Ni-base superalloys.

relax, and upon heating into the precipitation temperature range, additional contraction stresses accumulate due to precipitation. The combination of precipitation and relaxation stresses results in localized stress at the grain boundaries in *Stage D* that can lead to IG cracking.

4.4.2 Factors That Influence SAC Susceptibility

There are a number of factors that influence susceptibility to SAC. These include (i) composition, (ii) grain size, (iii) residual stress and restraint, and (iv) welding procedure. These are discussed in the following sections. It should be noted that susceptibility to SAC may involve a number of these factors.

4.4.2.1 Composition As discussed and illustrated in Figure 4.32, the susceptibility of superalloys to SAC is a strong function of the total amount (wt%) of precipitation strengthening elements in the alloy itself. Those alloys with higher total Al + Ti content are more susceptible to SAC than those that are lower in Al + Ti content. Note that all alloys with Al content higher than 3 wt% and alloys with 3–5 wt% Ti and more than 2 wt% Al are susceptible to SAC. This is related to the fact that as the Al + Ti content increases, (i) aging occurs more rapidly during PWHT (Fig. 4.33), (ii) the volume fraction of strengthening precipitate increases, and (iii) aging contraction stresses increase. An increasing volume fraction of γ' in the superalloys has the net effect of reducing ductility and increasing the local stresses by contraction during aging. This, in turn, lowers the ductility of the alloy and increases the tendency for cracking. By using niobium as the primary strengthening element, alloys such as Alloy 718 are effectively resistant to SAC due to the sluggish aging response of γ'' precipitation (Fig. 4.34) and the reduced contraction stresses that develop, as shown in Figure 4.35. It should be noted that alloys strengthened by γ'' may not be completely immune to SAC, but the slower precipitation reaction provides more time for the weld residual stresses to relax to a level where SAC is not possible.

The effect of other elements on SAC is not so clear. Hughes and Berry [88] showed that heats of René 41 with reduced levels of carbon are more resistant to SAC, while Koren *et al.* [91] found that low carbon content is detrimental to SAC resistance in Alloy 713C. It is proposed that the lower carbon content results in fewer carbides to pin the grain boundaries and prevent boundary migration in the HAZ. Thus, the lower-carbon alloys would be more prone to grain coarsening. As discussed in the next section, coarser grain size increases susceptibility to SAC.

Boron is added to many superalloys to improve stress rupture (creep) properties. Thamburaj *et al.* [92] found that higher levels of boron were related to the improved resistance of René 41 to SAC. Carlton and Prager [93] showed that the presence of oxygen was a prerequisite for SAC in René 41 and stated that oxygen segregation to grain boundaries will reduce grain boundary strength. They indicated that oxygen will have a similar effect with Alloy 718 and Waspaloy. There have been virtually no studies on the effect of the impurities sulfur and phosphorus on SAC in these alloys. While it is presumed that these impurities will have a negative effect, in practice, the low levels of S + P present in most superalloys make their practical influence negligible.

As noted previously, many of the alloys that are susceptible to SAC may also experience HAZ grain boundary liquation. If HAZ liquation cracks form, they can act as stress concentrators along the grain boundaries during PWHT. It is unclear how HAZ grain boundary liquation influences SAC. Elements such as Ti and Nb that are added as strengthening agents also form Ti- and Nb-rich MC-type carbides that are susceptible to constitutional liquation. Boron has high grain boundary affinity and can also promote grain boundary liquation.

4.4.2.2   Grain Size   Fine-grained materials increase the amount of grain boundary area and have been found to be more resistant to SAC than those that are coarse grained [88, 93]. Presumably, the increased grain boundary area of fine-grained alloys provides greater opportunity for uniform stress relaxation by grain boundary sliding. In addition, embrittling phases that may form at the grain boundaries are spread over a wider area forming a layer that is either thinner or discontinuous. It can also be argued that the fine grain size reduces the unit strain per grain boundary and thus stresses resulting from relaxation and/or aging are better accommodated in the structure, minimizing the strain localization at individual boundaries. These arguments are similar to those used to explain the beneficial effect of fine grain size on susceptibility to HAZ liquation cracking in Section 3.3.1. Use of fine-grained base materials for avoiding SAC may not always be an option since most Ni-base superalloys are used at elevated temperatures where coarser grain size provides better creep resistance.

4.4.2.3   Residual Stress and Restraint   It is widely believed that SAC is primarily the result of high residual stresses resulting from welding that are relaxed preferentially in the HAZ at the same time that the ductility of the HAZ is reduced by metallurgical (precipitation) reactions. In addition to welding residual stresses, it has also been suggested that stresses due to local thermal expansion and contraction during aging contribute to the stresses responsible for SAC [88, 90].

The level of weld residual stresses developed depend on heat input, component geometry, mechanical properties of the material, and elastic stiffness of the restraining elements. As with other materials, the residual stresses in heavily restrained superalloy weldments are considered to be on the order of the yield strength of these materials. In the HAZ, this would correspond to the yield strength of the alloy in the solution-annealed condition. The thermal stresses in complex assemblies are likely to be strongly influenced by component geometry and uniformity of heating. High thermal stresses can be generated by relatively small temperature differences in cases where the material is fully restrained against expansion. Since welds in turbine engine components are often severely restrained, the magnitude of stresses associated with stress relaxation during PWHT can be considerable.

Aging contraction stresses result from the precipitation of γ', which tend to increase with the volume fraction of γ' precipitation. Such contraction can lead to the development of significantly high stresses. Although the overall contraction of the component is important, the difference in aging contraction between the HAZ and the base metal may be of particular concern. This difference may lead to excessive localized straining in the HAZ and aggravate the tendency for SAC, particularly in the case where the base metal is in the fully aged condition. The stresses that accumulate by aging in the HAZ can be clearly seen from the work of Norton and Lippold [76] who showed stress increases on the order of 250–350 MPa (35–50 ksi) in simulated HAZs in Alloy 718 and Waspaloy that were aged for over 2 h (Fig. 4.35). Fawley and Prager [94] showed that for René 41, PWHT cracking could be avoided when the aging stresses were reduced by a more sluggish precipitation of γ'.

The propensity for SAC in a given alloy increases with the degree of weldment restraint. Even highly crack-resistant materials such as Alloy 718 can be susceptible to cracking during PWHT if the restraint level produces exceedingly large residual stresses [84, 87]. These residual stresses are then relaxed during PWHT, leading to cracking. The data of Norton and Lippold [76] showed that Alloy 718 exhibits a ductility curve similar to Waspaloy (Figure 4.36) but the minimum ductility for Waspaloy is much lower (~10% vs. 25%), resulting in higher susceptibility. If the ductility of Alloy 718 can be exhausted by high strains resulting from stress relaxation in the HAZ, cracking of this alloy is also possible. Thus, control of residual stresses during welding can be a good method for preventing SAC in superalloys that have moderate SAC susceptibility.

Many investigations have shown that the base metal should be soft (solution-annealed or overaged condition) to allow the stresses developed during welding and during PWHT to be relaxed more uniformly in the structure and not concentrated in the HAZ. Welds made on solution-annealed base metal have been shown to have considerably more resistance to SAC than those made on mill-annealed or fully aged metal [84]. Superior resistance to PWHT cracking in René 41 was obtained through slow cooling from the solution-annealing temperature, which led to coarse, overaged γ' precipitation and resulted in a softer base metal [92]. A two-step overaging treatment (solutionizing 1170°C (2140°F)/4 h/forced air cool and aging 1080°C (1975°F)/16 h/furnace cool to 1010°C (1850°F)/4 h/Ac) resulted in excellent resistance to PWHT cracking in Udimet 700 [90]. Again, the overaging of the γ' precipitates leads to an overall softening of the base metal and stabilizes these precipitates during reheating to the solution-annealing temperature during PWHT. The net effect is to reduce the local stresses that concentrate in the HAZ during PWHT.

4.4.2.4 *Welding Procedure*

Residual stresses can be reduced and metallurgical damage can be minimized by reducing the weld heat input. Wu and Herfert [86] showed that low weld heat input inhibits the precipitation of deleterious carbide films along the grain boundaries in René 41. Low heat input welding techniques may generally be useful as a partial solution to SAC, but it is unlikely that a complete solution can be obtained by simply "tweaking" welding parameters. Very low heat input can be obtained by using EB welding. But certain alloys develop HAZ liquation cracks as a result of EB welding, which may aggravate the degree of PWHT cracking during subsequent PWHT.

Preheating has been shown to be useful in reducing SAC, but the preheat temperatures may be extremely high. For example, in a study by Duvall and Doyle [95], Alloy 713C vanes were preheated to 538°C (1000°F) and held at this temperature while repair welds were made. Preheating and welding were carried out in an inert atmosphere. This procedure was found to substantially decrease the degree of both hot cracking (solidification and HAZ liquation cracking) and SAC. King *et al.* [96] found that high preheat temperatures, 705–955°C (1300–1750°F), successfully avoided cracking during welding and PWHT in high-strength, cast superalloy vanes. In the turbine engine industry, this technique is often referred to as "SWET" welding (superalloy welding at elevated temperature) [97].

The weld joint geometry can also be an important factor influencing SAC, since it can influence the restraint level in the weldment. Various studies [98–100] indicate that changes in weld bead contour influence the tendency for HAZ liquation cracking. Liquation cracks in EB welds are known to occur preferentially in the "nailhead" area of the weld and below [84]. These liquation cracks can then become initiation sites for SAC during subsequent heat treatment.

The use of lower-strength, more ductile filler metals such as Alloy 625 (a solid-solution-strengthened Ni-base alloy) was found to make the repair welds in Alloy 713C vanes resistant to SAC [84]. However, it should be recognized that the Alloy 625 weld metal cannot be substantially strengthened during PWHT.

4.4.2.5 *Effect of PWHT*

4.4.2.5 Effect of PWHT   Most nickel-base superalloys require a full solution anneal and aging treatment following welding in order to restore mechanical properties. Simple aging after welding is usually not sufficient since the weld metal and HAZ cannot be restored to full strength and there is the possibility of overaging the base metal (if welding is performed on fully aged material). In addition, the aging temperature in most alloys is not high enough to allow sufficient stress relief and severe cracking may occur.

As shown in Figure 4.31, rapid heating to the solution temperature may be effective in preventing SAC. This is possible since the material reaches the solution-annealing temperature where stress relaxation occurs and precipitation is suppressed. For small components or alloys with low or moderate (Ti+Al) content, such an approach may be possible. In such cases, both the weld residual stresses and stresses associated with precipitation are eliminated or avoided. In large components, such rapid heating approaches are usually not possible and can even be more damaging, since temperature gradients within the component can create large thermal stresses. Most PWHT of Ni-base superalloys is conducted in vacuum furnaces to avoid component oxidation, further restricting the rate at which the solution-annealing temperature can be reached.

A stepped-heating technique may be effective in some situations, particularly where weld residual stresses are not excessive. This technique involves slowly heating the component to about 500°C (930°F) and soaking at this temperature to reduce thermal gradients throughout the component and relieve some of the residual stress. The component is then rapidly heated through the crack-sensitive temperature range to the solution temperature. The success of such a technique is dependent on the original level of residual stress, the amount of residual stress that can be relieved during the low-temperature soak, and the rapidity at which the component can be heated through the precipitation range.

It has also been reported that a protective atmosphere during heat treatment is beneficial [91]. PWHT cracking was eliminated in high-purity dry argon, argon containing less than 0.5% oxygen, and vacuum atmospheres. The detrimental effect of an oxygen-rich atmosphere is thought to be due to the rapid diffusion of oxygen along grain boundaries and the consequent formation of oxides that are not able to resist plastic deformation during stress relaxation [93]. It is clear that oxygen is at best contributory and that the exclusion of oxygen from the heat treatment environment cannot eliminate SAC in most cases. For example, D'Annessa and Owens [101] showed that PWHT in a vacuum may eliminate cracking only in materials that have marginal susceptibility to SAC (such as Waspaloy and René 41), but is not effective in highly susceptible alloys.

Localized solution treatment at 1065°C (1950°F) for 5 min of repair welds in René 41 was found to prevent failure during subsequent aging [102]. The effects of the localized solution treatment were to (i) homogenize the weld metal, (ii) cause the aging precipitates to be more uniformly and finely distributed, and (iii) prevent the precipitation of carbides since fast cooling rates are associated with such treatments. However, the role of the various influences observed was not explained. In addition, care must be taken to avoid producing sufficiently high thermal stresses to leave high residual stresses on cooling.

4.4.3 Quantifying Susceptibility to Strain-age Cracking

There are a number of tests that have been used to quantify susceptibility to SAC. As with tests developed for other cracking phenomena, the techniques are of either the "self-restraint" or "simulative" (externally loaded) type. Generally speaking, the same techniques used for quantifying reheat cracking and discussed in Section 4.3.6 are used for SAC. A few of those tests are described here.

Prager and Shira [84] discuss a number of tests for evaluating susceptibility to SAC including the circular patch test, a "plug-weld" test, and a controlled heating rate test. The circular patch test has been widely used to study susceptibility to SAC, but it is a "go–no go" test. The test sample either cracks or it does not. In tests with René 41, they found that this test only identified the most susceptible heats of material. Heats with intermediate susceptibility, which may in fact be crack susceptible in practice, could not be readily identified. Similar problems are encountered with other self-restraint tests, as discussed in Chapter 9. Another disadvantage of the self-restraint tests is that the samples must often be sectioned in order to determine if cracking has occurred since the cracking may not be evident on the surface.

Because of the disadvantages of the self-restraint tests, a number of tests that either apply a stress or allow stress relaxation to occur have been developed. The controlled heating rate test is essentially a high-temperature tensile test where a sample is heated into the aging temperature range and then pulled to failure. Some data for René 41 from Prager and Shira [84] is shown in Figure 4.40. This test shows a minimum in ductility over a narrow temperature range. Materials with higher minimum ductility were deemed to be more resistant to SAC. Note that this test was used to evaluate base materials of different composition (and presumably grain size), but did not specifically test the weld HAZ.

Researchers at Haynes International developed a modified version of the controlled heating rate test using a Gleeble thermomechanical simulator [103, 104]. Thin-sheet material machined into reduced gage tensile samples is first heated to

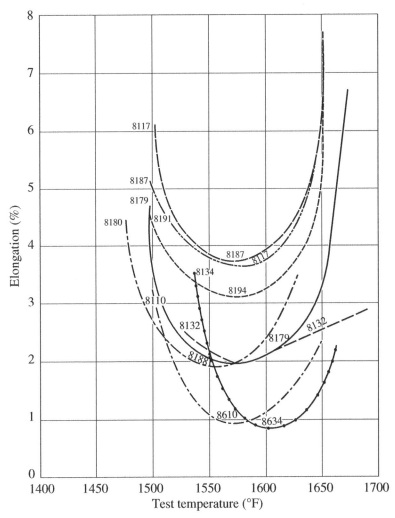

FIGURE 4.40   Controlled heating rate results for René 41 (From Ref. [84]. © WRC).

593°C at 56°C/s (1100°F at 100°F/s), then heated to 788°C at 17°C/s (1450°F at 30°F/s), and finally pulled to failure at 788°C at a rate of 1.6 mm/min (0.063 in./min). Total elongation is used to determine ductility. The test temperature of 788°C was selected as a compromise among a number of Ni-base superalloys based on minimum ductility in earlier testing. As such, it does not necessarily represent the absolute minimum for any of the materials tested. Typical results from this test are shown in Figure 4.41 for two different sample geometries [104]. A solid-solution-strengthened alloy (Hastelloy X) was added to the test matrix for comparison purposes.

The Welding Institute in the United Kingdom developed a constant load rupture test to evaluate samples in a variety of microstructure conditions, including simulated HAZs [82]. This test used a notched tensile sample that was heated to a predetermined

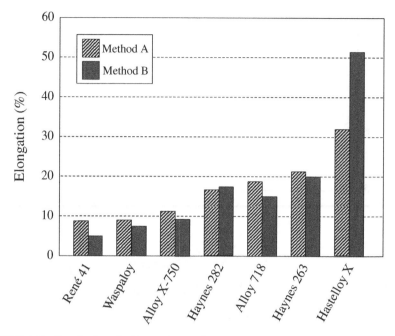

FIGURE 4.41   Results from the Haynes International controlled heating rate tensile test at 788°C (1450°F). Methods A and B refer to different sample geometries (From Ref. [104]. © AWS).

temperature and held under constant load until failure occurs or sample run out (10,000 min). The use of a notch in the sample allowed a critical stress intensity to be determined that could be used to predict failure. An example is shown in Figure 4.42.

Many of the other tests for SAC susceptibility used a Gleeble thermomechanical simulator [73]. Wu and Herfert [86] used the Gleeble to first simulate the HAZ in small René 41 tensile bars. A notch was then machined in the sample, and successive samples were loaded in constant tension at increasing stress levels (24–33 ksi) until failure occurred. The stress to cause failure was used as a measure of SAC susceptibility. Note that this test did not allow for any stress relaxation since the stress was held constant during the test. Duvall and Owczarski performed a HAZ simulation (including heating in the PMZ) and then used constant displacement at different PWHT temperatures to load the sample [89]. The sample was periodically cooled to room temperature to check for cracking. Using this technique, a C-curve cracking response was developed, such as the one for Waspaloy shown in Figure 4.43. Because this test used constant displacement, rather than constant stress, it allows for stress relaxation after the initial stress (displacement) is applied.

Franklin and Savage [81] used a technique similar to Duvall and Owczarski but included constant displacement control to study the effect of base metal pretreatment on SAC susceptibility in René 41. Their results on the base metal in the solution-annealed condition showed the effect of initial stress relaxation and then precipitation strengthening at longer hold times, as shown in Figure 4.44a. If the base metal

(a)   (b)

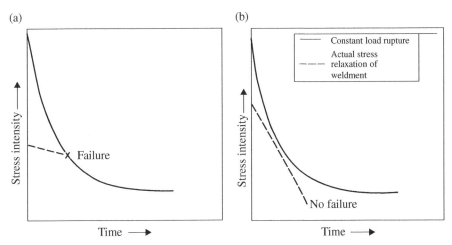

FIGURE 4.42 Constant load rupture test results used to predict weld failure based on stress relaxation (From Ref. [82]. © AWS).

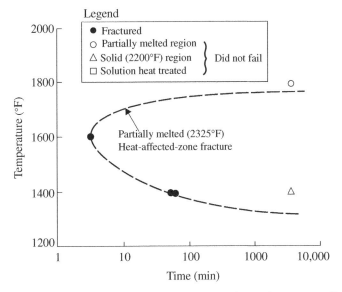

FIGURE 4.43 Time–temperature failure curve for Waspaloy under constant displacement testing (From Ref. [89]. © AWS).

was tested in the overaged condition, only stress relaxation occurs with no precipitation hardening at longer hold times, as shown in Figure 4.44b. They concluded that since precipitates would be dissolved in the high-temperature region of the HAZ, there was no advantage to overaging the base metal prior to welding.

Dix and Savage [105] developed a Gleeble-based test to evaluate the ductility response of the base metal as a function of displacement rate over a range of PWHT

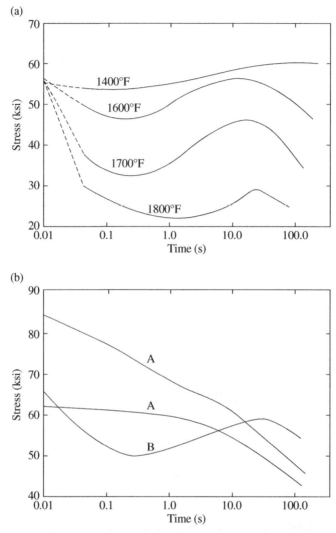

FIGURE 4.44 Stress relaxation behavior of René 41. (a) Effect of time and temperature on isothermal stress relaxation behavior of the solution-annealed condition and (b) behavior at 870°C (1600°F) for overaged (A) and solution-annealed (B) condition (From Ref. [81]. © AWS).

temperatures. These tests simply heated the sample to a given temperature and then pulled the sample to failure at a fixed displacement rate. Results for Inconel X-750 are shown in Figure 4.45. A ductility minimum is observed around 1600°F (870°C) at both displacement rates, and lower displacement rates result in lower ductility over a range of temperature. This reflects the fact that precipitation of γ' occurs much more rapidly at 870°C than at 700°C and that at the lower displacement rates, there is more time for precipitation to occur. It should be noted that the x-axis in Figure 4.45b

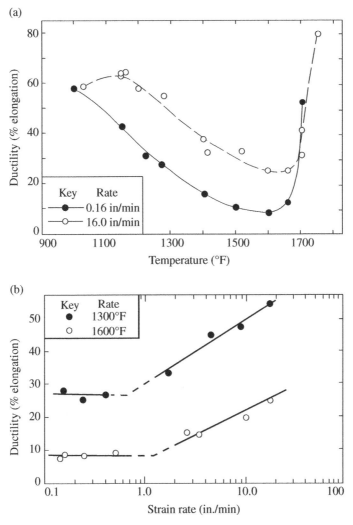

FIGURE 4.45   Strain-age cracking tests of Inconel X-750. (a) Effect of test temperature on ductility at two displacement rates and (b) effect of displacement rate and temperature on ductility (From Ref. [105]. © AWS).

indicates strain rate, while the term "displacement rate" is more appropriate for Gleeble-based testing since a temperature gradient exists along the gage section and the actual length of the isothermal hot zone in the sample is unknown, making it very difficult to determine the actual strain and strain rate.

There are a number of problems with the simulative tests for SAC that are described earlier. Many of them do not include an initial HAZ simulation thermal cycle. Since SAC almost always occurs in the high-temperature HAZ, it is necessary to precondition the sample microstructure using a thermal cycle representative of that

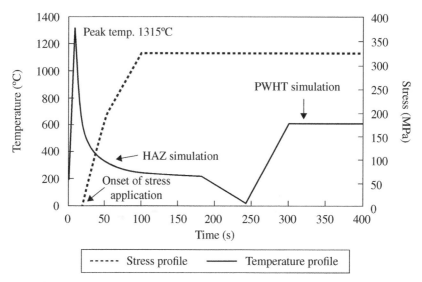

FIGURE 4.46   Lehigh stress relief cracking cycle (From Ref. [75]. © AWS).

experienced in the HAZ. This is a critical consideration since the effect of the base metal microstructure condition is effectively removed in this region and the HAZ prior to PWHT is essentially in the solution-annealed condition. Even if the base material is in the solution-annealed condition, the HAZ thermal cycle can still lead to grain growth and impurity segregation that can influence SAC susceptibility.

Another problem with some of these tests is that they do not allow for stress relaxation, which is a key element in the SAC mechanism. Samples tested under constant load (or stress) or constant extension rate do not allow relaxation to occur. Thus, the fixed displacement tests after a HAZ simulation thermal cycle are the most appropriate in terms of approximating actual conditions. Thus, of the simulative tests described earlier, that of Duvall and Owczarski [89] most closely simulates the actual conditions under which SAC occurs.

However, even this test ignores one key element critical to SAC. The tests that impose a HAZ thermal cycle to condition the microstructure prior to PWHT typically cool the sample back to room temperature under no load. This ignores the fact that under ordinary welding conditions, residual stress accumulates in the HAZ during cooling. Recognizing this, Gleeble-based tests developed by Balaguer *et al.* [74], Nawrocki *et al.* [75], and Norton and Lippold [76] have incorporated a displacement during cooling that results in a tensile residual stress at room temperature. The test used by Nawrocki *et al.* to evaluate stress relief cracking in low-alloy steels maintains this tensile stress on the sample throughout the PWHT cycle, as shown schematically in Figure 4.46. In practice, the tensile residual stress relaxes when the sample is heated to the PWHT temperature. This is the reason why tests using this constant load technique normally fail in relatively short times.

Recognizing this deficiency, the PWHT cracking technique developed by Norton and Lippold at OSU [76] maintains the sample in fixed displacement after cooling to room temperature. Upon heating to the PWHT temperature, the tensile stresses initially relax, and then, with time, the stress increases as contraction due to precipitation occurs. The rate and magnitude of this increase are indicative of SAC susceptibility, as shown in Figure 4.35. Note that the stress versus time curves shown in this figure represent the stress increase above the minimum relaxation stress after heating to the PWHT temperature.

Another problem encountered with Gleeble-based PWHT cracking tests is the large variation in time to failure when testing duplicate samples. In order to eliminate this variability and reduce testing time, Norton would terminate the test after a predetermined time (up to 4 h) and pull the sample to failure at the test temperature. Using this approach, ductility curves at specific times and temperatures can be obtained, as shown in Figure 4.36. The OSU PWHT cracking test is described in more detail in Chapter 9.

4.4.4 Identifying Strain-age Cracking

By definition, SAC only occurs in the HAZ of Ni-base superalloys usually strengthened by γ', $Ni_3(Ti, Al)$, precipitation. Strain-age cracks form during the PWHT cycle, so if cracks are present immediately following welding, they cannot have formed by a strain-age mechanism. Strain-age cracks are always IG and typically occur in close proximity to the fusion boundary. In some alloys, cracks may initiate and propagate in the PMZ. SAC is typically not associated with the fusion zone.

Because many of the alloys susceptible to SAC are also susceptible to HAZ liquation cracking, it may be difficult to distinguish these two forms of cracking using only optical metallography. As noted in Chapter 3, HAZ liquation cracks typically form along grain boundaries perpendicular to the fusion boundary and are limited to the narrow region over which liquid films are present. Strain-age cracks result from the accumulation of strain along HAZ grain boundaries and will form in response to stress relaxation and strain accumulation at the boundary. In many cases, this will result in cracks that run parallel to the fusion boundary.

Fractography may be required to verify SAC. Fracture is always macroscopically IG. At the microscopic level, the fracture surfaces may have either smooth or ductile IG features, as illustrated by the fractograph in Figure 4.38. In the latter case, the ductile dimples may be extremely fine, requiring high magnification in the SEM to resolve. Impurity segregation to grain boundaries is not a requirement for SAC, so the presence of impurities does not confirm or refute the presence of strain-age cracks.

4.4.5 Preventing Strain-age Cracking

SAC in the precipitation-strengthened Ni-base alloys occurs due to the local accumulation of strain and the concomitant hardening of the microstructure due to precipitation. Local strains develop due to both the relaxation of weld residual stresses and precipitation-induced stresses. Nonuniform heating during PWHT may also contribute some thermally induced stresses. The combination of straining and hardening due to precipitation is what leads to SAC.

To reduce or eliminate SAC, the following steps may be taken. A combination of these approaches may be necessary to avoid cracking:

- Minimize residual and thermally induced stresses by appropriate joint design and choice of welding process and materials. The use of lower weld heat input and smaller weld beads to minimize residual stress is often effective.
- Reduce the strength of the weld metal and/or base metal. The use of a filler metal with lower strength than the base metal and/or a solution-annealed base metal reduces residual stress in the HAZ. It has also been shown that overaging the base metal in γ'-strengthened alloys can reduce cracking susceptibility [90].
- During PWHT, heat as rapidly as possible through the precipitation temperature range (see Figure 4.31). This suppresses intragranular precipitation and allows residual stress to relax uniformly in the microstructure. This approach is limited by two factors: (i) the ability to heat rapidly based on component size and (ii) the kinetics of precipitation. As the (Ti+Al) content increases, the nose of the precipitation curve occurs at much shorter times (see Fig. 4.33), which does not allow the suppression of precipitation.
- Avoid partial melting along grain boundaries adjacent to the fusion boundary. Although it has not been conclusively shown that the presence of a PMZ increases susceptibility, the formation of HAZ liquation cracks clearly contribute to SAC since they can act as initiation points for cracking. Many of the Ni-base alloys are inherently susceptible to HAZ grain boundary liquation/cracking, so the only practical approach is to minimize the weld heat input and promote steeper HAZ temperature gradients, which reduce the region over which liquation occurs.
- Minimize HAZ grain size. Since SAC occurs due to strain accumulation at the grain boundary during the period of stress relaxation, a finer grain size in the HAZ will reduce the local strain on individual grain boundaries. This can also be accomplished by minimizing the grain size of the base metal, reducing heat input, or using high energy density (HED) processes that create steep temperature gradients in the HAZ. It should be noted that fine grain size in the base metal may not be desirable in the application of many Ni-base superalloys, since this can compromise the elevated temperature creep properties.
- Select alloys with lower Ti+Al content. This is an obvious solution that is often not practical since the alloy was selected to meet elevated temperature properties that require high Ti+Al content. Lower Ti+Al content will allow slower heating to the solution-annealing temperature without the onset of precipitation. The use of alloys that substitute Nb as a strengthening agent (such as Alloy 718) and form γ'', Ni_3Nb, greatly improves resistance to SAC.

4.5 LAMELLAR CRACKING

Lamellar cracking is a HAZ cracking phenomenon usually associated with plain-carbon or low-alloy steels. In the 1960s and 1970s, there was considerable research conducted on this form of cracking, also known as lamellar tearing. As will be described in this section, susceptibility to lamellar cracking is primarily controlled by steel cleanliness. Impurities such as sulfur and oxygen can promote the formation of intermetallic inclusions during steel processing that serve as the initiation sites for lamellar cracks. With the advent of "clean steel" technology in the 1980s, incidents of lamellar cracking have dramatically decreased. Another form of this cracking, sometimes referred to as "delamination cracking," has been observed in aluminum alloys.

Lamellar cracking results from the local decohesion in the HAZ that is associated with intermetallic stringers and/or a directional "texture" in the material that results from thermomechanical processing. In practice, lamellar cracking is almost always associated with rolled plate material, particularly in C–Mn steels in section thicknesses exceeding one inch. Most of the discussion in this section will concentrate on lamellar cracking in steels.

4.5.1 Mechanism of Lamellar Cracking

There are four factors that contribute to lamellar cracking. These include (i) low through-thickness ductility of the parent plate (often termed "short-transverse" ductility), (ii) the use of thick plates that increase restraint, (iii) high volume fraction of elongated stringers or inclusions along the rolling direction, and (iv) a weld joint geometry or welding practice that generates large through-thickness stresses.

Failure by lamellar cracking generally occurs in a region of the HAZ in C–Mn or low-alloy steels, which is just outside the transformed region, that is, an area heated to temperatures just below the lower critical (A_1) temperature. Crack initiation occurs locally and preferentially between the ferritic–pearlitic matrix and elongated stringers or inclusions.

The stringers and inclusions are remnants of the ingot solidification and subsequent deformation processes. For instance, sulfur is rejected along grain and subgrain boundaries during solidification. During the rolling process (to form plate), the sulfur, now in the form of manganese sulfide, is oriented via the deformation process along the rolling direction. If the sulfur content of the original ingot material was relatively high (>0.05 wt%), the "stringers" of MnS can be nearly continuous along distinct rolling bands in the longitudinal (or rolling) direction. At lower sulfur levels, the MnS becomes discontinuous, until at very low levels (<0.005 wt%) the MnS is distributed as discrete particles. Oxide and silicate stringers can form by the same mechanism and may also promote lamellar cracking if present in nearly continuous arrays.

An example of a lamellar crack in metallographic cross section is shown in Figure 4.47 from Threadgill [106]. The fusion boundary is indicated by the dotted line. Note that the lamellar cracks propagate in a "stairstep" fashion since they form at the stringer/matrix interfaces that run roughly parallel to the rolling direction. Individual lamellar cracks initiate and propagate by linking up with other cracks. Note that these cracks are at some distance from the fusion boundary and typically propagate through the untransformed region of the steel HAZ at temperatures below approximately 700°C (1290°F).

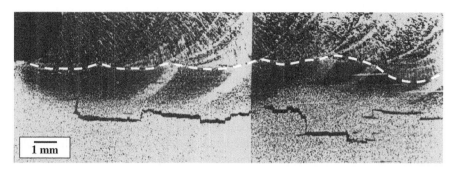

FIGURE 4.47   Lamellar cracking in a C–Mn steel (From Ref. [106]. © Maney).

Crack initiation and propagation are primarily along the stringer/matrix inter-face, resulting from decohesion at that interface. Once the crack forms, considerable plastic strain will be present at the crack tip, and ductile tearing may occur to connect the individual cracks. When these cracked regions are on different levels, the cracking will take on the characteristic "stairstep" appear-ance of lamellar cracking. Because of the nature of crack propagation, the fracture surface exhibits a very distinct "terraced" appearance where the terraces represent decohesion at the stringer/matrix interface and the ligaments joining these terraces represent regions of ductile tearing. A schematic of the decohe-sion and ductile tearing process is shown in Figure 4.48, as adapted from Farrar *et al.* [107].

Although the intermetallic inclusions persist in regions of the HAZ heated above the A_1 temperature, lamellar cracking does not occur in this region. Presumably, this is because the matrix strength decreases upon transformation to austenite and the intermetallic/matrix interface exhibits higher inherent ductility at temperatures above the A_1. The intermetallics do not completely dissolve at temperatures between the A_1 and the solidus (fusion boundary), but there may be some partial dissolution and modification of the interface that promote higher cohesive strength relative to the subcritical region of the HAZ.

Lamellar cracking will only occur when the welding stresses are high enough to exhaust the through-thickness ductility of the base metal and subcritical HAZ. An example of a configuration where high stresses can lead to lamellar cracking is shown in Figure 4.49 [106]. The combination of high welding-induced stress and restraint from the base plate results in lamellar cracking in the short-transverse tensile direction normal to the rolling direction.

A form of lamellar cracking is also observed in aluminum alloys and is some-times called delamination cracking or simply delamination. The underlying mecha-nism is different than for steels, since it is not normally associated with impurity stringers. Rather, aluminum alloys often have a strong texture in the rolling direction that imparts low through-thickness ductility. This form of cracking has been

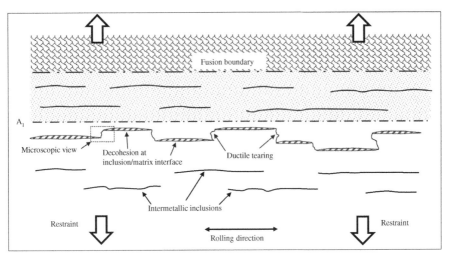

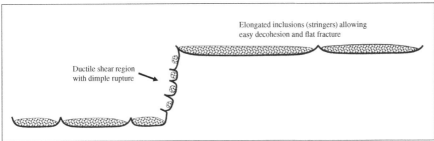

FIGURE 4.48 Schematic illustration of the decohesion and ductile tearing processes associated with lamellar cracking (Adapted from Ref. [107]).

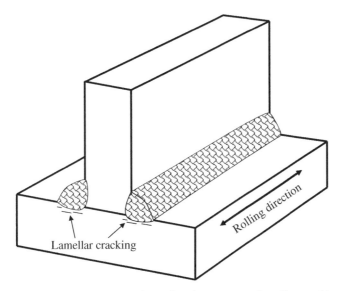

FIGURE 4.49 Example of weld configuration that promotes lamellar cracking (Adapted from Ref. [106]).

(a)

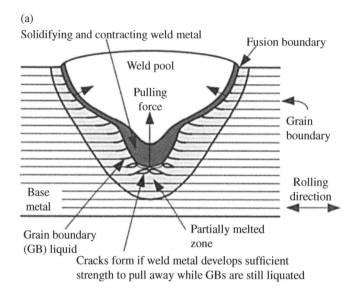

Cracks form if weld metal develops sufficient
strength to pull away while GBs are still liquated

(b)

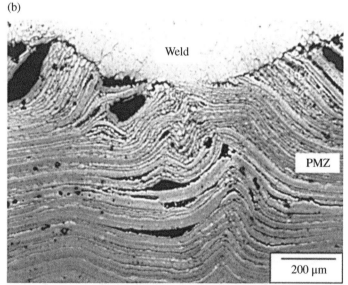

FIGURE 4.50   Delamination cracking in the HAZ of an aluminum alloy 7075 (From
Ref. [108]. © Wiley).

observed in many of the 2000-series (Al–Cu) alloys, particularly in those containing
lithium as an alloy addition. Such alloys include 2090 and 2195, the latter of which
is used extensively in aerospace and military applications. An example of delamination cracking in precipitation-hardened aluminum Alloy 7075 is shown in Figure 4.50
from Kou [108]. While the authors consider this delamination a form of HAZ liquation cracking, it is clear that much of this cracking occurs in the solid state.

4.5.2 Quantifying Lamellar Cracking

A number of test techniques have been used to quantify lamellar cracking in steels. Most of these tests are of the self-restraint type where welds are applied under high-restraint conditions to promote through-thickness stresses. Two examples of such tests are shown in Figures 4.51 and 4.52 [109, 110]. The Cranfield test consists of a test plate and support plate fixed at an angle between 45 and 60°. Multipass welding is performed at the intersection of these plates to generate high through-thickness stresses in the support plate. Welding conditions can be varied to investigate the effect of heat input, bead size, and other variables on the susceptibility to cracking. The "window" test developed at The Welding Institute uses a similar principle but with single-pass fillet welds applied between the restraint plate and the test plate. There is no direct quantification of susceptibility using these tests, and the test plates must be sectioned and inspected using metallographic techniques to identify cracks.

The Lehigh lamellar cracking test incorporates an external load on a welded cantilever beam in order to identify a critical weld restraint level (CWRL) to induce cracking. A schematic of the test fixture is shown in Figure 4.53 [111, 112]. The CWRL, reported as stress on the weld joint, was then correlated to susceptibility to lamellar cracking. Using this test, the effect of welding variables, number of passes, and material composition and condition could be quantified. As an example, the effect of the steel oxygen content on susceptibility to lamellar cracking is shown in Figure 4.54. Oxygen is thought to reduce the cohesive strength of the interface between the stringers (silicates and sulfides), thereby allowing void nucleation and growth at lower applied stresses [111].

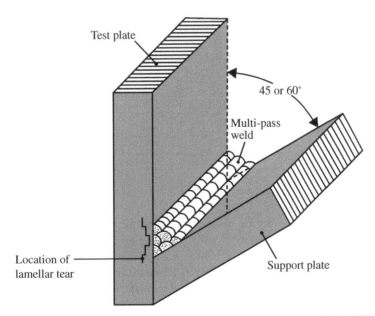

FIGURE 4.51   Cranfield test for lamellar cracking (From Ref. [109]. © AWS).

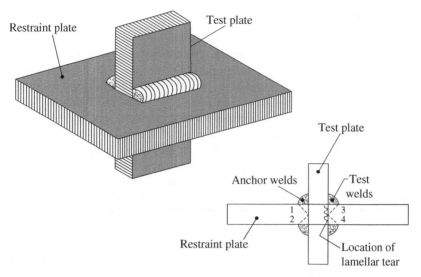

FIGURE 4.52   Window test for lamellar cracking (From Ref. [109]. © AWS).

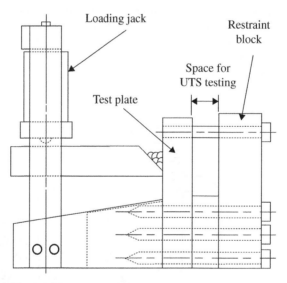

FIGURE 4.53   Lehigh lamellar cracking test fixture (From Ref. [111]. © AWS).

There are also a number of indirect (nonwelding) tests to quantify lamellar cracking. A simple short-transverse tensile test can give some indication of susceptibility. As ductility (determined by reduction in area) decreases, susceptibility to lamellar cracking increases since ductility in the short-transverse direction is dictated by the volume fraction and nature of inclusions. As inclusion content increases and the morphology becomes more "stringer-like," ductility will decrease. These tensile tests can be conducted at room temperature or elevated temperatures representative

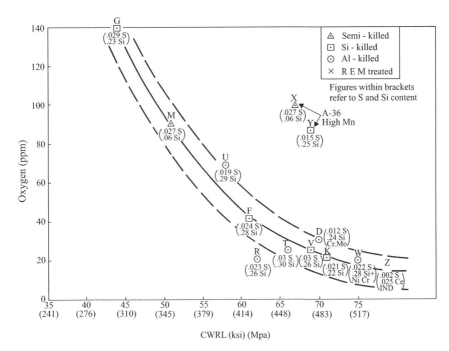

FIGURE 4.54 Effect of oxygen content on lamellar cracking using the Lehigh test (From Ref. [111]. © AWS).

of the subcritical HAZ. In plate thicknesses below 1 in. (~25 mm), tensile testing in the short-transverse direction may be difficult because of the restriction on tensile sample length.

Dickinson *et al.* [109] developed a compression test to evaluate through-thickness ductility and compared results of this test to the CWRL of the Lehigh test. In the limited testing that was conducted, the correlation was very good, as shown in Figure 4.55. Relative to testing in tension, this test has the advantage that shorter test samples can be used allowing evaluation of plate thicknesses less than 1 in.

4.5.3 Identifying Lamellar Cracking

Lamellar cracking in steels is one of the easiest forms of weld cracking to identify. As shown in Figures 4.47 and 4.48, cracks normally exhibit an irregular, "stairstep" appearance since they initiate along inclusion stringers oriented in the rolling direction of the plate and propagate between stringers by ductile overload. Lamellar cracks normally form in the subcritical region of the HAZ, although propagation into the intercritical region (or beyond) may be possible. Lamellar cracking in steels is typically not observed in the CGHAZ.

Steel plate exceeding 1 in. (~25 mm) in thickness is most susceptible. The most susceptible steels have apparent nonmetallic inclusions distributed in a linear fashion along the rolling direction. These can be easily identified in metallographic cross

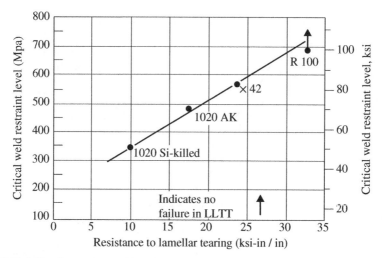

FIGURE 4.55 Comparison of lamellar cracking susceptibility using the Lehigh test and compression test (From Ref. [109]. © AWS).

section in either the etched or unetched (as-polished) conditions. These inclusions are most often sulfides or silicates that develop a directional orientation along the rolling direction during hot rolling. Steels susceptible to lamellar cracking often have high levels of sulfur and oxygen.

The fracture surface usually consists of a number of plateaus or terraces. Ductile tearing (dimpled rupture) is typical in the segments connecting the plateaus, while the fracture mode on the plateau (representing the stringer/matrix interface) may vary from flat fracture to microscopic dimpled rupture. An example of a fracture surface where cracking occurred at the stringer/matrix interface is shown in Figure 4.56 [113]. Note that there is a very high density of inclusions on this fracture surface.

Lamellar cracking usually occurs during welding due to both intrinsic and extrinsic welding stresses. In some cases, it has been observed after a certain time delay suggesting that <u>hydrogen embrittlement</u> may contribute to the cracking mechanism. This would require sufficient time for hydrogen to diffuse from the weld metal to the region where lamellar cracking occurs.

4.5.4 Preventing Lamellar Cracking

As mentioned previously, reports of lamellar cracking have diminished significantly since the widespread introduction of "clean steel" technology in the 1980s. Lower levels of sulfur and oxygen in steels greatly reduce the formation of intermetallic inclusions in the form of stringers along the rolling direction. Hence, selection of "clean" steels is the most effective method for preventing lamellar cracking.

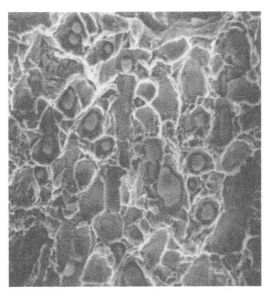

FIGURE 4.56 Fracture surface of lamellar crack typical of the terraced region of the crack with high fraction of inclusions (From Ref. [113]. © TWI).

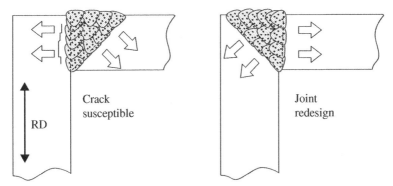

FIGURE 4.57 Example of joint redesign to avoid lamellar cracking.

Some steels may contain additions of rare earth elements to control inclusion morphology. These steels contain elements such as cerium that react with sulfur to produce a spheroidized sulfide, thus avoiding the formation of elongated sulfide stringers.

Lamellar cracking is unique to plate material because of the inclusion morphology that develops during rolling. Substituting castings or forgings for plate material, where appropriate, is also an effective method for avoiding lamellar cracking. In particular, the forging of steels tends to break up the inclusion structure that gives rise to cracking.

Weld joint designs that minimize through-thickness stresses normal to the susceptible microstructure in the rolling direction are also an effective method. An example of this approach is shown in Figure 4.57 where a corner joint has been redesigned to

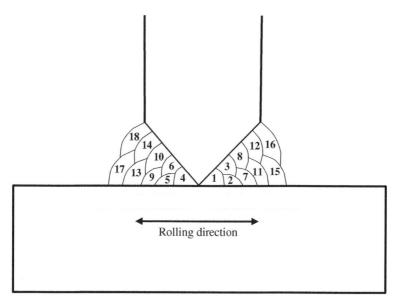

FIGURE 4.58   Balancing weld stresses by weld sequence and bead placement (Adapted from Ref. [106]).

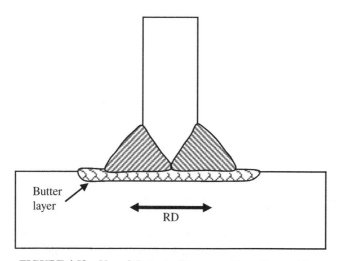

FIGURE 4.59   Use of "buttering" to prevent lamellar cracking.

reduce stresses in the through-thickness direction. Another approach is to alter the weld pass sequence and the size of the individual weld beads to reduce the weld stresses that lead to cracking. An example of this is shown in Figure 4.58 for a highly restrained attachment weld.

Another technique that is quite effective for eliminating lamellar cracking is the application of a "butter" layer on the susceptible material, as shown in Figure 4.59. This butter layer effectively isolates the susceptible steel from the attachment weld such that the welding stresses in the plate are essentially nil. The butter layer that is applied must have sufficient ductility and should be immune from other forms of cracking, such as liquation cracking or DDC. While this approach significantly increases fabrication costs, it is a very reliable method to avoid lamellar cracking in situations where other techniques are not possible.

Finally, even though it is not clear to what degree hydrogen contributes to lamellar cracking in steels, the use of low-hydrogen practice when using processes such as shielded metal arc welding (SMAW) may be appropriate. This includes proper storage of welding electrodes and use of preheat.

4.6   COPPER CONTAMINATION CRACKING

Copper Contamination Cracking (CCC) is a liquid metal embrittlement (LME) phenomenon that has been observed in the weld HAZ of steels and cobalt-based alloys. This form of cracking is categorized as solid-state cracking since the HAZ remains in the solid state and the copper (or copper alloy) melts and penetrates the HAZ (or weld metal) grain boundaries.

CCC may be misinterpreted as HAZ liquation cracking, since the failure mode is IG and the failure location is in a region of the HAZ heated above 1000°C (1832°F). CCC differs from HAZ liquation cracking in three important ways:

- Cracking is usually observed in a region of the HAZ slightly removed from the fusion line. It may occur 2–3 mm away depending on the HAZ thermal history.
- Although cracking is IG, these cracks are not continuous to or across the fusion boundary and generally are restricted to a specific temperature regime in the HAZ. Depending on the applied stresses, they commonly run parallel to the fusion boundary.
- There is no apparent heat-to-heat difference in cracking susceptibility within an alloy system.

In many of the documented cases of weld CCC, the source of the copper is from weld fixturing used by most fabricators due to the high thermal conductivity (i.e., heat sinking) of copper and copper-base alloys. Other sources of copper are contact tips for wire feeders, nozzles for delivery of shielding gas, and copper tooling.

4.6.1   Mechanism for Copper Contamination Cracking

LME is a phenomenon that has been studied for many years in a variety of material systems. Systems including Al–Ga, Al–Sn, Al–In, Fe–Hg, Ti–Cd, and Cu–In have been shown to exhibit LME. In general, three criteria must be met for LME to occur:

- Low mutual solubility between the liquid and solid metals.
- No intermetallic formation between the solid and liquid couple.

- A barrier to plastic flow intragranularly. This results in a high stress concentration at the crack tip along the grain boundary that promotes crack propagation by the LME mechanism.

In all these systems, the basic principle is the same: the lower-melting-temperature material melts and penetrates the grain boundary of the higher-melting-temperature material. From a welding standpoint, the only system of practical significance is the Fe–Cu system. Another system of practical importance that may show some suscepti-bility to LME is the Fe–Zn system, since there is wide usage of galvanized (Zn-coated) steels in the automotive industry. Some cracking due to zinc embrittlement during resistance spot welding has been reported, but is not well documented. It should be noted that at the melting temperature of Zn (420°C), the steel will be bcc ferrite. A list of systems that have been reported to exhibit LME is provided in Table 4.4.

In Fe-based systems that transform to austenite at elevated temperature, the presence of copper can lead to LME. Since pure copper melts at 1083°C (1982°F), only regions of the HAZ heated above this temperature are potentially susceptible to CCC. The molten copper then penetrates the austenite grain boundaries and results in subsequent embrittlement if suitable restraint is present. A schematic of the CCC mechanism is shown in Figure 4.60.

TABLE 4.4 Systems susceptible to liquid metal embrittlement

Parent metal	Melting point (°C)	LME metal	Melting point (°C)
Al	660	Ga	30
		Sn	232
		In	157
Fe	1535	Hg	−39
		Zn	420
		Cu	1083
Ti	1660	Cd	321
Cu	1083	In	157

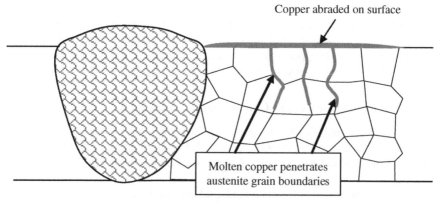

FIGURE 4.60 Schematic of the copper-contamination cracking mechanism.

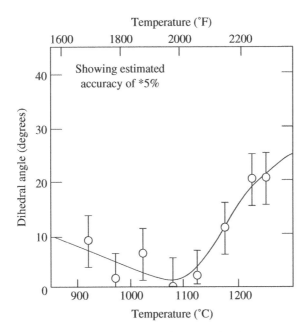

FIGURE 4.61   Effect of wetting angle (dihedral angle) of molten copper as a function of temperature (From Ref. [114]. © AWS).

This mechanism requires that the liquid copper have a high affinity for the solid boundary and can penetrate the boundary rapidly. This, in turn, requires that liquid copper is able to effectively wet the boundary, as dictated by the contact, or dihedral, angle between the liquid and solid substrate. As in other liquid–solid systems, the ratio of solid–liquid to solid–solid surface energies can be used to describe the wetting characteristics, as shown in Figure 2.45.

The plot of dihedral angle versus temperature in Figure 4.61 shows that for mild steels, wetting is most effective near the melting point of copper and becomes progressively less effective at higher temperatures. This explains why CCC in the HAZ of steels tends to be localized in the HAZ and does not extend to the fusion boundary.

Consideration of boundary wetting also explains why materials that are austenitic in the molten copper temperature range are most susceptible to CCC. Molten copper does not wet ferrite–ferrite or ferrite–austenite boundaries as effectively as austenite–austenite boundaries, and thus, materials such as ferritic stainless steels are not susceptible to CCC. It is important to recognize that most structural steels are austenitic over the melting temperature range of copper and copper-base alloys and, thus, are susceptible to CCC.

4.6.2   Quantifying Copper Contamination Cracking

There have been several tests used to evaluate susceptibility to CCC. Holbert *et al.* [115] used a form of the circular patch test to evaluate the cracking tendency in some austenitic stainless steels. This approach resulted in CCC in the HAZ, but could not provide

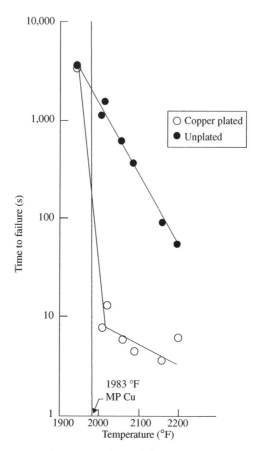

FIGURE 4.62   Stress rupture test for copper-contamination cracking using Type 304 stainless steel (From Savage *et al.* [114]. © AWS).

any ranking with respect to susceptibility. Savage *et al.* [114, 116] used both the spot Varestraint test and a constant load stress rupture test to quantify cracking in a variety of alloys. For both test techniques, copper was plated onto the surface of the sample. For the stress rupture tests, samples were initially loaded at a constant stress and then held at different temperatures until rupture occurred. An example of the results of this technique for Type 304 stainless steel is shown in Figure 4.62. Note that the time to failure drops dramatically upon heating above the melting point of the copper.

Using the spot Varestraint test (see Chapter 9), a large number of Fe-based and Ni-based alloys and two Co-based alloys were evaluated [114]. Tests were conducted using two levels of applied strain (approximately 1.0 and 3.5%). Samples were then examined for cracking and compared to companion samples that had not been copper plated. Based on these tests, alloys were either listed as susceptible or not susceptible, as shown in Table 4.5. Note that the Ni-base alloys are resistant to CCC despite an austenitic (fcc) microstructure. The two ferritic stainless steels (Types 430 and 446) are resistant since they are ferritic (bcc) at the melting temperature of copper.

TABLE 4.5   Materials susceptible or resistant
to copper-contamination cracking based on the
spot Varestraint test[a]

Susceptible	Resistant
Carbon and HSLA steels	*Stainless steels*
HY-80	Type 430
AISI 4130	Type 446
AISI 4340	
Auto body stock	*Ni-base alloys*
	Alloy 718
Stainless steels	Alloy 750
Type 304	René 41
Type 316	Waspaloy
Type 321	Hastelloy X
Type 347	
Type 410	*Aluminum alloys*
A-286	6061-T6
Co-based alloys	
L-605	
Haynes 188	

[a]From Ref. [114].

4.6.3   Identifying Copper Contamination Cracking

CCC is normally quite easy to identify using metallographic and advanced character-ization techniques, including SEM analysis. Cracking is normally in the HAZ but at some distance from the fusion boundary, as illustrated by the photomicrographs in Figure 4.63 for Type 304L [115] and Co-base Alloy 188 [117]. CCC may be confused with HAZ liquation cracking, but the separation from the fusion boundary is what normally distinguishes CCC. When CCC is observed in metallographic sections in the as-polished condition, the presence of copper is usually evident along the grain boundaries.

CCC can also be identified by analysis of the fracture surface. Since CCC results from infiltration of molten copper along the grain boundaries, the fracture surface will exhibit high levels of copper when SEM/XEDS analysis of the fracture surface is conducted.

4.6.4   Preventing Copper Contamination Cracking

Preventing CCC is usually quite straightforward if the cracking mechanism is prop-erly identified and the source of copper is found. In most instances, this source will be fixturing, gas shielding nozzles, or other copper-base equipment that the parts to be welded have come in contact with. The copper must be abraded onto, or otherwise applied to, the surface in a bulk form. Copper added as an alloying element does not cause CCC. For example, the "weathering steels" and some high-strength steels may contain up to 1 wt% Cu to improve general corrosion behavior or to form Cu-rich precipitates, but these steels are not susceptible to CCC.

(a)   (b)

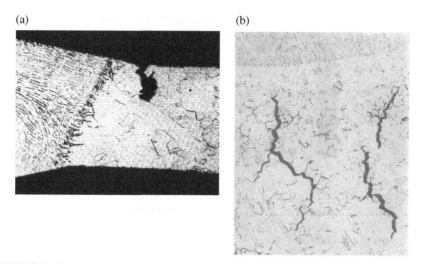

FIGURE 4.63   Examples of copper-contamination cracking. (a) Type 304 L (From Ref. [115]. © AWS) and (b) Co-base Alloy 188 (From Ref. [117]. © AWS).

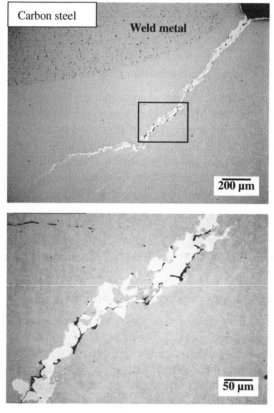

FIGURE 4.64   Copper infiltration along grain boundaries in the HAZ of a carbon steel.

In some instances, CCC has occurred where Cu has been inadvertently electro-chemically plated onto the surface of the steel. It can also occur if susceptible materials are exposed to molten copper. An example is shown in Figure 4.63, where a carbon steel ladle was used to extract molten copper samples for analysis. These ladles failed after a short period of time due to copper infiltration in both the HAZ and weld metal (Figure 4.64).

Selection of alloys that are ferritic at elevated temperatures is an effective method to avoid cracking. Molten copper does not readily wet ferrite–ferrite or ferrite–austenite boundaries and, hence, prevents penetration of molten copper along the boundaries. Ferritic stainless steels, such as Type 430, and duplex stainless steels are resistant to CCC. Ni-base alloys are also resistant, despite their austenitic (fcc) microstructure, in part due to the extensive solubility of Cu in Ni. Austenitic stainless steels are generally susceptible to CCC, but their susceptibility is reduced if some ferrite forms along the austenite grain boundaries in the HAZ.

REFERENCES

[1] Bengough GD. A study of the properties of alloys at high temperatures. J Inst Metals 1912;VII:123–174.

[2] Rhines FN, Wray PJ. Investigation of the intermediate temperature ductility minimum in metals. Trans ASM 1961;54:117–128.

[3] Mintz B, Abu-Shosa R, Shaker M. Influence of deformation induced ferrite, grain boundary sliding, and dynamic recrystallization on hot ductility of 0.1–0.75 wt% C steels. Mater Sci Technol 1993;9:907–914.

[4] Hemsworth B, Boniszewski T, Eaton NF. Classification and definition of high temperature welding cracks in alloys. Metal Constr Brit Weld J 1969;16 (1):5–16.

[5] Cordea JN, Kammer PA, Martin DC. Causes of fissuring in Ni-base and stainless steel alloy weld metals. Weld J 1964;43 (11):481s–491s.

[6] Haddrill DM, Baker RG. Microcracking in austenitic weld metal. Brit Weld J 1965;12 (8):411–418.

[7] Lippold JC, Ramirez AJ. Elevated temperature grain boundary embrittlement and ductility-dip cracking of Ni-base weld metals. Proceedings of the Conference on Vessel Penetration, Inspection, Cracking, and Repairs, NRC; September 29–October 2, 2003; Gaithersburg, MD.

[8] Nissley NE, Collins MG, Guaytima G, Lippold JC. Development of the strain-to-fracture test for evaluating ductility-dip cracking in austenitic stainless steels and Ni-base alloys. Weld World 2002;46 (7/8):32–40.

[9] Lippold JC, Nissley NE. Further investigations of ductility-dip cracking in high chromium, Ni-base filler metals. Weld World 2007;51 (9/10):24–30.

[10] Dupont JN, Lippold JC, Kiser SD. *Welding Metallurgy and Weldability of Nickel Base Alloys*. Hoboken, NJ: Wiley and Sons, Inc; 2009. p 138–142.

[11] Yamaguchi S, Kobayashi H, Matsumiya T, Hayami S. Effect of minor elements on hot workability of nickel-base superalloys. Metals Technol 1979;6 (5):170–175.

[12] Zhang YC, Nakagawa H, Matsuda F. Weldability of Fe–36%Ni alloy (report V). Trans JWRI 1985;14 (2):119–124.

[13] Zhang YC, Nakagawa H, Matsuda F. Weldability of Fe–36%Ni alloy (report VI). Trans JWRI 1985;14 (5):125–134.

[14] Ramirez AJ, Lippold JC. High temperature cracking in nickel-base weld metal, Part 2: Insight into the mechanism. Mater Sci Eng A 2004;380:245–258.

[15] Ramirez AJ, Lippold JC. New insight into the mechanism of ductility dip cracking in Ni-base weld metals. In: Böllinghaus T, Herold H, editors. *Hot Cracking Phenomena in Welds*. Berlin/Heidelberg: Springer; 2005. p 19–41.

[16] Nishimoto K, Saida K, Okauchi H. Microcracking in multipass weld metal of alloy 690, Part 1: Microcracking susceptibility in reheated weld metal. Sci Technol Weld Join 2006;11 (4):455–461.

[17] Nishimoto K, Saida K, Okauchi H, Ohta K. Microcracking in multipass weld metal of alloy 690, Part 2: Microcracking susceptibility in reheated weld metal. Sci Technol Weld Join 2006;11 (4):462–470.

[18] Nishimoto K, Saida K, Okauchi H, Ohta K. Microcracking in multipass weld metal of alloy 690, Part 3: Microcracking susceptibility in reheated weld metal. Sci Technol Weld Join 2006;11 (4):471–479.

[19] Noecker II FF, Dupont JN. Metallurgical investigation into ductility dip cracking in Ni-based alloys, Part I. Weld J 2009;88 (1):7s–20s.

[20] Noecker II FF, Dupont JN. Metallurgical investigation into ductility-dip cracking in Ni-based alloys, Part II. Weld J 2009;88 (3):62s–77s.

[21] Young GA, Capobianco TE, Penik MA, Morris BW, McGee JJ. The mechanism for ductility dip cracking in nickel–chromium alloys, Weld J 2008;87 (2):31s–43s.

[22] Ramirez AJ, Lippold JC. High temperature cracking in nickel-base weld metal, Part 1: Ductility and fracture behavior. Mater Sci Eng A 2004;380:259–271.

[23] Matsuda F. Weldability of Fe–36% Ni alloy, II: Effect of chemical composition on reheated hot cracking in weld metal. Trans JWRI 1984;13 (2):241–247.

[24] Collins MG, Lippold JC. An investigation of ductility-dip cracking in nickel-base filler metals, Part 1. Weld J 2003;82 (10):288s–295s.

[25] Collins MG, Lippold JC. An investigation of ductility-dip cracking in Ni-base filler metals, Part 2. Weld J 2003;82 (12):348s–354s.

[26] Collins MG, Ramirez AJ, Lippold JC. An investigation of ductility-dip cracking in Ni-base filler metals, Part 3. Weld J 2004;83 (2):39s–49s.

[27] Lippold JC, Nissley NE. Ductility dip cracking in high-chromium, Ni-base filler metals. In: Böllinghaus T, Herold H, Cross CE, Lippold JC, editors. *Hot Cracking Phenomena in Welds II*. Berlin/Heidelberg: Springer; 2008. p 409–426.

[28] Nissley NE, Lippold JC. Ductility-dip cracking susceptibility of Ni-based weld metals, Part 1: Strain-to-fracture testing. Weld J 2008;87 (10):257s–264s.

[29] Nissley NE, Lippold JC. Ductility-dip cracking susceptibility of Ni-based weld metals, Part 2: Microstructural characterization. Weld J 2009;88 (6):131s–140s.

[30] Alexandrov BT, Hope AT, Sowards JW, McCracken S, Lippold JC. Weldability studies of high-Cr, Ni-base filler metals for nuclear applications. Weld World 2011;55 (3/4):65–76.

[31] Nissley NE. Intermediate temperature grain boundary embrittlement in Ni-base weld metals [PhD Dissertation]. Columbus, OH: Ohio State University; 2006.

[32] Matsuda F. Hot crack susceptibility of weld metal. Proceedings of 1st US–Japan Symposium on Advances in Welding Metallurgy; 7–8 June, 1990, San Francisco, CA. American Welding Society, Miami, FL. 1990. p 19–36.

[33] Lippold JC, Lin W. A methodology for quantifying heat-affected zone liquation cracking susceptibility. Unpublished research [unpublished PhD Thesis]. Columbus, OH: Edison Welding Institute; 1994.

[34] Kikel JM, Parker DM. Ductility-dip cracking susceptibility of filler metal 52 and alloy 690. In: *Trends in Welding Research V*. Metals Park, OH: ASM International; 1999. p 757–762.

[35] Nissley NE, Lippold JC. Development of the strain-to-fracture test for evaluating ductility-dip cracking in austenitic alloys. Weld J 2003;82 (12):355s–364s.

[36] Matsuda F, Nakagawa H. Some fractographic features of various weld cracking and fracture surface with scanning electron microscope. Trans JWRI 1977;6 (1):81–90.

[37] Meitzner CF. *Stress Relief Cracking in Steel Weldments*. WRC Bulletin No. 211. New York: Welding Research Council; 1975.

[38] Dhooge A, Vinckier A. Reheat cracking – a review of recent studies. Int J Pres Ves Pip 1987;27 (4):239–269.

[39] Dhooge A, Vinckier A. Reheat cracking – a review of recent studies (1984–1990). Weld World 1992;30 (3/4):44–71.

[40] Tamaki K, Suzuki J, Nakaseka Y, Tajiri M. Effects of carbides on reheat cracking sensitivity. Trans JWS 1984;15 (1).

[41] Tamaki K, Suzuki J. Effect of chromium and molybdenum on reheat cracking sensitivity of steels. Trans JWS 1983;14 (2):124–127.

[42] Tamaki K, Suzuki J. Combined influences of phosphorus, chromium and molybdenum on reheat cracking of steels. Trans JWS 1984;15 (2):117–124.

[43] Tamaki K, Suzuki J, Tajiri M. Effect of vanadium and titanium on reheat cracking sensitivity. Trans JWS 1984;15 (1):17–24.

[44] Tamaki K, Suzuki J. Reheat cracking test on high strength steels by modified implant test. Trans JWS 1983;14 (2):25–30.

[45] McMahon Jr CJ, Shin J. Comparison of stress relief cracking in A508, C12 and A533B pressure vessel steels. Metal Sci 1984;8 (18).

[46] Watanabe T, Savage WF. A study of reheat cracking in weld heat affected zone of high strength steel. Trans Natl Res Inst Metals 1984;26 (4).

[47] Kikuchi T, Nakao Y. Effect of impurity elements on the reheat cracking in the weld zones of steel. Proceedings of 4th International JWS Symposium; September 1982; Osaka, Japan.

[48] Hippsley CA, Knott JF, Edwards BC. A study of stress-relief cracking in 2.25Cr–1Mo steel – I. The effects of P segregation. Acta Metall 1980;28:869–885.

[49] Hippsley CA, Knott JF, Edwards BC. A study of stress-relief cracking in 2.25Cr–1Mo steel – II. The effects of multi-component segregation. Acta Metall 1982;30:641–654.

[50] Shin J, McMahon CJ. Mechanisms of stress-relief cracking in a ferritic steel. Acta Metall 1984;32 (9):1535–1582.

[51] Lundin CD, Khan KK. *Fundamental Studies of the Metallurgical Causes and Mitigation of Reheat Cracking in 1.25Cr–0.5Mo and 2.25Cr–1Mo Steels*. WRC Bulletin No. 409. New York: Welding Research Council; 1996.

[52] Meitzner CF, Pense AW. Stress-relief cracking in low-alloy steel weldments. Weld J 1969;48 (10):431s–440s.

[53] Vinckier AG, Pense AW. *A Review of Underclad Cracking in Pressure Vessel Components.* WRC Bulletin No. 197. New York: Welding Research Council; 1974.

[54] Haure J, Bocquet P. Fissuration sous les revêtments inoxyables des pièces pour cuvées sous pression (Cracking below stainless steel cladding under tension). Convention No. 6210-75/3/303; September 1975; Creusot Loire.

[55] Brear JM, King BL. An assessment of the embrittling effects of certain residual elements in two nuclear pressure vessel steels (A533B, A508). Philos. Trans. R Soc London, A 1980;295:291.

[56] Lundin CD, Khan K, Gill T. PWHT/reheat/stress rupture cracking and heat-affected zone toughness in Cu-precipitation hardenable steel A710. Welding Research Council Progress Report No. 46; 1991. p 31–51.

[57] Thomas RD Jr. HAZ cracking in thick sections of austenitic stainless steels, Part 1. Weld J 1984;63 (9):24–32.

[58] Thomas Jr RD. HAZ cracking in thick sections of austenitic stainless steels, Part 2. Weld J 1984;63 (9):355s–368s.

[59] Curran RM, Rankin AW. Welding Type 347 stainless steel for 1100°F turbine operation. Weld J 1955;34 (3):205–213.

[60] Christoffel RJ. Cracking in Type 347 heat-affected zone during stress relaxation. Weld J 1962;41 (6):251s–256s.

[61] Christoffel RJ. Notch-rupture strength of Type 347 heat-affected zone. Weld J 1960;39 (7):315s–320s.

[62] Younger RN, Haddrill DM, Baker RG. Post-weld heat treatment of high-temperature austenitic steels. JISI 1963;201:693–698.

[63] Thomas Jr RD, Messler Jr RW. *Welding Type 347 Stainless Steel: An Interpretive Report.* WRC Bulletin No. 421. New York: Welding Research Council; 1997.

[64] Van Wortel H. Control of relaxation cracking in austenitic high temperature components. *NACE 2007*, 11–15 March, 2007, Nashville, TN. p 2216–2228.

[65] Kiso T, Seshimo I, Okazaki T. Cracking in welds of heavy wall nickel alloy piping during fabrication. *NACE 2011*, 13–17 March, 2011, Houston, TX.

[66] Nishimoto K, Matsunaga T, Tanaka T, Okazaki T. Effect of bismuth on reheat cracking susceptibility in Type 308 FCAW weld metal. Weld World 1998;41:220–235.

[67] Dupont JN, Lippold JC, Kiser SD. *Welding Metallurgy and Weldability of Nickel Base Alloys.* Hoboken, NJ: Wiley and Sons, Inc; 2009. p 364–367.

[68] Glossop BA, Eaton NF, Boniszewski T. Reheat cracking in Cr–Mo–V steel weldments. Metal Constr Brit Weld J 1969;1 (2s):68–73.

[69] Debray W. Reheat cracking in low alloy steels. *Metal Constr* February 1976; 8(2):74–77.

[70] Pense AW, Galda EJ, Powell GT. Stress relief cracking in pressure vessel steels. Weld J 1971;50 (8):374s–378s.

[71] Tamaki K, Suzuki J. RHC test on high strength steel by a modified implant test. Trans JWS 1983;14 (2):33–38.

[72] Younger RN, Baker RG. Heat-affected zone cracking in welded high-temperature austenitic steels. JISI 1960;196:188–194.

[73] *Gleeble® Systems: Defining a New Era in Thermal-Mechanical Physical Simulation and Testing.* Poestenkill, NY: Dynamic Systems Inc. Available at http://gleeble.com. Accessed July 1, 2014.

[74] Balaguer JP, Wang Z, Nippes EF. Stress relief cracking of copper containing HSLA steel. Weld J 1989;68 (4):121s–131s.

[75] Nawrocki JG, Dupont JN, Robino CV, Puskar JD, Marder AR. The mechanism of stress-relief cracking in a ferritic alloy steel. Weld J 2003;82 (2):25s–35s.

[76] Norton SJ, Lippold JC. Development of a Gleeble-based test for postweld heat treatment cracking susceptibility. Trends in Welding Research. Proceedings of 6th Annual Conference; 2003, 15–19 April 2002, Calloway Gardens, GA. ASM International. p 609–614.

[77] Wilson RM, Burchfield WF. Nickel and high-nickel alloys for pressure vessels. Weld J 1956;35 (1):32s–40s.

[78] Kou S. *Welding Metallurgy*, 1st edn. Hoboken, NJ: Wiley Interscience, Inc.

[79] Berry TF, Hughes WP. A study of the strain-age cracking characteristics in welded René 41 – Phase II. Weld J 1969;46 (11):505s–513s.

[80] Baker RG, Newman RP. Cracking in welds. Metal Constr Brit Weld J 1969;1:1–4.

[81] Franklin JG, Savage WF. Stress relaxation and strain-age cracking in René 41 weldments. Weld J 1974;53 (9):380s–387s.

[82] McKeon D. Reheat cracking in high nickel alloy heat-affected zones. Weld J 1971;50 (5):201s–205s.

[83] Nakao Y. Study on reheat cracking of Ni-base superalloy, Waspaloy. Trans JWS 1988;19 (1):66–74.

[84] Prager M, Shira CS. *Welding of Precipitation-Hardening Nickel-Base Alloys*. Welding Research Council Bulletin 128. New York: Welding Research Council; 1968.

[85] Younger RN, Baker RG. Heat-affected zone cracking in welded austenitic steels during heat treatment. Brit Weld J 1961;8 (12):579–587.

[86] Wu KC, Herfert RE. Microstructural studies of René 41 simulated weld heat-affected zones. Weld J 1967;46 (1):32s–38s.

[87] Weiss S, Hughes WP, Macke HJ. Welding evaluation of high temperature sheet materials by restraint patch testing. Weld J 1962;41 (1):17s–22s.

[88] Hughes WP, Berry TF. A study of the strain-age cracking characteristics in welded René 41-Phase I. Weld J 1967;46 (8):361s–370s.

[89] Duvall DS, Owczarski WA. Studies of postweld heat-treatment cracking in nickel-base alloys. Weld J 1969;48 (1):10s–22s.

[90] Duvall DS, Owczarski WA. Heat treatments for improving the weldability and formability of Udimet 700. Weld J 1971;50 (9):401s–409s.

[91] Koren A, Roman N, Weisshaus I, Kaufman A. Improving the weldability of Ni-base superalloy 713C. Weld J 1982;61 (11):348s–351s.

[92] Thamburaj R, Goldak JA, Wallace W. The influence of chemical composition on postweld heat treatment cracking in René 41. SAMPE Quarterly 1979;10:6–12.

[93] Carlton JB, Prager M. *Variables Influencing the Strain-Age Cracking and Mechanical Properties of René 41 and Related Alloys*. Welding Research Council Bulletin 150. New York: Welding Research Council; 1970. p 13–23.

[94] Fawley RW, Prager M. *Evaluating the Resistance of René 41 to Strain-Age Cracking*. Welding Research Council Bulletin 150. New York: Welding Research Council; 1970. p 1–12.

[95] Duvall DS, Doyle JR. Repair of turbine blades and vanes. ASME Publication 73-GT-44; 1973. ASME, New York, NY.

[96] King RW, Hatala RW, Hauser HA. Welding of superalloy turbine hardware. Metals Eng Q 1970;10:55–58.

[97] Flowers G, Kelley E, Grossklaus W, Barber J, Grubbs G, Williams L. Elevated-temperature, plasma-transferred arc welding of nickel-base superalloy articles. US patent 6,084,196. July 4, 2000.

[98] Lucas MJ, Jackson CE. The welded heat affected zone in nickel-base alloy 718. Weld J 1970;49 (2):46s–54s.

[99] Adam P. Welding of high-strength gas turbine alloys. In: *High Temperature Alloys for Gas Turbines*. London: Applied Science Publishes; 1978. p 737–768.

[100] Arata Y. Fundamental studies on electron beam welding of heat-resistant superalloys for nuclear plants (Report 4). Trans JWRI 1978;7:41–48.

[101] D'Annessa AT, Owens JS. Effects of furnace atmosphere on heat treatment cracking of René 41 weldments. Weld J 1973;52 (12):568s–575s.

[102] Lepkowski WJ, Monroe RE, Rieppel PJ. Studies on repair welding age-hardenable nickel-base alloys. Weld J 1960;39 (9):392s–400s.

[103] Rowe MD. Ranking the resistance of wrought superalloys to strain-age cracking. Weld J 2006;85 (2):27s–34s.

[104] Metzler DA. A Gleeble®-based methodology for ranking the strain-age cracking susceptibility of Ni-based superalloys. Weld J 2008;87 (10):249s–256s.

[105] Dix AW, Savage WF. Factors influencing strain-age cracking in Inconel X-750. Weld J 1971;50 (6):247s–252s.

[106] Threadgill PL. Avoiding HAZ defects in welded structures. Metals Mater July 1985;1985:422–429.

[107] Farrar JCM, Dolby RE, Baker RG. Lamellar tearing in welded structure steels. Weld J 1969;48 (7):274s–282s.

[108] Kou S. *Welding Metallurgy*. 2nd ed. Hoboken, NJ: Wiley-Interscience; 2003. p 327.

[109] Dickinson DW, Ferguson BL, Ries GD. Prediction of lamellar tearing susceptibility by mechanical testing. Weld J 1980;59 (11):343s–348s.

[110] Jubb JEM. *Lamellar Tearing*. Welding Research Council Bulletin 168. New York: Welding Research Council; 1971.

[111] Ganesh H, Stout RD. *Material variables affecting lamellar tearing susceptibility in steels*. Weld J 1976;55 (11):341s–355s.

[112] Ganesh H, Stout RD. Effect of welding variables on lamellar tearing susceptibility in the Lehigh test. Weld J 1977;56 (3):78s–87s.

[113] Elliot DN. A fractographic examination of lamellar tearing in multi-run fillet welds. Metal Constr Brit Weld J 1969;1 (2s):50–63.

[114] Savage WF, Nippes EF, Mushala MC. Liquid metal embrittlement of the heat-affected zone by copper contamination. Weld J 1978;57 (8):237s–245s.

[115] Holbert RK, Dobbins AG, Bennet RK. Copper contamination cracking in thin stainless steel sheet. Weld J 1987;66 (8):38–44.

[116] Savage WF, Nippes EF, Mushala MC. Copper-contamination cracking in the weld heat-affected zone. Weld J 1978;57 (5):145s–152s.

[117] Matthews SJ, Maddock MO, Savage WF. How copper surface contamination affects weldability of cobalt superalloys. Weld J 1972;51 (5):326–328.

5

HYDROGEN-INDUCED CRACKING

5.1 INTRODUCTION

The presence of hydrogen in the weld metal or HAZ may lead to a form of cracking known as hydrogen-induced cracking (HIC). This form of cracking is also commonly referred to as "hydrogen-assisted cracking" (HAC) or "cold cracking" since it occurs at or near room temperature after the weld has cooled. The loss of ductility associated with the presence of hydrogen is often referred to as hydrogen embrittlement. This form of cracking is most often associated with steels, but the presence of sufficient hydrogen can lead to cracking or embrittlement in other materials.

Although hydrogen is present at some trace level in virtually all materials, it is the introduction of hydrogen during the welding process that allows hydrogen to be present at some threshold level to promote cracking. Atomic hydrogen is very mobile in the microstructure even at room temperature, allowing it to diffuse to regions of stress concentration and susceptible microstructure.

While HIC normally occurs almost immediately upon cooling to room temperature, it may also occur after a delay. This form of HIC is referred to as "delayed" cracking. This suggests that an incubation time is necessary for the hydrogen to diffuse, and accumulate, in a location where cracking occurs after a threshold level of hydrogen is reached.

Despite the fact that hydrogen cracking and its variants have been studied for over 60 years, there is still no universally accepted mechanism. This is due in part to the fact that hydrogen is very difficult to detect using analytical tools. Unlike other trace

Welding Metallurgy and Weldability, First Edition. John C. Lippold.
© 2015 John Wiley & Sons, Inc. Published 2015 by John Wiley & Sons, Inc.

elements (P, S, B, O) that lead to cracking or embrittlement, hydrogen is effectively undetectable within the microstructure. Computational modeling tools have improved the understanding of hydrogen behavior in metals, and significant progress has been made in predicting embrittlement.

In the absence of a precise hydrogen embrittlement mechanism, methods have been devised to avoid HIC in most structural materials, particularly steels. Modification and control of composition, welding process conditions, and micro-structure can effectively be used to avoid this form of weld cracking. In practice, HIC should be avoidable in most situations if the appropriate procedures are applied.

5.2   HYDROGEN EMBRITTLEMENT THEORIES

HIC is most often associated with welding due to the fact that hydrogen is introduced into the weld pool during the welding process. High temperatures in the welding arc lead to the dissociation of hydrogen gas, water vapor, and hydrogen-bearing compounds, resulting in the formation of atomic hydrogen that can readily be absorbed into the molten weld pool and diffuse into the surrounding metal matrix. There are three major sources for hydrogen during welding: (i) moisture in the elec-trode covering, flux, shielding gas, and ambient environment; (ii) decomposition products of cellulosic-type electrode coatings and combustion products of oxyfuel gas welding; and (iii) contaminants containing hydrocarbons (grease, oil, cutting fluid, etc.) on the surface of the filler metal and base plate prior to welding.

The generation of hydrogen in the weld metal and then diffusion into the HAZ to cause cracking during welding are shown in Figure 5.1 [1]. Hydrogen in atomic form is quite mobile in the molten weld pool and diffuses rapidly into the surrounding HAZ. As shown in Figure 5.2, there is considerable solubility of hydrogen in iron

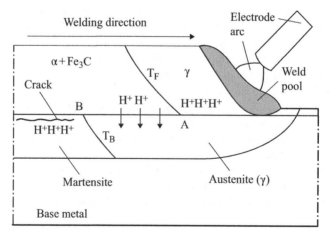

FIGURE 5.1   Hydrogen diffusion from weld metal to HAZ during welding (From Ref. [1]. © Wiley).

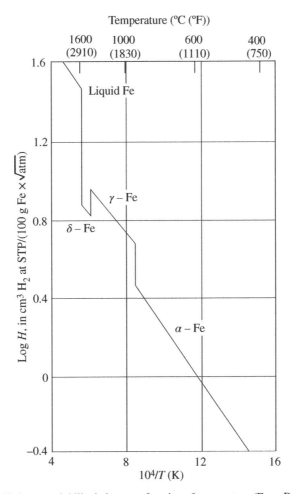

FIGURE 5.2   Hydrogen solubility in iron as a function of temperature (From Ref. [2]. © ASM).

at elevated temperatures, and its diffusivity is quite high [2]. As the weld metal and HAZ cool to room temperature, the diffusivity of hydrogen remains quite high and hydrogen is able to diffuse some distance into the HAZ. Diffusion data for both ferritic (bcc) and austenitic (fcc) materials is shown in Figure 5.3 from Coe [3]. Even at temperatures below 100°C, hydrogen is quite mobile in the ferritic microstructure. This explains why welds can suffer from delayed cracking, since even at room temperature hydrogen can diffuse at an appreciable rate.

Because of widespread problems with HIC, associated both with welding and service exposure, the mechanism of HIC has been studied since the 1940s. A number of theories have evolved, but there is still not a unified mechanism. This is probably because the behavior of hydrogen is different among materials, attributed to the differences in crystal structure, strengthening mechanisms, diffusion characteristics, and other variables. Among all the engineering materials, HIC is most prevalent in

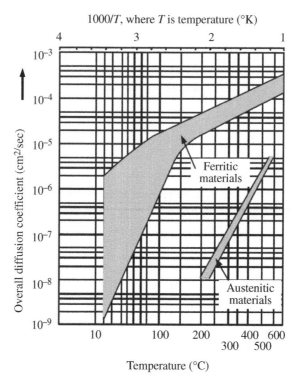

FIGURE 5.3 Diffusion coefficient of hydrogen in ferrite and austenite as a function of temperature (From Ref. [3]. © Wiley).

steel because of its complex transformation behavior (ease of formation of hard martensite) and also because relatively low levels of hydrogen can induce cracking. No single mechanism can comprehensively explain all the phenomena associated with HIC, and it is possible that several mechanisms work cooperatively to induce cracking. Some of the more popular theories that have evolved over the years are summarized in the following sections.

5.2.1 Planar Pressure Theory

The planar pressure theory was proposed by Zapffe and Sims [4]. This theory related cracking to the development of hydrogen bubbles in the microstructure based on diffusion of atomic hydrogen to defect sites such as grain boundaries and interfaces. Once the hydrogen accumulates at these sites, it combines to form molecular hydrogen (H_2) bubbles with high internal pressure. If this pressure exceeds the elastic strength (yield strength) of the material, these internal voids can grow and link together. Eventually, the linkage of these voids leads to the formation of cracks by a plastic deformation or cleavage mechanism. The presence of voids on the fracture surface that were observable using an optical microscope probably led Zapffe and Sims to propose internal pressure as a driving force for hydrogen cracking. With the advent of electron microscopy, high-resolution characterization of the initiation phase of HIC did not support this theory. Although the planar pressure theory is not applicable to HIC, it may explain a phenomenon known as "hydrogen blisters" where long-term exposure to hydrogen (or an environment that generates hydrogen) can lead to surface decohesion appearing as blisters.

5.2.2 Surface Adsorption Theory

The surface adsorption theory was proposed by Petch [5, 6]. It was postulated that the surface free energy of an atomically clean metal is lowered by hydrogen adsorbed onto the internal crack surface. The fracture stress is then lowered by the square root of the surface energy, according to Griffith's thermodynamic criterion. Under these conditions, cracks can propagate under low applied stress. The surface adsorption theory has good experimental support, but there has never been a satisfactory mechanistic explanation of the behavior of hydrogen at the atomistic level at the crack tip.

5.2.3 Decohesion Theory

The decohesion theory by Troiano [7] proposed that high triaxial stresses that are present at a stress concentration, such as a crack tip, attract dissolved atomic hydrogen and locally reduce the cohesive strength. If the critical hydrogen concentration is reached, a small crack forms and propagates into the region with high hydrogen concentration. Further crack growth must await a localized increase in hydrogen concentration in front of the crack. This crack initiation and propagation process repeats until either the critical level of hydrogen is not present or the level of triaxial stress decreases. This process is shown schematically in Figure 5.4.

The decohesion theory was also supported by Oriani [8–10]. He proposed that the maximum cohesive force between atoms is lowered by the dissolved hydrogen, which accumulates at trap sites, such as voids, grain boundaries, and interfaces. If the local stress exceeds the cohesive force, which is lowered by dissolved hydrogen, the atomic bonds can be broken and cracking occurs. The effect of trap sites on hydrogen embrittlement assumes that hydrogen can accumulate at specific locations in the microstructure and the "strength" of the trap site determines under which conditions hydrogen may participate in the process [11, 12]. For example, the interface between certain precipitates and the matrix may represent very strong trap sites, preventing hydrogen from escaping into the matrix. Savage *et al.* [13] used the decohesion theory to explain the intermittent release of hydrogen bubbles at the crack tip that they observed while monitoring crack propagation. Presumably, as the crack propagates, hydrogen in trap sites is released at the crack tip.

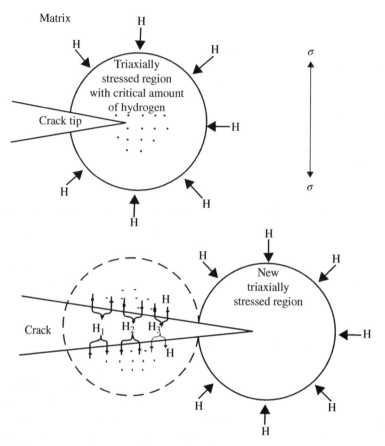

FIGURE 5.4   Schematic of the decohesion theory as proposed by Troiano [7]. © ASM.

5.2.4   Hydrogen-Enhanced Localized Plasticity Theory

Several researchers, including Sofronis [14–16], Birnbaum and Sofronis [17], and Lynch [18], have contributed to the development of the hydrogen-enhanced local-ized plasticity (HELP) theory. Their mechanism is supported by experimental observations and theoretical calculations over a range of temperatures and strain rates. Their fundamental consideration is that hydrogen in solid solution facilitates dislocation motion and increases local plasticity. This increases the amount of deformation that occurs in a local region adjacent to the fracture surface. According to this theory, the fracture process is not characterized by embrittlement but rather a highly localized plastic failure process.

The HELP model assumes that hydrogen solute acts to shield the elastic inter-actions between obstacles and dislocations. The dislocation mobility is therefore enhanced due to the reduction in the interaction energy between the elastic stress components. This phenomenon was confirmed by experimental observation in various engineering materials [19–23]. As a result, if the hydrogen distribution is inhomogeneous, the intrinsic material flow stress varies as a function of location with local plastic flow occurring at low stresses and higher hydrogen concentrations.

5.2.5 Beachem's Stress Intensity Model

A model that related stress intensity at the crack tip and hydrogen concentration with fracture behavior was proposed by Beachem at the Naval Research Laboratory in the 1970s [24]. He used a wedge-loaded specimen in a hydrogen environment to create a decreasing stress intensity factor (K) at the crack tip as the crack extends. Changes in fracture mode as a function of stress intensity factor were assessed based on experimental observation (fractography). The model suggested that the microscopic deformation ahead of crack tip is aided by the concentrated hydrogen dissolved in the lattice ahead of the crack tip. A decrease in the stress intensity factor results in a gradually decreasing microscopic plasticity, as shown in Figure 5.5. It was therefore concluded that the fracture behavior depends on the combination of hydrogen concentration and stress intensity factor at the crack tip. When the stress intensity factor is low and hydrogen concentration is high, the fracture mode is intergranular (IG), which is the most energetically favorable process as it involves the least amount of plastic deformation compared to microvoid coalescence (MVC) and quasi-cleavage (QC) fracture modes. At higher stress intensity and lower hydrogen concentration, QC

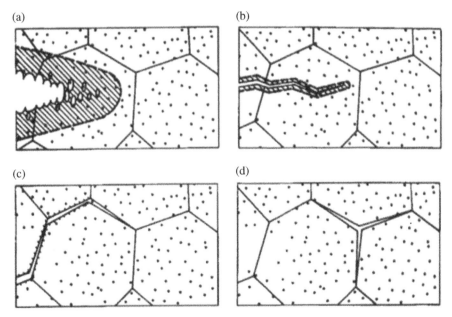

FIGURE 5.5 Microscopic fracture modes observed as a function of decreasing stress intensity factor and concomitant decreasing cracking rate. (a) High K, MVC, (b) intermediate K, QC, (c) low K, IG, and (d) IG cracking with an assist from hydrogen pressure (From Ref. [24]. © Springer).

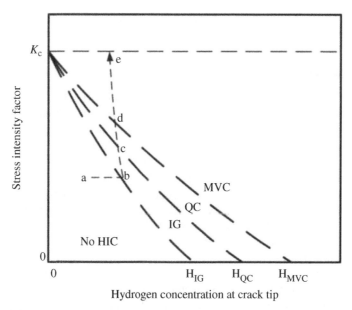

FIGURE 5.6   Combined effect of stress intensity factor and hydrogen concentration at the crack tip on the fracture mode (From Ref. [24]. © Springer).

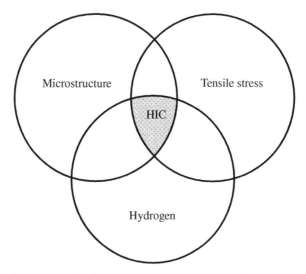

FIGURE 5.7   Interaction of hydrogen, microstructure, and tensile restraint to promote HIC.

and MVC modes replace IG, as shown in Figure 5.6. As the crack propagates, it is possible for the fracture mode to shift from one mode to another as a function of either stress intensity or hydrogen concentration. Beachem's model has been widely adopted and validated by fractographic analysis of hydrogen cracking test specimens conducted by other researchers [25–28].

Although a unified HIC mechanism still does not exist, it is generally agreed that the occurrence of HIC in steels requires the simultaneous presence of a threshold level of hydrogen, a susceptible microstructure of high hardness, and tensile restraint, as shown schematically in Figure 5.7. In addition, the temperature must be in the range from –100 to 200°C (–150 to 390°F). If one of these four contributing factors is eliminated, HIC can be avoided. From a welding standpoint, the most common approaches to avoid HIC are to control microstructure by composition and weld thermal cycle control and to minimize the uptake of hydrogen. These approaches will be discussed in more detail in the following sections.

5.3   FACTORS THAT INFLUENCE HIC

Hydrogen cracking in steels may occur in both the weld metal and the HAZ, although it is most prevalent in the HAZ due to the combination of microstructure and tensile restraint that exists in this region of the weld. The composition and microstructure of the weld metal can usually be controlled to minimize or eliminate hydrogen cracking. For example, filler metals of slightly undermatched strength with ferritic and/or bainitic microstructure can be used to weld high-strength steels. As indicated previously, four conditions must be fulfilled for HIC to occur, namely, (i) a threshold hydrogen level, (ii) susceptible microstructure, (iii) high restraint level, and (iv) near ambient temperature.

5.3.1   Hydrogen in Welds

Hydrogen can be introduced into the weld from several sources. These include (i) hydrogen in the original base material, (ii) moisture in electrode coatings and fluxes, (iii) organic contaminants (grease or oil), (iv) hydrogen in the shielding gas, and, on occasion, (v) humidity from the atmosphere. Hydrogen from these sources may be assimilated into the molten weld pool since the liquid has a much higher solubility for hydrogen than the solid (particularly iron and aluminum). After solidification, the diffusivity of the hydrogen in the microstructure is extremely high, and the hydrogen diffuses down a concentration gradient into the HAZ.

The actual measurement of the level of hydrogen in materials has been the subject of much research effort (and considerable controversy). It is still difficult to assign a "safe" hydrogen level to welds and HAZs in engineering materials. The threshold level of hydrogen that promotes cracking is dependent on the material, its microstructure, and the stress state. For example, high-strength martensitic steels require less diffusible hydrogen than low-strength ferritic steels to cause cracking. As a general rule, every effort should be made to minimize or eliminate the sources of hydrogen.

There is also a difference between the total hydrogen content that is measured and the fraction of this total that is available to diffuse and contribute to HIC. The microstructure may contain many sites where hydrogen can be trapped—grain boundaries and particle/matrix interfaces—and restrict its diffusion in the matrix [12]. Total hydrogen content is a measure of all the hydrogen that is present in the material and can be obtained using special analytical techniques. This includes hydrogen that is free in the matrix and that which is "trapped." In general, it is only the free, or diffusible, hydrogen that can contribute to HIC. Special methods must be used to measure the diffusible hydrogen, and they generally rely on allowing hydrogen to diffuse out of the material over time under conditions where the hydrogen evolution can be measured.

In most cases, it is desirable to measure the actual weldment hydrogen content rather than rely on an estimate. Hydrogen is reasonably difficult to measure since it is so mobile and can diffuse from the sample at room temperature. For that reason, rather elaborate and detailed procedures have been developed for determining hydrogen, as described in ANSI/AWS A4.3-93.

Using this procedure, samples are prepared by welding using the desired process and consumable in a copper fixture in order to accelerate cooling to room temperature. Immediately following welding, the sample is plunged into an iced bath and held at 0°C (32°F) until hydrogen analysis is performed. If properly followed, this procedure will retain most of the hydrogen.

Two analysis methods have been widely used. The first is known as the mercury displacement method and involves submersing the sample in a mercury bath held at a constant 45°C (113°F). The sample is left in the bath for up to a minimum of 72 h. As the hydrogen diffuses out of the sample, it displaces mercury in a calibrated tube known as a eudiometer. Mercury is chosen because it is molten at room temperature and has very low solubility for hydrogen. After 72 h, the amount of mercury displaced is measured, and the volume of hydrogen calculated and then converted to units of ml/100 g. This is a measure of diffusible hydrogen. Environmental issues with the use of mercury have severely curtailed the use of this technique in recent years. A simpler and more direct method for measuring hydrogen is to place the sample in a gas chromatograph. This instrument measures hydrogen directly and in a relatively short period (minutes). Use of a chromatograph is described in ASTM E260. The obvious advantages over the mercury displacement method are offset by the expense of the equipment.

Total hydrogen content can be measured using combustion techniques, such as the LECO method [29]. This technique is also useful for measuring the levels of carbon and nitrogen. In materials that contain stable trap sites, the total hydrogen measurement may not be a good estimate of the diffusible hydrogen that contributes to HIC.

The welding process selected and the consumable used both have a major influence on the amount of hydrogen that can be assimilated into the weld metal and HAZ during welding. Figure 5.8 shows the hydrogen levels that are possible for a variety of processes. Weld hydrogen levels are measured in the units of ml/100 g [30].

As shown on the x-axis in this plot, actual hydrogen levels can range from as low as 2 to over 30 ml/100 g. Total hydrogen levels less than 10 ml/100 g are considered low and normally present little risk to HIC (unless the weld metal or HAZ hardness is very high). Levels between 10 and 20 generally require special precautions. Above 20, HIC may be very difficult to avoid under high restraint conditions.

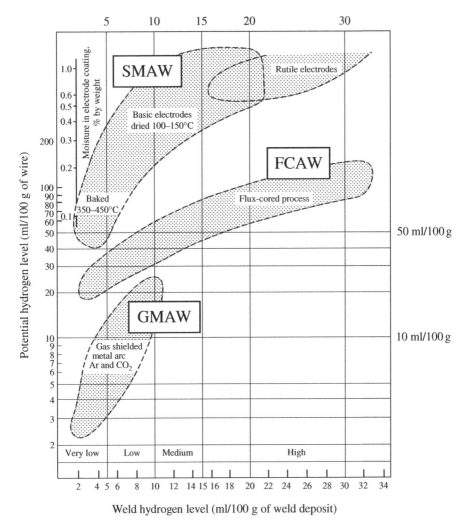

FIGURE 5.8   Effect of welding process selection on hydrogen content (From Ref. [30]. ©
Elsevier).

The flux-shielded processes (shielded metal arc welding (SMAW), FCAW, and
SAW) are the most susceptible to hydrogen pickup, since moisture in the flux is the
largest contributor to hydrogen generation. SMAW and FCAW consumables can produce a wide range of hydrogen levels based on the nature of the flux or slag system that
is used. In particular, note that rutile electrodes (such as E6010 and E7010) produce high
levels of weld hydrogen. Conversely, processes that use shielding gas for protection,
such as GMAW and GTAW, have low potential for hydrogen in the weld metal.

Hydrogen introduced into the weld through the arc or molten weld pool must
diffuse to the adjacent HAZ in order for the HAZ to be susceptible to HIC.
Because the base metal hydrogen content is generally low, the weld metal becomes

the primary source of hydrogen. As shown in Figure 5.3, hydrogen diffusivity in the solid is a function of temperature and microstructure. Diffusion of hydrogen in ferritic (bcc) materials is much faster than in austenite (fcc) because of the more open crystal lattice structure of bcc versus fcc. In general, diffusion of hydrogen in ferritic steels is faster by over 3 orders of magnitude (1000 times) relative to austenitic alloys. This explains, in part, why steels that are austenitic at or near room temperature are more resistant to HIC. The wide band of hydrogen diffusivity below approximately 150°C reflects the wide range of microstructures that can be present in steels—ferrite, pearlite, bainite, and martensite. Diffusion rates in bainite and martensite are typically the highest.

5.3.2   Effect of Microstructure

In C–Mn and low-alloy steels, a wide variety of weld microstructures can exist, depending on both the specific composition of the steel and the welding process/ parameters. In general, hard and/or brittle microstructures are the most susceptible to hydrogen cracking. These microstructures generally form due to rapid cooling from above the upper critical temperature (A_3) and are normally martensitic or bainitic.

Fast cooling rates are the result of low heat input welding processes, low preheat and/or interpass temperatures, and thick section sizes. The effect of cooling rate from the upper critical temperature to room temperature on the microstructure generated in a plain-carbon (AISI 1040) and low-alloy steel (AISI 4340) is demonstrated by the continuous cooling transformation (CCT) diagrams shown in Figure 5.9 [31]. On both diagrams, three cooling curves have been superimposed, which represent high (A), medium (B), and low (C) cooling rates from above the A_3 temperature. This is representative of the range that might be expected in the HAZ during welding. For most steels, the cooling rate in the range from 800 to 500°C (1470–930°F) is the most critical since this is the range over which austenite transforms to lower-temperature transformation products. The time to cool through this range, designated t_{8-5}, is often used as a measure of the cooling rate.

For the AISI 1040 steel, even the fastest cooling rate will not produce a fully mar-tensitic structure, and the slower cooling rates (B and C) result in microstructures consisting primarily of ferrite and pearlite. In contrast, the AISI 4340 steel forms 100% martensite at both the high and medium cooling rates and a mixture of bainite and martensite at the slowest cooling rates. This demonstrates the important effect of hardenability on the microstructure that forms in the HAZ. For hardenable steels, such as 4340, martensite will form easily in the HAZ and cannot be avoided by con-trol of heat input or preheat/interpass temperature. The HAZ microstructure of the 1040 steel, with lower hardenability, can be greatly influenced by control of cooling rate, and HIC susceptibility can be reduced by avoiding martensite and bainite in this steel. For the hardenable steels, such as 4340, HIC must be controlled by reducing H content in the HAZ since microstructure control during welding is not possible.

The relative susceptibility of different microstructure types observed in steels is provided in Table 5.1. Untempered martensite is the most susceptible, while ferrite and pearlite, acicular ferrite, and austenite are the most resistant. To some

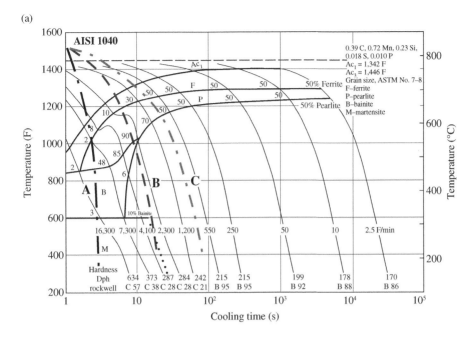

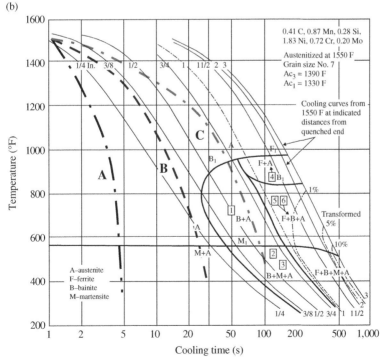

FIGURE 5.9 Continuous cooling transformation diagrams. (a) AISI 1040 and (b) AISI 4340 (From Ref. [31]. © ASM).

TABLE 5.1   Relative ranking of HIC susceptibility according to microstructure

Microstructure	
Untempered martensite Untempered martensite and bainite Bainite Tempered martensite Ferrite and pearlite Acicular ferrite Austenite	**Decreasing susceptibility to HIC**

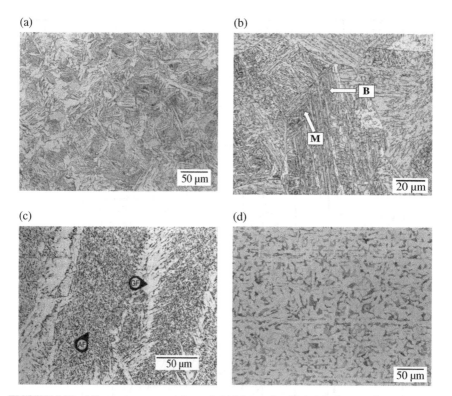

FIGURE 5.10   Microstructure types in steel. (a) Martensite, (b) bainite/martensite, (c) acicular ferrite with grain boundary ferrite, and (d) ferrite and pearlite.

extent, hardness can be used as a measure of the HIC susceptibility of the microstructure. In general, microstructures with hardness levels below 35 HRC (~350 VHN) are resistant to HIC in most conditions. Examples of microstructure types found in carbon steels are provided in Figure 5.10.

Composition, in conjunction with cooling rate, ultimately controls the microstructure and hardness of the steels. Carbon content has the strongest influence on hardness, while other alloy additions (together with carbon) influence the hardenability of the

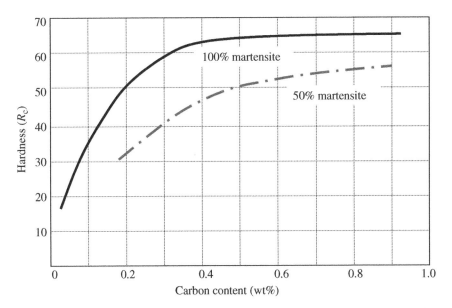

FIGURE 5.11 Relationship between hardness and carbon content.

steel. Hardenability is a concept that predicts the ease of forming martensite in the microstructure. Since martensite is the most susceptible microstructure to HIC, the ability to predict whether it will form is an important factor in determining susceptibility to HIC.

The hardness of martensite is dictated almost entirely by the carbon content of the steel, as illustrated in Figure 5.11. Under conditions where 100% martensite forms, the hardness is predicted by the top curve. Note that at even relatively low carbon content (0.2 wt%), the maximum hardness is on the order of 50 HRC. This explains why many steels that are resistant to HIC have carbon levels at or below approximately 0.1 wt%. Hardness can also be controlled by reducing the amount of martensite that forms, as shown by the lower curve in Figure 5.11 for 50% martensite. This can be achieved by controlling the cooling rate or reducing the hardenability of the steel.

Carbon equivalent (CE) formulae are used to predict susceptibility to HIC. In general, the higher the value of CE, the more susceptible the steel is to HIC. Unfortunately, there is not a single CE formula that can be used for all steels. The IIW CE formula, CE_{IIW}, is the one that is most widely used because it can be generally applied to most plain-carbon and C–Mn steels:

$$CE_{IIW} = C + \frac{Mn}{6} + \frac{Cu + Ni}{15} + \frac{Cr + Mo + V}{5}$$

However, many other CE formulae exist that may be more appropriate for specific alloy systems, as listed in Table 5.2. Yurioka and Suzuki [32] have grouped these formulae into categories based on the carbon and alloying element content

TABLE 5.2   Carbon equivalent formulae for steels[a]

Group	Formula (values in wt%)	References
A	$CE_{IIW} = C + Mn/6 + (Cu + Ni)/15 + (Cr + Mo + V)/5$	1
	$CE_{WES} = C + Si/24 + Mn/6 + Ni/40 + Cr/5 + Mo/4 + V/14$	2
	$CE_{Stout} = C + Mn/6 + Cu/40 + Ni/20 + (Cr + Mo)/10$	3
B	$CE_{DNV} = C + Si/24 + Mn/10 + (Ni + Cu)/40 + Cr/5 + Mo/4 + V/10$	4
	$P_N = C + Si/20 + Mn/10 + Cu/20 + Cr/30 + Mo/20 + 4P/3$	5
C	$P_{CM} = C + Si/30 + Mn/20 + Cu/20 + Ni/60 + Cr/20 + Mo/15 + V/10 + 5B$	6
	$CE_{PLS} = C + Si/25 + Mn/16 + Cu/16 + Ni/60 + Cr/20 + Mo/40 + V/15$	7
	$CE_{HSLA} = C + Mn/16 + Ni/50 + Cr/23 + Mo/7 + Nb/5 + V/9$	8
D	$CE_N = C + A(C)[Si/24 + Mn/6 + Cu/15 + (Cr + Mo + Nb + V)/5 + 5B]$	9
	where, $A(C) = 0.75 + 0.25 \tan h[20(C - 0.12)]$	

[a]From Ref. [32].

Group A: medium-carbon (>0.16 wt%) steels, not for low-alloy steels.

Group B: medium- to low-carbon and medium- to low-alloy steels, stronger effect of carbon than Group A.

Group C: low-carbon and low-alloy steels including HSLA steels.

Group D: range of carbon and alloy content, special factor for carbon.

[1] Dearden J. O'Neill H. A guide to the selection and welding of low-alloy structural steels, Trans Inst Weld 1940;3:203–214.

[2] Kihara H, Suzuki H, Kanatani Y.Studies on weld hardening of steel (Report 3), Journal of Japanese Welding Society, 1958, 27(1):36–42.

[3] Stout RD, Vasudevan R, Pense AW. A field weldability test for pipeline steels, Weld J 1976;55(4):89s–94s.

[4] Hannerz NE. The influence of silicon on the mechanical properties and the weldability of mild and high tensile structural steels, IIW Doc. IX-1169-80; 1980.

[5] Tanaka J, Kitada T. Study on the fillet weld cracking, JWS 1972;41(8):915–924.

[6] Ito Y, Bessho K. Cracking parameter of high strength steels related to heat affected zone cracking, JWS 1968;37(9):983–991.

[7] Duren C, Niederhoff K. Hardness in the heat affected zone of pipeline girth welds. *Proceedings of the Third International Conference on Welding and Performance of Pipelines*; November 1986; London. The Welding Institute.

[8] Graville BA. *Proceedings of the Conference on Welding of HSLA structural steels*; November 1976; Rome. American Society for Metals and Associazione di Metallurgia.

[9] Yurioka N, Suzuki H, Oshita S. Determination of necessary preheating temperature. Weld J 1983;62(6):147s–153s.

of the various steels. In general, the CE_{IIW} formula tends to be less predictive of hardenability as carbon content decreases and alloy content increases. In Section 5.6, it will be shown how the concept of CE is used to determine welding procedures and/or preheat temperature to prevent HIC during welding.

5.3.3   Restraint

Restraint is a term often used to describe the stress and strain state of a weld resulting from a variety of factors. High restraint is usually a necessary entity for HIC. In general, restraint is the hardest variable to control and measure with respect to HIC. Most predictive relationships for hydrogen cracking simply relate restraint to material thickness and possibly joint design. Other factors that contribute to weld restraint are base material and weld metal strength and external fixturing. In many cases, it is not the overall (global) restraint that is important, but rather the local tensile stresses that are present at stress concentration points, such as the toe of the weld. Since hydrogen is so mobile in the structure, it is possible for hydrogen to accumulate at a region of high tensile stress.

In the case of HIC associated with welds, residual stresses may subject regions of the HAZ to near-yield stress levels. The presence of geometric defects such as undercut, excessive overbead, excessive drop-through (underbead), and other stress concentrators can lead to locally high stress levels. In SMAW, slag intrusions at the toe of the weld may act as stress concentrations points. Almost all HIC initiates at stress concentrations in the weld, as illustrated in Figure 5.12 [33]. As will be described later in this chapter, most tests for HIC incorporate a notch or other stress concentration feature. It is also important that these stress concentrations are very close to the CGHAZ where the most susceptible microstructure and highest hydrogen levels are expected.

(a)

(b)

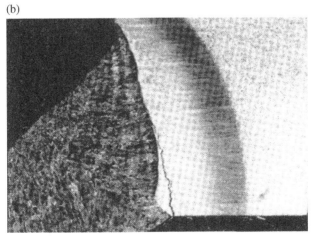

FIGURE 5.12 Examples of hydrogen cracking initiating at stress concentration points (From Ref. [33]. © Maney).

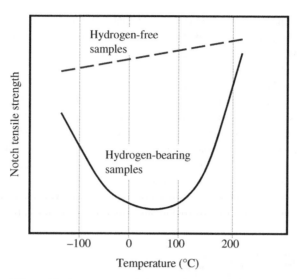

FIGURE 5.13 Effect of hydrogen on notch bar tensile strength in hydrogen (Adapted from Ref. [33]).

5.3.4 Temperature

The susceptibility of steels to hydrogen cracking is very dependent on temperature. This dependence is illustrated in Figure 5.13. At temperatures above 200°C (390°F), hydrogen cracks are less likely to form since the hydrogen can readily diffuse and the inherent resistance to fracture (fracture toughness) of the microstructure increases. At very low temperatures (< −50°C), hydrogen mobility decreases and its effect is minimized, although the decrease in the influence of hydrogen is often masked by the inherently low resistance to fracture in certain microstructures (e.g., martensite). In steels, the ductile-to-brittle transition temperature is normally in the range from 20 to −100°C.

As discussed previously, not all the hydrogen that is present in the material contributes to HIC. Hydrogen is present in both "diffusible" and "nondiffusible" forms. Only the diffusible hydrogen contributes to HIC, since it is able to diffuse to high-stress regions. Nondiffusible hydrogen is "trapped" in the structure and is not able to diffuse. Trapping sites can include grain boundaries and interfaces between particles and the matrix.

5.4 QUANTIFYING SUSCEPTIBILITY TO HIC

A number of methods and test techniques have been developed to quantify susceptibility to HIC. One indirect method for assessing the potential susceptibility to HIC is to measure the hardenability of the material. Historically, this has been done using the Jominy end quench method. There are many direct tests for HIC, some of which have been described previously for quantifying other types of cracking such as reheat

cracking. As with tests for other cracking mechanisms, both self-restraint and externally loaded tests have been developed. Among the self-restraint tests, the controlled
thermal severity (CTS) test, the Y-groove (or Tekken) test, and the gapped bead-on-
plate (G-BOP) test are the most popular. Among the externally loaded tests, the
implant test, tensile restraint cracking (TRC) test, and augmented strain cracking test
are most often used. Each of these approaches will be briefly described. Many of these
test techniques are described in AWS standard B4.0 [34] and ISO 17642-3:2005(E).
A review of hydrogen cracking tests has been published by Kannengiesser and
Böllinghaus that discusses the use and application of these tests [35].

5.4.1 Jominy End Quench Method

The Jominy end quench method is a simple, straightforward procedure for quantifying hardenability [36, 37]. As shown in Figure 5.14, a cylindrical sample 4 in. long
and 1 inch in diameter is austenitized in a furnace and then placed in a fixture where

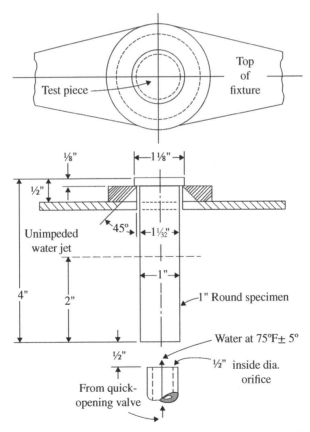

FIGURE 5.14 Schematic illustration of the Jominy end quench fixture (From Ref. [38].
© WRC).

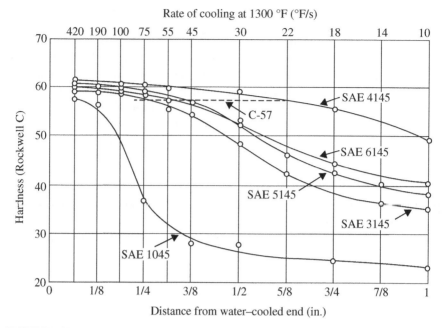

FIGURE 5.15   Jominy curves for a number of steels with 0.45 wt% C (From Ref. [38]. © WRC).

water is immediately sprayed on one end. This results in a cooling rate gradient along the length of the sample. A flat is then ground along the axis of the pin, and hardness measurements are made at 1/16 in. intervals from the quenched end. The hardness versus distance plot is used to determine the hardenability of the steel. It is also possible to determine the microstructure along the length of the pin by polishing and etching. If the cooling rate along the pin can be measured or estimated, the Jominy technique can also be used to construct a CCT diagram.

Typical Jominy hardenability curves for steels with similar carbon content (0.45 wt% C) are shown in Figure 5.15 [38]. Note that the hardness of all these steels is equivalent at the quenched end, reflecting the effect of carbon on hardness when all the steels are fully martensitic. The hardness of the 1045 steel drops off rapidly since it only forms martensite under very rapid cooling conditions. The other steels are more hardenable since they contain alloy additions that allow martensite to form at lower cooling rates. The 4145 steel has the highest hardenability since it forms martensite at even the slowest cooling rates. It is also (potentially) the steel that would be most susceptible to HIC.

Examples of Jominy end quench hardenability plots for a plain-carbon (1050) and low-alloy (4340) steel are shown in Figure 5.16 [38]. The plain-carbon steel has the potential for high hardness at high cooling rates but low hardenability since its hardness (and % martensite) drops off rapidly as the cooling rate decreases. The 4340 steel exhibits effectively no change in hardness with cooling rate and is considered a very hardenable steel. Not surprisingly, special precautions must be taken with 4340 (and similar hardenable steels) to prevent HIC during welding.

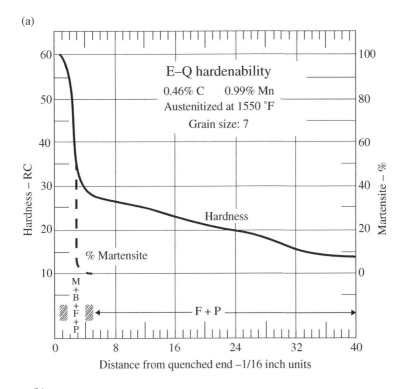

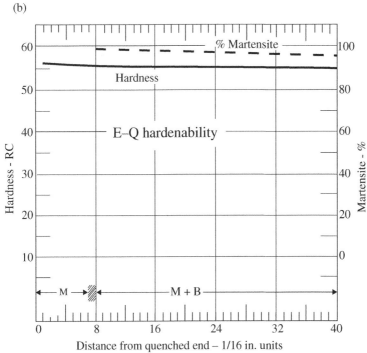

FIGURE 5.16 Jominy hardenability data for two steels. (a) AISI 1050 and (b) AISI 4340 (From Ref. [38]. © WRC).

5.4.2 Controlled Thermal Severity Test

The CTS test was developed at The Welding Institute [39]. The specimen arrangement in this test consists of one bottom plate and one top plate, which is the steel to be investigated. The top plate is bolted to the bottom plate, and two anchor fillet welds are made to provide restraint. Two test welds made under controlled conditions are deposited on opposite sides, as illustrated in Figure 5.17 [39]. Tensile restraint is created in the fillet weld metal and HAZ of the top plate. The small notch (10 mm in depth) machined in the top plate provides stress concentration at the weld metal/HAZ interface and tends to promote crack initiation at this location. After welding, the test welds are sectioned for metallographic examination to determine if hydrogen cracking has occurred. The section in Figure 5.12b is representative of a crack in a CTS sample.

Although originally designed to evaluate HAZ HIC, it is possible that the CTS test can also evaluate susceptibility of weld metal cracking. However, it was later found that the tensile restraint provided by the CTS test may be lower as compared to the restraint present in actual welded structures [40]. As a result, the requirements to avoid hydrogen cracking, such as preheat temperature, can be underestimated if

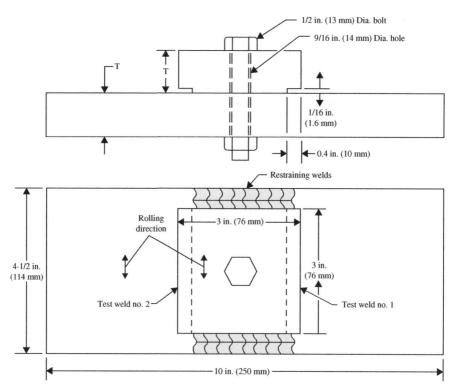

FIGURE 5.17 Schematic illustration of the controlled thermal severity test (From Ref. [35]. © AWS).

determined based on the CTS test results. For example, the critical level of preheat required to prevent cracking may be higher than predicted by the CTS results. The CTS test was used to develop the British Standard, BS 5155, that is described in Section 5.6.

5.4.3 The Y-Groove (Tekken) Test

The Y-groove, or Tekken, test was developed by Kihara *et al.* [41]. In the original test, low-hydrogen filler metal is used to deposit two restraining welds on each end of the weld groove prepared from the steel of interest. A single test weld is then deposited with the filler metal of interest in the restrained assembly, as shown in Figure 5.18 [35]. Note that the weld joint is in the shape of a "Y"; hence, this test is more popularly known as the Y-groove test. The design of the joint results in a notch

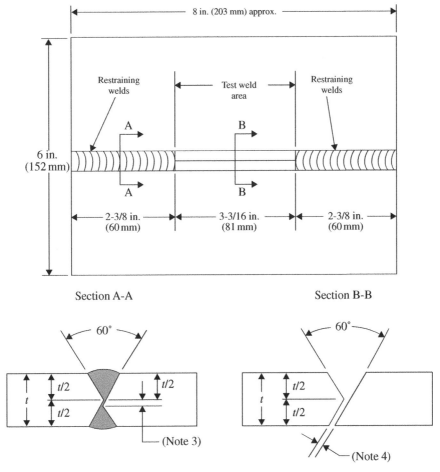

FIGURE 5.18   Schematic illustration of the Y-groove (or Tekken) test (From Ref. [35]. © Springer).

at the root of the weld. Because of the angle of this notch, a hydrogen crack may propagate into either the weld metal or HAZ. After placement of the test weld, the weld is sectioned after no less than 72 h to account for delayed cracking effects. Metallographic inspection at several locations is used to determine if cracking has occurred.

Because of its design, the Tekken test produces higher stress concentration than the CTS test and allows the crack to propagate in either the weld metal or the HAZ. The Tekken test has also been used to determine the critical welding parameters to avoid cracking [42, 43] or the effect of composition on cracking susceptibility [44]. However, sample preparation requirements for the Tekken test can be quite stringent, and machining tolerances and sample surface roughness must be carefully controlled. Variations in sample preparation can affect crack nucleation and propagation conditions resulting in a high level of scatter in the test results [45]. In particular, the placement of the restraining welds on each end of the sample can lead to distortion of the test weld joint geometry. One approach to avoid this is to use a water-jet cutting method to prepare the Y-groove out of a single plate [42]. This then eliminates the need for the restraining welds and decreases the overall cost of sample preparation.

Cross sections from the Y-groove test are shown in Figure 5.19 [35]. In one case, the weld metal strength exceeds (overmatches) the base metal strength, and cracking initiates in the CGHAZ at the weld root. In the case where the weld strength undermatches the base metal, cracking occurs in the weld metal. This demonstrates the versatility of the Y-groove test in that it can be used to assess both weld metal and HAZ HIC.

5.4.4 Gapped Bead-on-Plate Test

The G-BOP test was developed by Graville and McParlan to evaluate HIC in the weld metal [46, 47]. The test configuration consists of two blocks with a plate thickness of 50 mm (~2.0 in.), which are clamped together. One block has a machined recess through the thickness, so a gap is created between the two blocks when they are

(a)   (b)

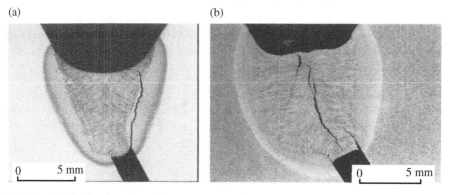

FIGURE 5.19   Sections from Y-groove test. (a) Weld metal strength overmatch and (b) weld metal strength undermatch (From Ref. [35]. © Springer).

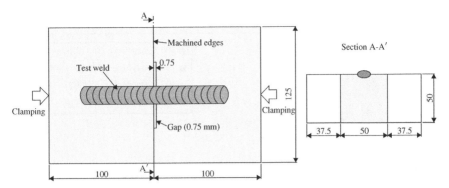

FIGURE 5.20   Schematic illustration of the gapped bead-on-plate (G-BOP) test (From Ref. [35]. © Springer).

clamped together, as illustrated in Figure 5.20. The welding consumable of interest is deposited directly over the machined gap resulting in a tensile stress along the length of the weld. This creates a high stress concentration at the root of the weld due to the presence of the gap. After at least 24 h of the completion of welding, the clamping force is removed and the assembly is placed in a furnace or heated locally with a torch to produce a heat tint on the surface of any crack that had formed. The blocks are then separated, and the fraction of heat-tinted weld cross section is determined to quantify the level of HIC.

The G-BOP test was further modified by Hart [48] by using a "buttering" procedure prior to the actual test weld to lower the dilution of the filler metal by the base metal. The dilution is controlled within approximately 10%. This modification makes the G-BOP test suitable to evaluate the weld metal hydrogen cracking susceptibility of multipass welds, where limited dilution exists.

In a controlled study using a variety of solid wire and flux-cored consumables, Marianetti [49] evaluated the variables associated with the G-BOP test and found that gap width in the range from 0.375 to 1.625 mm (0.015–0.065 in.) had little effect on the cracking susceptibility and that clamping force between 2,000 and 10,000 pounds also had only a small effect. He developed a procedure that resulted in very good reproducibility with variation less than 10% among identical tests. This procedure is described in more detail in Chapter 9.

The G-BOP test has been used extensively to determine the preheat temperature to avoid cracking and investigate the effect of filler metal composition on the cracking susceptibility [49–51]. Examples of G-BOP results for two flux-cored consumables with different strength levels (E70T-5 and E110T-5) are shown in Figure 5.21. Note that this test is very effective at identifying a minimum preheat temperature for avoiding HIC.

5.4.5   The Implant Test

The implant test was developed in the 1960s by Henri Granjon at the Institut de Soudure (French Welding Institute) [52]. The "implant" is essentially a cylindrical sample notched on one end that was inserted into a clearance hole in the center of a specimen plate, with

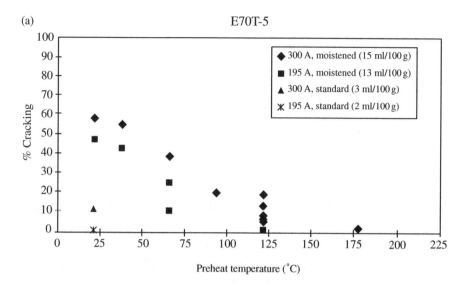

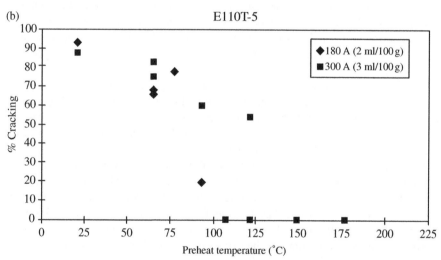

FIGURE 5.21   G-BOP results for two flux-cored electrodes. (a) E70T-5 and (b) E110T-5 (From Ref. [49]).

the top surface of the notched implant specimen flush with that of the specimen plate. A weld bead was then deposited on the top of the specimen plate directly over the notched sample and the hole, creating a HAZ in the cylindrical sample and locating the notch within the HAZ, as illustrated in Figure 5.22. The early sample developed by Granjon incorporated a single notch that was used as a stress concentration to cause cracking to occur in the HAZ instead of the weld metal. After welding, the complete assembly was placed in a fixture and loaded in tension. The time to failure for a series of tests performed

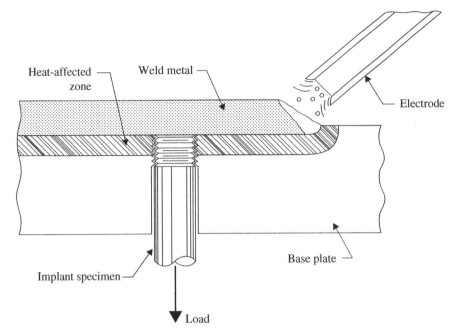

FIGURE 5.22   Schematic of the implant test (From Ref. [35]. © Springer).

at various stress levels was recorded. The applied stress was then plotted against the time to failure, as shown in Figure 5.23 [34]. In the relationship, the maximum stress under which the implant specimen did not fail after 24 h loading was defined as the lower critical stress, which is an important index to rate the HAZ HIC susceptibility.

The major disadvantage associated with the original implant test was considerable scatter in time-to-failure data because the notch could not be reliably located in the same HAZ region in a series of tests of the same material and testing conditions. To resolve this, Sawhill *et al.* [53] proposed a modified implant specimen, in which a helical notch was used instead of the single circumferential notch, as shown in Figure 5.22. This was easily accomplished by simply machining standard threads on the end of the sample. The modification ensured that the entire HAZ is traversed by the helical notch as shown in Figure 5.24, which represents a cross section of an implant specimen after welding. The crack thus initiates and propagates in the most susceptible HAZ microstructure, which is usually within the CGHAZ region.

In order to provide comparative results among different steels, an embrittlement index (I) was defined to rate the HIC susceptibility, as shown in the following equation, where NTS represents the notch tensile stress of the base metal and LCS is the lower critical stress, as illustrated in Figure 5.23. As the embrittlement index increases, susceptibility to HIC increases:

$$I = \frac{NTS - LCS}{NTS}$$

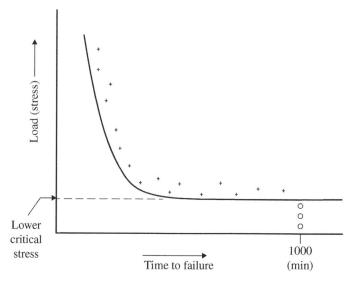

Note: Solid dots = failure.
Open dots = no failure.

FIGURE 5.23 Typical load (stress) versus time-to-failure relationship for the implant test (From Ref. [34]. © AWS).

FIGURE 5.24 Modified implant test specimen cross section showing helical notch (From Ref. [53]. © AWS).

Others have simply used the ratio of LCS to NTS to indicate susceptibility to HAZ HIC. An example of how LCS/NTS can be used to evaluate the cracking susceptibility of pipeline steels is shown in Figure 5.25 from the work of Signes and Howe [54]. Although there is some scatter in the data, the general trend of increasing susceptibility to HIC (lower LCS/NTS values) with increasing CE is apparent. Other

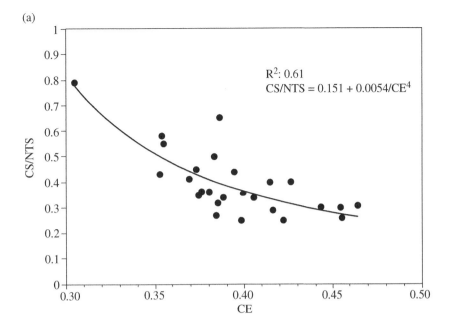

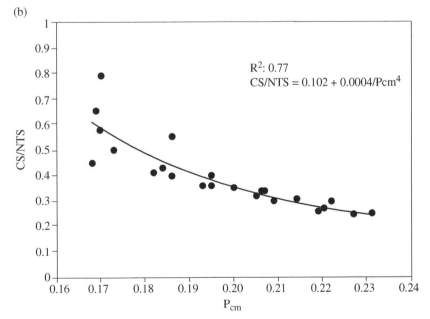

FIGURE 5.25   Relationship between LCS/NTS and carbon equivalent values for a number of pipeline steels. (a) CE_{IIW} and (b) P_{CM} (From Ref. [54]. © AWS).

(a)

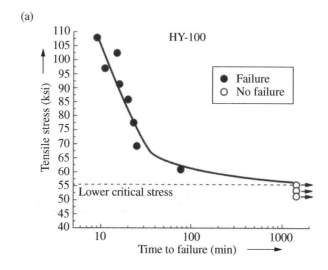

(b)

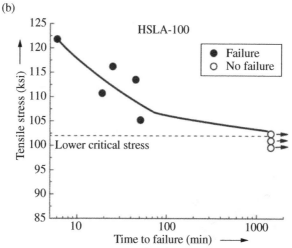

FIGURE 5.26 Implant test results for naval steels. (a) HY-100 and (b) HSLA-100 (From Ref. [25]. © AWS).

results for high-strength naval steels from the work of Yue and Lippold [25] are shown in Figure 5.26. The steel with the highest carbon content and CE (HY-100) exhibits the lowest LCS and would have higher susceptibility to HIC.

In summary, the modified implant test has a number of advantages over other laboratory test methods for evaluating HAZ HIC susceptibility of high-strength steels. First of all, the testing procedures have already been standardized in AWS standard B4.0 [34] and ISO 17642-3:2005(E) [55]. This increases the likelihood of reproducible results of tests conducted at different laboratories. Secondly, the test can be conducted using a wide range of welding processes and consumables. It can also be used to study the effect of preheat on eliminating HIC. This test is described in more detail in Chapter 9.

5.4.6 Tensile Restraint Cracking Test

The TRC test was developed by Suzuki and Nakamura [56] to investigate the hydrogen-induced root cracking in butt weld joints. The schematic of the TRC test is shown in Figure 5.27. After welding is completed, a constant tensile load is applied on the specimen until root cracking failure occurs. The time required for crack initiation and propagation is dependent on the tensile stress applied. By varying the tensile load, the time to failure will change, and a plot of tensile stress versus failure time can be generated, as shown in Figure 5.28 [32]. This test allows welding conditions to be varied and the influence of preheat determined. Similar to the implant test, the TRC test identifies a critical stress below which HIC does not occur.

The TRC test was modified by Matsuda *et al.* [57, 58] to apply longitudinal tensile stress to the test weld metal, as opposed to the transverse stress applied in Figure 5.27. This modified method was therefore named longitudinal bead-tensile restraint cracking (LB-TRC) test. Similar to the G-BOP test, the weld was deposited over a gap that increases the stress concentration at the root of the weld. The LB-TRC test has proven to be a reliable test method to evaluate weld metal HIC [59, 60].

In the TRC test and its modifications, the tensile stress state, geometry, and dimensions of the test specimens can be adjusted to simulate actual welding conditions

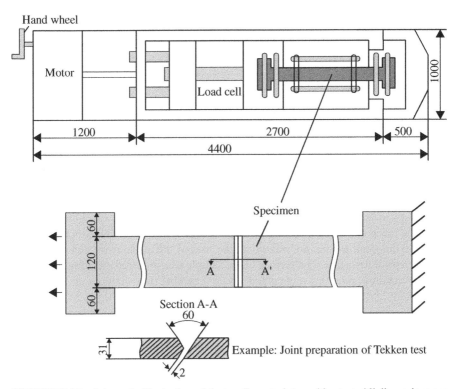

FIGURE 5.27 Schematic illustration of the tensile restraint cracking test. All dimensions are in mm. (From Ref. [35]. © Springer).

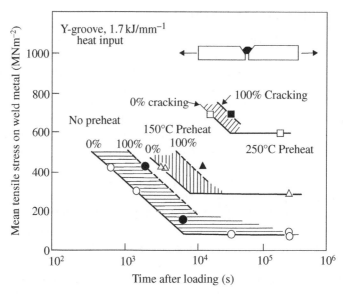

FIGURE 5.28   Relationship between applied tensile stress and time to failure with different preheating temperatures (From Ref. [32]. © Maney).

and potentially produce laboratory test results that translate to actual welded components. However, because the sample geometry produces a high stress concentration in the weld metal, the TRC and LB-TRC tests are not very effective for evaluating HAZ HIC.

5.4.7   Augmented Strain Cracking Test

The augmented strain cracking test was developed by Savage *et al.* [61, 62] in the 1970s at Rensselaer Polytechnic Institute. The test was designed to induce hydrogen cracking in small laboratory-scale specimens and directly observe crack initiation and propagation in the test specimens. The specimen was produced by a single-pass bead-on-plate welding process. Immediately after welding, the test weldments were quenched in ice water and then transferred to a bath of dry ice and alcohol in order to reduce hydrogen diffusion and minimize the loss of diffusible hydrogen out of the specimen prior to testing. The surface of the sample was prepared metallographically prior to loading into the augmented strain cracking test fixture, as shown in Figure 5.29.

The loading apparatus can provide either a constant stress or a constant strain. The constant stress on the test weldments was provided by the four-point bending arrangement. The constant plastic strain was applied by using a radiused die block pushing against the test weldment during the test process. The approximate value of the constant strain (ε) can be calculated by the relationship $\varepsilon = t/2R$, where t is the thickness of the specimen and R is the radius of curvature of the die block. The constant strain applied on the outer surface of the test weldments can be varied either

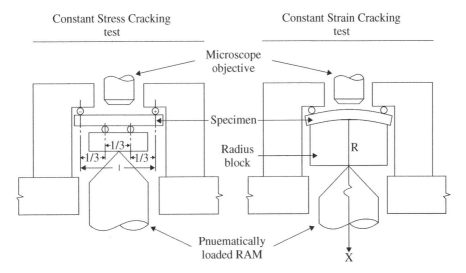

FIGURE 5.29   The augmented strain cracking test (From Ref. [61]. © AWS).

by using a die block with an appropriate radius of curvature or by changing the test specimen thickness.

A microscope was incorporated into the test setup to observe the crack initiation and propagation. A film of immersion oil was spread over the polished surface in order to make direct observation of the evolution of hydrogen. By direct observation, the rate of the evolution of hydrogen bubbles from surface cracks was measured, and a relationship of evolution rate as a function of loading time under different applied stresses was obtained, as illustrated in Figure 5.30. It was found that increasing the applied stress increases the diffusion rate of hydrogen to the region of triaxial stress concentration just ahead of the crack tip, implying hydrogen diffusion is stress induced. These findings provided strong evidence to support Troiano's decohesion theory [7]. Attempts were also made to monitor crack initiation and propagation using acoustic emission, but these were largely unsuccessful.

5.5   IDENTIFYING HIC

HIC or HAC occurs immediately after welding has been completed or after a delay time. In general, if cracking does occur, it will be within 72h of the completion of welding. Because of this delayed cracking possibility, many weld inspection standards require at least a 48h waiting period following welding before final inspection can be performed.

Since high tensile stress and stress concentration are key factors that promote cracking, cracks normally initiate at weld discontinuities associated with the weld geometry. In most cases, a natural notch effect is created at the weld root or toe where the weld metal meets the base metal. Such a notch, even though not sharp, will

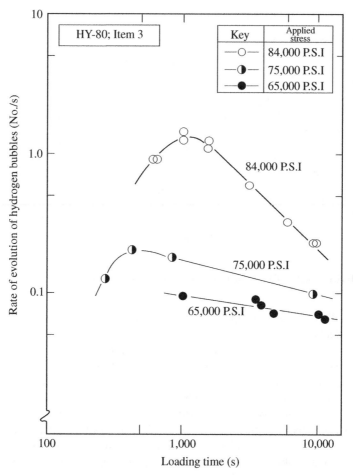

FIGURE 5.30 Rate of evolution of hydrogen bubbles in the Augmented Strain Cracking test as a function of loading time at different applied stresses (From Ref. [61] . © AWS).

concentrate stress. Two examples of stress concentration leading to cracking were shown previously in Figure 5.12.

HIC may occur in either the weld metal or HAZ. Since there is a strong correlation between HIC susceptibility and hardness, cracking is most prevalent in the region of highest hardness. For most steels, this occurs in the CGHAZ just adjacent to the fusion boundary. In the vast majority of cases, HIC initiates at a free surface at the fusion boundary and initially propagates into the CGHAZ. This occurs because all three conditions for hydrogen cracking (Fig. 5.7) are satisfied—stress concentration, high hardness, and hydrogen that has diffused from the molten weld metal. It is possible for hydrogen cracking to occur in other locations if the necessary conditions are satisfied. For example, underbead cracking during cladding may be a form of HIC if sufficient stresses are present.

HIC may propagate in either an IG or transgranular manner. If it is IG, the crack path is along prior austenite grain boundaries. IG fracture is favored at lower stress intensities (see Figs. 5.5 and 5.6) and if there is high impurity segregation at the grain boundaries. In most cases, crack propagation is transgranular and, as discussed in the following text, can exhibit different fracture surface morphologies depending on the stress state at the crack tip and the hydrogen concentration.

In some cases, such as with low-alloy steels, it may be difficult to distinguish between HIC and reheat cracking. Reheat cracks are not present in the as-welded structure, so cracking present immediately following welding is most likely HIC. Reheat cracks are always IG and propagate along prior austenite grain boundaries in very close proximity to the fusion boundary. Hydrogen cracks are typically not so closely aligned with the fusion boundary and tend to wander within the CGHAZ, as shown in Figure 5.12a. In general, the most compelling distinguishing feature between reheat cracks and hydrogen cracks is the tendency for hydrogen cracks to propagate in a transgranular mode or a mixed IG/transgranular mode. Examples of both IG and transgranular HIC are provided in Figure 5.31.

Fracture surface analysis in the SEM is a great aid in identifying HIC. Transgranular fracture displays either a QC or ductile rupture mode. These fracture morphologies in addition to IG HIC are shown in Figure 5.32. The mode described as "MVC" is essentially the same as ductile dimple rupture where voids form at discrete locations in the microstructure (such as a particle/matrix interface) and link together to form a crack surface with a dimpled appearance. QC fracture is unique to hydrogen embrittlement and represents a mixture of true cleavage and ductile rupture with fracture along crystallographic planes within the grain but with some microductility. Hydrogen-assisted IG fracture occurs along prior austenite grain boundaries and, as mentioned previously, may be difficult to distinguish from other forms of solid-state IG fracture, such as reheat or stress relief cracking.

Additional discussion on failure mode and fracture interpretation pertinent to HIC is included in Chapter 8.

5.6   PREVENTING HIC

Of all the weldability issues associated with welded structures, more effort has been concentrated on the control and prevention of HIC than all the others combined. The prevention of HIC requires that one of the major factors that contribute to cracking be eliminated. These include hydrogen, microstructure, restraint, and temperature.

Reducing hydrogen uptake during welding requires careful control and handling of base metals and consumables. In processes using coated electrodes or fluxes, baking these electrodes or fluxes in the temperature range from 250 to 450°C followed by storage at a temperature above 100°C is highly recommended. These precautions remove the moisture that can later dissociate in the weld to form hydrogen. The use of "low-hydrogen" electrodes that do not contain compounds that dissociate to form hydrogen is also recommended. For example, cellulosic electrodes will generate hydrogen in the welding arc. Cleanliness is

(a)

(b)

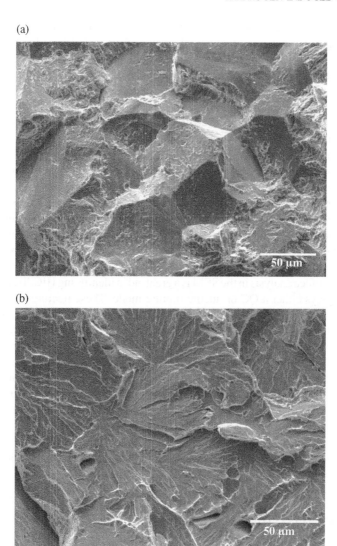

FIGURE 5.31   Examples of (a) intergranular and (b) transgranular HIC (Courtesy of Xin Yue).

also critical. Paint, rust, grease, and other lubricants can all act as sources of hydrogen on either the base material or the filler material. Proper cleaning procedures should be used prior to welding.

The microstructure of the weld metal and HAZ has a strong influence on cracking susceptibility. In the weld fusion zone, microstructural control is most often implemented by selection of filler metal compositions that produce more resistant microstructures (see Table 5.1). In the HAZ, HIC-resistant microstructures can be produced in some steels by reducing the cooling rate, usually through the use of higher weld heat input, preheat, or a combination of these. The effectiveness of microstructure

(a)

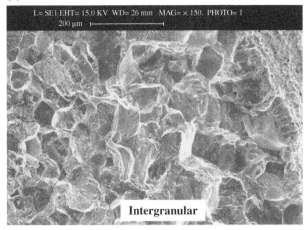

(b)

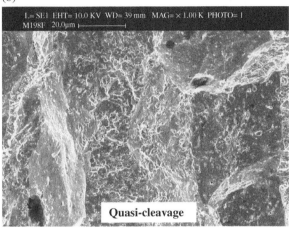

(c)

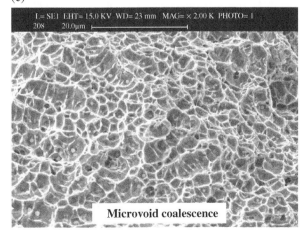

FIGURE 5.32 Fracture modes associated with HIC. (a) Intergranular, (b) quasicleavage, and (c) microvoid coalescence.

control by weld thermal conditions decreases as the CE of the steel increases. For multipass welding, weld schedules may also be developed, which effectively temper the underlying HAZ. For example, the so-called "temperbead" techniques have been developed to temper the HAZ in large structures where postweld heat treatment is not an option.

Weld restraint is often the most difficult of the variables related to hydrogen cracking to control. Attention should be paid to weld joint designs and welding practices that minimize the level of residual stresses, particularly in thick-section weldments. The use of preheat and interpass temperature control may help reduce residual stress effects. Performing a stress relief heat treatment immediately follow-ing welding is the most effective method to eliminate residual stress effects. In addition to the global restraint conditions that generate high tensile residual stresses, local stress concentration can be very damaging. Eliminating stress concentrations by proper weld design or removing them by grinding or other means following welding can be very effective in preventing cracking.

The temperature range for HIC is usually considered to be from −100 to 200°C (−150 to 390°F), and, thus, all welds will naturally cool into this range. Controlling cooling rates above this range may allow sufficient time for hydrogen to diffuse out of the structure. In severe cases, it may be possible to hold the structure above 200°C to allow hydrogen to diffuse out and then cool to room temperature. In some harden-able steels, the microstructure may be fully (or partially) austenitic at this temperature, which will effectively prevent HIC and allow the hydrogen level to decrease prior to full transformation upon cooling.

Not surprisingly, many predictive/preventative tools have been developed over the years for avoiding HIC. These have been reviewed in some detail by Yurioka and Suzuki [32] and can be grouped into two general categories, (i) cooling time methods and (ii) stress criterion methods. The cooling time methods typically include factors for composition (hardenability), hydrogen concentration, and joint restraint.

One cooling time method developed by Itoh and Bessho [63] determines a critical cooling time from 300 to 100°C (572–212°F) using a cracking index (CI), P_W, where

$$t_{100} = 1.4 \times 10^4 \left(P_W - 0.28 \right)^2$$

$$P_W = P_{CM} + \frac{H_{GL}}{60} + \frac{R_F}{400,000}$$

P_{CM} is the CE (see Table 5.2), H_{GL} is the diffusible hydrogen content determined by the glycerin method, and R_F is the joint restraint factor. P_W was derived from the results of the Tekken test on steels of varying strength levels. Using this relationship, an actual cooling time exceeding t_{100} will avoid HIC. This is presumably because the increased cooling time reduces the hardness of the HAZ and/or allows more hydrogen to diffuse from the structure.

Yurioka *et al.* [64] used a similar approach in conjunction with the Tekken test to determine a critical cooling time from 300 to 100°C. Their criterion used a CI that included CE_N (Table 5.2); a stress concentration factor, K_t; and the yield strength of the weld metal, σ_w:

$$t_{100} = \exp\left(68.05CI^3 - 181.77CI^2 - 163.80CI - 41.65\right)$$

$$CI = CE_N + 0.15 \log H_{GL} + 0.70 \log \left(0.0017{\cdot}K_t{\cdot}\sigma_w\right)$$

Because both of these methods were developed based on the Tekken (Y-groove) test that produces very high stress concentration at the root of the weld, they tend to be overly conservative, particularly in plain-carbon steels. The implication of this conservative approach is that it is possible to avoid HIC with faster cooling times (lower preheat), but exceeding the critical cooling time will almost certainly assure resistance to HIC.

There are also methods that use the critical stress level to predict susceptibility to HIC [65, 66]. While the level of tensile stress and stress intensity are key elements in the hydrogen cracking mechanism, they are much more difficult to control and quantify. As a result, direct methods that predict preheat or prescribe welding conditions that control cooling rate are the most widely used to prevent HIC.

There are four of these methods that have been included in the ISO/TR 17844: Welding—Comparison of standardised methods for the avoidance of cold cracks. These include the following:

1. CE method (derived from British Standard BS 5135)
2. CET method (derived from a German Standard and based on the Y-groove test)
3. CE_N method (derived from Japanese Standard JIS B8285 and based on the Y-groove test)
4. AWS method (derived from AWS D1.1: Structural Welding Code for Steel)

All of these methods are similar in that they allow the determination of a preheat temperature that will be sufficient to prevent HIC as a function of CE and heat input. Two of these methods, the CE method and the AWS method, are described here.

5.6.1 CE Method

The CE method constitutes the British Standard BS 5135. It was first included as a British Standard in 1973 and has been successfully used for many years to avoid HIC in a range of steels. It was developed primarily for SMAW, or manual metal arc welding (MMAW), as it is called in Europe. Much of the data was gathered using the CTS test described in Section 5.4.2. This method contains a number of charts that allow for determination of the preheat temperature that will avoid HIC. Two of these charts

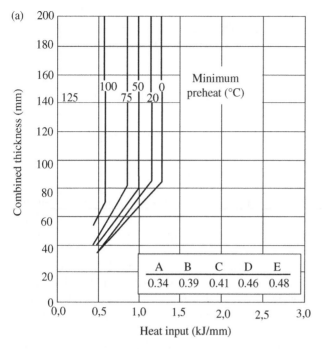

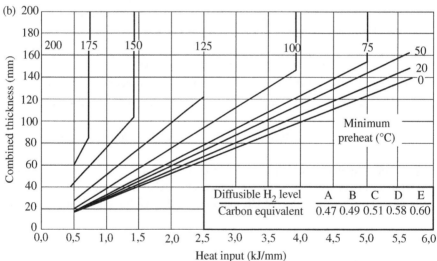

FIGURE 5.33 Preheat charts for use with the CE method. (a) Low CE and (b) intermediate CE (From Ref. [30]. © Elsevier).

are shown in Figure 5.33 for C–Mn steels with relatively low and intermediate CE. In order to use the charts, the following steps must be followed:

1. Determine the CE for the material using the CE_{IIW} formula (Table 5.2).
2. Estimate or measure the diffusible hydrogen content of the weld deposit and rate the hydrogen level as A, B, C, D, or E according to the following.

Diffusible hydrogen (ml/100 g)	CE axis
>15	A
10–15	B
5–10	C
3–5	D
<3	E

3. Calculate the heat input (in kJ/mm) using

$$\text{Heat input} \left(\text{kJ}/\text{mm}\right) = \frac{V \times I \times 60}{1000 \times TS \left(\text{mm}/\text{min}\right)}$$

4. Based on the weld heat input, plate thickness, hydrogen level, and CE, determine the preheat temperature using the appropriate chart.

The standard contains tables to determine heat input based on the "run length" for a given SMAW electrode. The effect of consumable type and process (GTAW, GMAW, and FCAW) is taken into consideration by the selection of the approximate hydrogen level (if the exact value is not known). For example, flux-cored electrodes are usually assigned levels B to D, while solid wires are considered level D or E.

The chart in Figure 5.33b is also for a C–Mn steel but for higher CE levels than in Figure 5.33a. This chart also allows for a wider range of heat input than the chart in Figure 5.33a. The methodology is the same as described earlier. The following is an example of how this chart can be used:

Material composition (wt%): 0.35C, 0.85Mn, 0.22Si, and 0.10Cr
Plate thickness: 100 mm
Welding process: FCAW
Consumable: E81T-5

Welding conditions: Current = 400 A, voltage = 11 V, and TS = 200 mm/min

1. $CE_{IIW} = 0.35 + 0.85/6 + 0.10/5 = 0.51$
2. Diffusible hydrogen level = 8 ml/100 g (based on measurement)—category C
3. Heat input = 1.32 kJ/mm
4. For 100 mm-thick plate, minimum preheat = 150°C

5.6.2 AWS Method

The AWS method involves either controlling weld heat input and associated weld cooling rate to control hardness or controlling preheat to reduce the hydrogen content in the weld deposit. The first step is to determine which of these control methods should be used. This is done using the chart in Figure 5.34 to determine the "zone" in which the material lies:

Zone 1. Cracking is unlikely due to the low carbon content but may occur with high hydrogen or high restraint. The hydrogen control method should be used to determine preheat in this zone.

Zone 2. The hardness control method should be applied to determine minimum energy input required to achieve the desired cooling rate. If the heat input required is not practical, the hydrogen control method should be used.

Zone 3. The hydrogen control method should always be used in this zone.

Note that the CE formula used for the AWS method is slightly different from CE_{IIW} since it contains a term for Si:

$$CE_{AWS} = C + \frac{Mn + Si}{6} + \frac{Ni + Cu}{15} + \frac{Cr + Mo + V}{5}$$

The cooling rate method should be used for compositions that fall within Zone 2 in Figure 5.34. Based on the CE of the steel, the cooling rate should be estimated

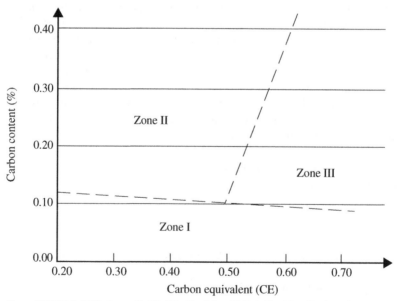

From AWS D1.1-2000, Appendix XI, $CE = C + (Mn + Si)/6 + (Cr + Mo + V)/5 + (Ni + Cu)/15$

FIGURE 5.34 AWS method chart for determining zone classification of steel based on carbon content and carbon equivalent (From AWS D1.1, Structural Welding Code for Steel. © AWS).

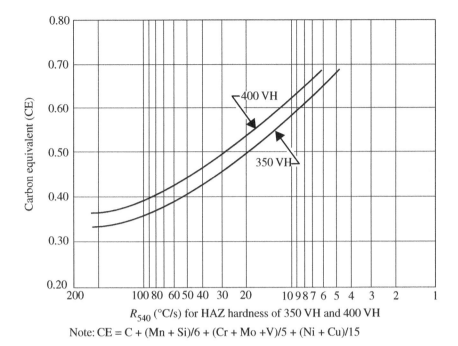

FIGURE 5.35   AWS method chart for estimating cooling rate based on carbon equivalent (From AWS D1.1, Structural Welding Code for Steel. © AWS).

using the chart in Figure 5.35. For highly restrained or high hydrogen welds, the lower hardness level curve (350 VHN) should be used. Once the cooling rate is determined, charts such as those shown in Figure 5.36 can be used to estimate the welding heat input required to avoid HIC.

It should also be noted that the charts shown in Figure 5.36 and additional charts found in AWS D1.1 and ISO/TR 17844 were developed for SAW fillet welds. For fillet welds made using other processes, the following multiplication factors should be applied to the heat inputs determined using these charts.

Welding process	Multiplication factor
SMAW	1.5
GMAW	1.25
FCAW	1.25

For example, if the same steel in the example for the CE method is considered using the AWS method, it would fall within Zone 2 and the cooling rate (hardness control) method would be applied. Based on the CE (~0.55), a cooling rate of 12°C/s would be required to achieve a hardness level of 350 VHN. Assuming section thicknesses of 0.5 in. (~12.5 mm), the chart in Figure 5.36a would estimate a heat input of approximately 50 kJ/in. Using the multiplication factor for FCAW (1.25), the

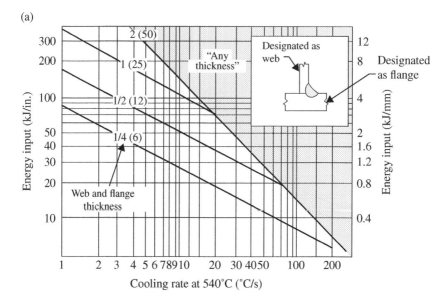

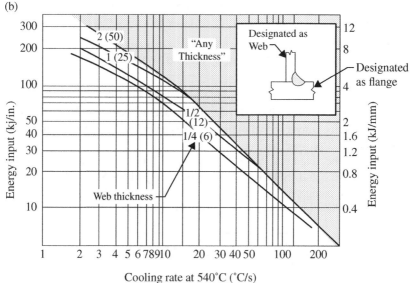

FIGURE 5.36 Charts used to estimate heat input based on cooling rate and section thickness. (a) Equal web and flange thickness and (b) 1 inch flange with variable web thickness (From AWS D1.1, Structural Welding Code for Steel. © AWS).

recommended minimum heat input to achieve the desired hardness of 350 VHN would be 62 kJ/in. (~2.5 kJ/mm).

The hydrogen control method is recommended for situations where the steel falls within Zone 1 and high restraint and hydrogen are anticipated and all cases with

Zone 3. Applying the AWS method for these steels, a different CE must be used, designated as P_{CM}:

$$P_{CM} = C + \frac{Si}{30} + \frac{Mn}{20} + \frac{Cu}{20} + \frac{Ni}{60} + \frac{Cr}{20} + \frac{Mo}{15} + \frac{V}{10} + 5B$$

The combination of P_{CM} and hydrogen level is then used with Table 5.3 to establish a susceptibility index (A–G) for the weldment. The hydrogen level is rated from low to high, H1 to H3. These ratings are defined as follows:

- H1: Extra low hydrogen. These consumables produce a diffusible hydrogen content below 5 ml/100 g. These would include (i) low-hydrogen electrodes taken immediately from hermetically sealed containers, dried at 700–800°F for 1 h, and used within 2 h and (ii) clean solid wires used for GTAW or GMAW.
- H2: Low hydrogen. These consumable will yield a diffusible hydrogen content between 5 and 10 ml/100 g. Examples would be (i) low-hydrogen electrodes taken immediately from hermetically sealed containers and used within 4 h and (ii) SAW with dry flux.

TABLE 5.3   Estimating the susceptibility index using the AWS method[a]

	Susceptibility index grouping as function of hydrogen level "H" and composition parameter, P_{cm} (see XI6.2.3) Susceptibility index[2] grouping Carbon equivalent = P_{cm}^1				
Hydrogen level, H	<0.18	<0.23	<0.28	<0.33	<0.38
H1	A	B	C	D	E
H2	B	C	D	E	F
H3	C	D	E	F	G

[a]From AWS D1.1, Structural Welding Code for Steel. Courtesy of American Welding Society.

$$^1 P_{cm} = C = \frac{Si}{30} + \frac{Mn}{20} + \frac{Cu}{20} + \frac{Ni}{60} + \frac{Cr}{20} + \frac{Mo}{15} + \frac{V}{10} + 5B$$

[2]Susceptibility index—$12P_{cm} + \log_{10} H$

[3]Susceptibility index groupings, A through G, encompass the combined effect of the composition parameter, P_{cm}, and hydrogen level, H, in accordance with the formula shown in Note 2.

The exact numerical quantities are obtained from the note 2 formula using the stated values of P_{cm} and the following values of H, given in ml/100 g of weld metal (see XI6.2.2, a, b, c):

H1—5; H2—10; H3—30

For greater convenience, susceptibility index groupings have been expressed in the table by means of letters, A through G, to cover the following narrow ranges:

$A = 3.0$; $B = 3.1–3.5$; $C = 3.6–4.0$; $D = 4.1–4.5$; $E = 4.6–5.0$; $F = 5.1–5.5$; $G = 5.6–7.0$

These groupings are used in Table XI-2 in conjunction with restraint and thickness to determine the minimum preheat and interpass temperature.

TABLE 5.4 Determining minimum preheat and interpass temperature based on susceptibility index, restraint level, and plate thickness[a]

Restraint level	Thickness[b] (in.)	Minimum preheat and interpass temperatures for three levels of restraint (see XI6.2.4) Minimum preheat and interpass temperature (F) Susceptibility index grouping						
		A	B	C	D	E	F	G
Low	<3/8	<65	<65	<65	<65	140	280	300
	3/8–3/4	<65	<65	65	140	210	280	300
	3/4–1-1/2	<65	<65	65	175	230	280	300
	1-1/2–3	65	65	100	200	250	280	300
	>3	65	65	100	200	250	280	300
Medium	<3/8	<65	<65	<65	<65	160	280	320
	3/8–3/4	<65	<65	65	175	240	290	320
	3/4–1-1/2	<65	65	165	230	280	300	320
	1-1/2–3	65	175	230	265	300	300	320
	>3	200	250	280	300	320	320	320
High	<3/8	<65	<65	<65	100	230	300	320
	3/8–3/4	<65	65	150	220	280	320	320
	3/4–1-1/2	65	185	240	280	300	320	320
	1-1/2–3	240	265	300	300	320	320	320
	>3	240	265	300	300	320	320	320

[a]From b D1.1, Structural Welding Code for Steel. Courtesy of American Welding Society.
[b]Thickness is that of the thicker part welded.

- H3: Hydrogen not controlled. This covers all cases where the hydrogen is not controlled as described in H1 and H2. Actual hydrogen levels may be as high as 30 ml/100 g.

Once the susceptibility index is determined, Table 5.4 can be used to select minimum preheat and interpass temperature to prevent HIC. To use this table, the material thickness must be known, and some judgment as to the restraint level must be made. The following general guidelines for determining restraint level are taken from AWS D1.1:

Low restraint. This level describes fillet and groove welds in which reasonable freedom of movement exists, such as a freestanding structure.

Medium restraint. Situations where the members being welded are already attached to structural work and there is reduced freedom of movement.

High restraint. Almost no freedom of movement exists, such as in repair welds or when welding very thick material.

When selecting the restraint level, it is advisable to be conservative and select a level higher than is apparent, unless there is prior knowledge or expertise that suggests otherwise.

As an example, assume a low-alloy steel with composition Fe–0.20C–0.2Si–0.8Mn–1.0Cr–0.5Mo. Using Figure 5.34, this steel would be in Zone 3 (CE = 0.67), and the hydrogen control method would be specified. Using Table 5.3 and assuming a low-hydrogen condition, the P_{CM} for this steel (0.33) would result in a susceptibility index of "E." Assuming the section thickness is on the order of 0.5 in. (12.5 mm) and the restraint level is moderate, Table 5.4 would recommend a minimum preheat of 240°F (115°C) to avoid HIC.

Within ISO/TR 17844, there is a comparison of the four methods (CE, CET, CE_N, and AWS) based on a range of steel compositions and section thickness. Not surprisingly, there is some disagreement among these methods when estimating preheat to prevent HIC. Some of this disagreement results from the test techniques used to develop these approaches. For example, the CE method is based largely on CTS test results, while the CET and CE_N methods were based on Y-groove testing. Since the Y-groove test produces higher restraint than the CTS test, preheat predicted by CET and CE_N methods tends to be higher than the CE method. In general, the AWS method tends to be the most conservative (higher preheat values) and the CE method the least conservative. These methods should be used as guidelines, particularly when welding high-carbon and/or high-hardenability steels.

REFERENCES

[1] Kou S. *Welding Metallurgy*. Hoboken, NJ: Wiley-Interscience; 2003. p 411.

[2] Lesnewich A. Influence of welding on steel weldment soundness. In: *ASM Handbook*, Vol. 6. Materials Park, OH: ASM International; 1993, p 408–415.

[3] Coe FR. *Welding Steels without Hydrogen Cracking*. Cambridge, UK: Welding Institute; 1973.

[4] Zapffe C, Sims C. Hydrogen embrittlement, internal stress and defects in steel. Trans AIME 1941;145:225–271.

[5] Petch NJ, Stables P. Delayed fracture of metals under static load. Nature 1952;169:842–843.

[6] Petch NJ. The lowering of fracture-stress due to surface adsorption. Philos Mag 1956;1:331–337.

[7] Troiano AR. The role of hydrogen and other interstitials in the mechanical behavior of metals. Trans ASM 1960;52:54–80.

[8] Oriani R, Josephic P. Equilibrium aspects of hydrogen-induced cracking of steels. Acta Metall 1974;22:1065–1074.

[9] Oriani R, Josephic P. Equilibrium and kinetic studies of the hydrogen-assisted cracking of steel. Acta Metall 1977;25:979–988.

[10] Oriani R. Hydrogen embrittlement of steels. Ann Rev Mater Sci 1978;8:327–357.

[11] Pressouyre G, Bernstein IM. A kinetic trapping model for hydrogen-induced cracking. Acta Metall 1979;27:89–100.

[12] Pressouyre G. Trap theory of hydrogen embrittlement. Acta Metall 1980;28:895–911.

[13] Savage WF, Nippes EF, Szekeres ES. Hydrogen induced cold cracking in a low alloy steel. Weld J 1976;55 (9):276s–283s.

[14] Sofronis P, Birnbaum HK. Hydrogen induced shear localization of the plastic flow in metals and alloys. Eur J Mech 2001;20:857–872.

[15] Sofronis P. The influence of mobility of dissolved hydrogen on the elastic response of a metal. J Mech Phys Solids 1995;43:1385–1407.

[16] Sofronis P, Birnbaum HK. Mechanics of the hydrogen-dislocation-impurity interactions—I. Increasing shear modulus. J Mech Phys Solids 1995;43:49–90.

[17] Birnbaum HK, Sofronis P. Hydrogen-enhanced localized plasticity: a mechanism for hydrogen-related fracture. Mater Sci Eng 1994;176:191–202.

[18] Lynch S. Environmentally assisted cracking: overview of evidence for an adsorption-induced localised-slip process. Acta Metall 1988;36:2639–2661.

[19] Shih D, Robertson I, Birnbaum HK. Hydrogen embrittlement of α titanium: in-situ TEM studies. Acta Metall 1988;36:111–124.

[20] Robertson I, Birnbaum HK. An HVEM study of hydrogen effects on the deformation and fracture of nickel. Acta Metall 1986;34:353–366.

[21] Bond G, Robertson I, Birnbaum HK. The influence of hydrogen on deformation and fracture processes in high-strength aluminum alloys. Acta Metall 1987;35:2289–2296.

[22] Robertson I, Birnbaum HK. Effect of hydrogen on the dislocation structure of deformed nickel. Scripta Metall 1984;18:269–274.

[23] Michler T, Naumann J. Microstructural aspects upon hydrogen environment embrittlement of various bcc steels. Int J Hydrogen Energy 2010;35:821–832.

[24] Beachem C. A new model for hydrogen-assisted cracking (hydrogen "embrittlement"). Metall Trans 1972;3:441–455.

[25] Yue X, Lippold JC. Evaluation of heat-affected zone hydrogen-induced cracking in Navy steels. Weld J 2013;92 (1):20s–28s.

[26] Vasudevan R, Stout RD, Pense AW. Hydrogen-assisted cracking in HSLA pipeline steels, Weld J 1981;60 (9):155s–168s.

[27] Savage WF, Nippes EF, Sawhill Jr JM. Hydrogen induced cracking during implant testing of alloy steels. Weld J 1976;55 (12):400s–407s.

[28] Gedeon SA, Eagar TW. Assessing hydrogen-assisted cracking fracture modes in high-strength steel weldments. Weld J 1990;69 (6):213s–220s.

[29] Available at http://www.leco.com/products/analytical-sciences/oxygen-nitrogen-hydrogen-analyzers. Accessed July 4, 2014.

[30] Bailey N, Coe FR, Gooch TG, Hart PHM, Jenkins N, Pargeter RJ. *Welding Steels without Hydrogen Cracking*. Cambridge, UK: Woodhead Publishing Limited; 1993.

[31] American Society for Metals. *Atlas of Isothermal Transformation and Cooling Transformation Diagrams*. Metals Park, OH: ASM International; 1977.

[32] Yurioka N, Suzuki H. Hydrogen assisted cracking in C–Mn and low alloy steel weldments. Int Mater Rev 1990;35:217–249.

[33] Threadgill PL. Avoiding HAZ defects in welded structures. Metals Mater 1985;July:422–429.

[34] American Welding Society. *AWS B4.0: Standard Methods for Mechanical Testing of Welds*. Miami, FL: American Welding Society; 2007.

[35] Kannengiesser T, Boellinghaus T. Cold cracking tests: an overview of present technologies and applications. Weld World 2013;57 (1):3–37.

[36] Jominy WE, Boegehold AL. A hardenability test for carburizing steel. Trans ASM 1938;26:574–606..

[37] ASTM A255-10: Standard Methods for Determining Hardenability of Steel.

[38] Stout RD. *Weldability of Steels*. 4th ed. New York: Welding Research Council; 1987.

[39] Cottrell C. Controlled thermal severity cracking test simulates practical welded joints. Weld J 1953;32 (6):257s–272s.

[40] Campbell W. Experiences with HAZ cold cracking tests on a C-Mn structural steel. Weld J 1976;55 (5):135s–143s.

[41] Kihara H, Suzuki H, Nakamura H. Weld cracking tests of high strength steels and electrodes. Weld J 1962;41 (1):36s–48s.

[42] Alexandrov B *et al.* Cold cracking in weldments of steel S 690 QT. Weld World 2005;49:64–73.

[43] Alcantara NG, Rogerson JH. A prediction diagram for preventing hydrogen-assisted cracking in weld metal. Weld J 1984;63 (4):116s–122s.

[44] Reddy GM, Mohandas T, Sarma D. Cold cracking studies on low alloy steel weldments: effect of filler metal composition. Sci Technol Weld Join 2003;8:407–414.

[45] Wingrove L. *An appraisal of the Tekken test* [PhD Dissertation]. Wollongong: University of Wollongong; 1986.

[46] McParlan M, Graville BA. Hydrogen cracking in weld metals. Weld J 1976;55 (4):95s–102s.

[47] McParlan M, Graville BA. Development of the G-BOP test for weld metal cracking. IIW Document IX-922-75: International Institute of Welding, Paris; 1975.

[48] Hart PM. Resistance to hydrogen cracking in steel weld metals. Weld J 1986;65 (1):14s–22s.

[49] Marianetti C. *The development of the G-BOP test and the assessment of weld metal hydrogen cracking* [MS Thesis]. Columbus, OH: The Ohio State University; 1998.

[50] Chakravarti AP, Bala SR. Evaluation of weld metal cold cracking using the G-BOP test. Weld J 1989;68(1):1s–8s.

[51] Atkins G, Thiessen D, Nissley N, Adonyi Y. Welding process effects in weldability testing of steels. Weld J 2002;81 (4):61s–68s.

[52] Granjon H. The implant method for studying the weldability of high strength steels. Metal Constr Brit Weld J 1969;1:509–515.

[53] Sawhill Jr J, Dix AW, Savage WF. Modified implant test for studying delayed cracking. Weld J 1974;53(12):554s–559s.

[54] Signes E, Howe P. Hydrogen-assisted cracking in high-strength pipeline steels. Weld J 1988;67(8):163s–170s.

[55] ISO 17642-3:2005. *Destructive tests on welds in metallic materials—Cold cracking tests for weldments—Arc welding processes—Part 3: Externally loaded tests.* 1st edn: International Organization for Standardization, Geneva, Switzerland; 2005.

[56] Suzuki H, Nakamura H. Effect of external restraint on root cracking in high strength steels by TRC testing. IIW Doc. IX-371-63: International Institute of Welding, Paris; 1963.

[57] Matsuda F, Nakagawa H, Shirozaki K. Evaluation of the cold cracking susceptibility of weld metal in high strength steels using the longitudinal bead-TRC test. Weld Int 1988;2:229–233.

[58] Matsuda F, Nakagawa H, Shirozaki K. The longitudinal bead-TRC test for cold crack susceptibility of weld metal for high strength steels. Weld Int 1988;2:135–139.

[59] Shinozaki K, Ke L, North TH. Hydrogen cracking in duplex stainless steel weld metal. Weld J 1992;71(11):387s–396s.

[60] Shinozaki K, North TH. Effect of oxygen on hydrogen cracking in high-strength weld metal. Metall Trans 1990;21:1287–1298.

[61] Savage WF, Nippes EF, Homma H. Hydrogen induced cracking in HY-80 steel weldments. Weld J 1976;55(4):368s–376.

[62] Savage WF, Nippes EF, Tokunaga Y. Hydrogen induced cracking in HY-130 steel weldments. Weld J 1978;57(4):118s–126s.

[63] Itoh Y, Bessho K. Cracking parameter of high strength steels related to HAZ cracking. JWS 1968;37 (9):983–989.

[64] Yurioka Y, Suzuki H, Ohshita S. Determination of necessary preheating temperature. Weld J 1983;62(6):147s–153s.

[65] Satoh K, Terasaki T, Ohkuma Y. Relationship between critical stress of HAZ cracking and residual diffusible hydrogen content. JWS 1979;48 (4):248–252.

Karppi RAJ, Nevasmaa P. Contribution of comparison of methods for determining welding procedures for the avoidance of hydrogen cracking. IIW Doc. IX-1673-92: International Institute of Welding, Paris; 1992.

6

CORROSION

6.1 INTRODUCTION

Corrosion is often associated with welded structures, since the microstructure, properties, and composition of the weld metal and HAZ may be quite different than those of the base metal. Corrosion takes a number of forms, as described in the following section, and may result in general (uniform), localized, or microstructure-specific attack.

Often, the corrosion rate associated with welds is much higher than the base metal. The reason for this is usually a combination of the effect of microstructure and residual stress. Highly stressed regions surrounding welds may result in accelerated corrosion relative to the base metal. For example, the spot welds in automotive steels are normally the first place where corrosion attack takes place in car bodies.

There are a number of standardized tests that have been developed to quantify susceptibility to corrosion in different environments. Many of these are accelerated tests that allow long-term corrosion behavior to be estimated using short-term laboratory tests. Most of these tests can be conducted using welded samples. Corrosion control of welded structures is often the critical factor in determining the ultimate life, or fitness for service, of a structure.

Welding Metallurgy and Weldability, First Edition. John C. Lippold.
© 2015 John Wiley & Sons, Inc. Published 2015 by John Wiley & Sons, Inc.

6.2 FORMS OF CORROSION

Corrosion of engineering alloys is manifested in a variety of forms, most of which are macroscopic (visible with the naked eye) and others require microscopic examination. Fontana and Green [1] defined eight forms of corrosion, many of which can be linked to degradation and eventual failure of welded structures, as listed below. This classification was based on the visual appearance of the corrosion. A ninth form of corrosion that has been determined to be distinct from these other forms is termed microbiologically induced corrosion (MIC):

1. General
2. Galvanic
3. Crevice
4. Selective leaching
5. Erosion
6. Pitting
7. Intergranular
8. Stress assisted
9. Microbiologically induced

The general characteristics of the first five of these corrosion forms are briefly reviewed in the following sections. Corrosion resulting from pitting, intergranular attack (IGA), and stress-assisted mechanism are covered in more detail since these forms are most pertinent to welded structures.

6.2.1 General Corrosion

This is the most common form of corrosion encountered and is by far the most costly (over 10 billion dollars a year is spent in the United States alone to repair or prevent uniform attack). Uniform attack, or general corrosion, is characterized by a chemical or electrochemical reaction at the metal surface that results in the uniform formation of a corrosion product. In steels, this corrosion product is manifested as "rust," the oxidation of iron to form Fe_2O_3 (hematite) and Fe_3O_4 (magnetite).

General corrosion is primarily a function of composition, rather than microstructure. In cases where the weld metal and base metal compositions are similar (such as with autogenous and homogenous welds), the welds may be only slightly more susceptible to general corrosion than the surrounding base material. Increased corrosion rates in welded structures are often associated with residual stresses that may accelerate attack in certain environments.

General corrosion rates vary widely among materials. Much general corrosion data has been gathered in a variety of environments, including atmospheric, soil, and more aggressive solutions. For example, there is considerable general corrosion data available for chloride-bearing environments such as seawater. A comparison of atmospheric corrosion rates for different materials is provided in Table 6.1 [2].

TABLE 6.1   General corrosion rates of metals and alloys[a]

Material	Rate, mils/year (mm/year)
Aluminum	0.032 (0.0008)
Copper	0.047 (0.0012)
Lead	0.017 (0.0004)
Tin	0.047 (0.0012)
Nickel	0.128 (0.0032)
Monel (70N–30Cu)	0.053 (0.0013)
Zinc	0.202 (0.0051)
AISI 1020 steel	0.48 (0.0120)
Low-alloy steel (1% Cr)	0.09 (0.0023)
Type 304 stainless steel	Nil

[a]From Ref. [2].

Damage resulting from general atmospheric corrosion is usually prevented by the use of coatings or inhibitors (such as paint) or the selection of corrosion-resistant materials (such as aluminum alloys, stainless steels, and Ni-base alloys for resistance under normal atmospheric conditions).

6.2.2   Galvanic Corrosion

A chemical potential difference exists between dissimilar metals when they are both in contact with a corrosive and/or conductive medium. If these metals are, in turn, in electrical contact in the presence of this same medium, the potential difference results in a flow of current between the two. By convention, the material in which a corrosion reaction (oxidation) occurs is defined as the *anode* and the material where reduction occurs is defined as the *cathode*. When two materials are connected electrically in a corrosive environment, corrosion occurs in the anodic material.

A galvanic series of some commercial alloys and pure metals in seawater is provided in Table 6.2. Metals that are anodic (active) will corrode preferentially when in contact with those that are cathodic (noble) in a corrosive medium. This medium may range from pure water (or steam) to highly corrosive acids. The rate of attack of a cathodic material relative to the anode is related to the difference in potential and the area that is exposed to the environment. Thus, from this table, magnesium would corrode rapidly if coupled to steel in seawater, while little attack would be expected in a cupronickel coupled to an austenitic stainless steel (18Cr–8Ni).

The galvanic effect can be used to provide "cathodic protection" in certain environments. For example, steel ship hulls are often cathodically protected by coupling to a massive magnesium block, which will corrode preferentially and protect the ship hull during seawater exposure.

Controlling the relative differences in solution potential among the base metal, HAZ, and weld metal is the key to avoiding galvanic attack, as illustrated in Figure 6.1.

TABLE 6.2   Galvanic series for metals in seawater

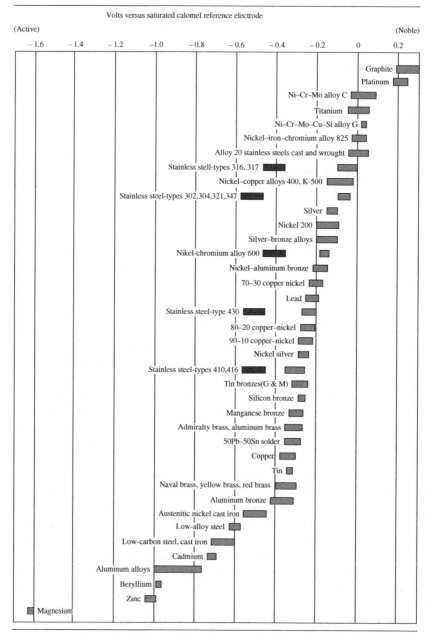

Volts versus saturated calomel reference electrode

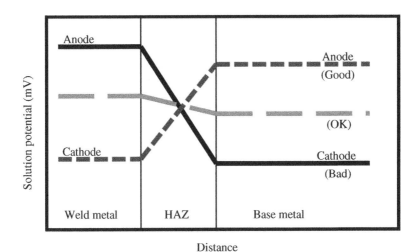

FIGURE 6.1   Galvanic couples and solution potentials.

In the galvanic couple, the anode (active metal) will corrode preferentially to the cathode (noble metal). If the anode area is small relative to the cathode, this attack can be quite rapid. Thus, in environments where corrosion is a concern, it is important to insure that the base metal is anodic to the weld metal and HAZ.

In the case where the weld metal is anodic and the differential in solution potential is large, very rapid attack of the weld metal is possible. For most filler metal/base metal combinations, this is not a problem since the difference in potential is usually small. However, when selecting filler metals for dissimilar combinations or when the filler metal is of significantly different composition than the base metal, it is recommended that the relative potentials of the materials are determined. This is particularly important when the welded structure is exposed to aqueous environments. The filler metals used for many common aluminum alloys are often very different in composition than the base metal and may require special attention to galvanic effects in certain environments.

6.2.3   Crevice Corrosion

Intense localized corrosion can occur at crevices and notches on metal surfaces. Crevice corrosion is usually associated with mechanical fasteners such as bolts or rivets but has resulted from the presence of weld discontinuities. Such crevices can result from lack of penetration defects, slag intrusions (e.g., at the weld toe), and cracks, which are open to the surface of the weld.

To function as a crevice corrosion site, the crevice must be wide enough to permit entry of the liquid corrosive medium but narrow enough to allow a stagnant concentration to form within the crevice. As a result, defects that are only open a few thousandths of an inch at the surface are the most susceptible. The austenitic stainless steels tend to be particularly susceptible to crevice corrosion in seawater.

6.2.4 Selective Leaching

Selective leaching is the removal of one element from an alloy by corrosion processes. The most common example of this is the selective removal of zinc from brass alloys, commonly known as dezincification. Weld regions do not appear to be any more susceptible than the base material to this form of corrosive attack.

6.2.5 Erosion Corrosion

This form of attack is basically general corrosion that is accelerated by the relative motion of the corrosive medium. Most metals and alloys are susceptible to this form of corrosive attack. Metals that are relatively soft and readily damaged or worn by mechanical abrasion, such as copper, are usually the most susceptible to erosion corrosion. The weld region in many materials may be more susceptible to this form of attack than the base material due to the softening (annealing) of the structure resulting from the weld thermal cycle. This effect is usually quite subtle.

Erosion corrosion has been observed in power generation applications where fluids or steam are being pumped at high velocities using rotors or impellers. Both the rotors and piping systems may be susceptible to erosion corrosion.

In some situations, welding can be used to mitigate erosion corrosion. For example, the application of "hard-facing" filler metals can locally increase hardness and reduce or eliminate erosion associated with softer materials. In other cases, the use of filler metals with higher corrosion resistance can have a similar effect.

6.2.6 Pitting

Pitting is an extremely localized attack that is manifested by holes, or pits, in the metal surface. Pitting is a particularly insidious form of corrosion since it is difficult to detect until the structure has been severely attacked. Pits usually grow in the direction of gravity, only rarely forming on vertical surfaces or growing upward from the bottom of horizontal surfaces. As shown in Figure 6.2, there may be little observable damage on the surface of the structure, while in the subsurface the corrosion attack may be substantial. Pitting is an autocatalytic process whereby anodic metal dissolution occurs within the pit, creating a local corrosion cell. For pitting in seawater, this anodic reaction results in formation of an acidic (pH 1.5–2.0) solution at the bottom of the pit that results in rapid dissolution of the metal.

In general, damage by pitting can be avoided by proper alloy selection. For instance, plain-carbon steel is more resistant to pitting in some environments than stainless steel (18Cr–8Ni). In situations where corrosion-resistant (general corrosion) materials are necessary, the selection of Mo-bearing stainless steels (Type 316), duplex stainless steels, or nickel-base alloys is recommended (e.g., Hastelloy).

Pitting is influenced primarily by the composition of the metal and does not appear to prefer weld regions over the surrounding base metal unless segregation of critical alloying elements has occurred. Elemental partitioning during weld solidification may enhance localized pitting attack (due to microscopic inhomogeneity).

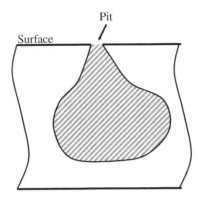

FIGURE 6.2   Illustration of pitting attack.

TABLE 6.3   Effect of alloying and impurity elements on pitting corrosion of stainless steels

Element	Effect on resistance
Chromium	Increases
Nickel	Increases
Molybdenum	Increases
Tungsten	Increases
Silicon	Decreases, except with Mo
Titanium and niobium	Decreases
Sulfur	Decreases
Carbon	Decreases, especially when sensitized
Nitrogen	Increases

As a class of materials, the stainless steels and aluminum alloys tend to be the most susceptible to damage by pitting. For the stainless steels, resistance to pitting can be influenced by variation in alloying additions. Table 6.3 lists the effect of various alloying and impurity elements on the pitting corrosion resistance of stainless steels.

For the stainless steels, resistance to pitting can be quantified using the pitting resistance equivalent (PRE) based on the content of Cr, Mo, W, and N:

$$PRE = Cr + 3.3(Mo + 0.5W) + 16N$$

Note the potent effect of nitrogen with regard to increasing resistance to pitting. Nitrogen is intentionally added to duplex stainless steels to improve pitting corrosion resistance as well as to promote a balanced austenite + ferrite (duplex) microstructure. The PRE values of some austenitic and duplex stainless steels are provided in Table 6.4. Materials with a PRE greater than 32 are resistant to pitting in seawater, while resistance to hydrogen sulfide (H_2S) requires a PRE above 40.

The critical pitting temperature (CPT) can also be used to quantify pitting resistance. Figure 6.3 shows the effect of Mo additions to stainless steel on the CPT [3].

TABLE 6.4 Pitting resistance equivalent (PRE) values for stainless steels

Alloy	UNS	Alloy type	PRE$_N$
304L	S30403	Austenitic	20
316L	S31603	Austenitic	25
317L	S31703	Austenitic	31
254SMO	S31254	Superaustenitic	41
AL-6XN	N08367	Superaustenitic	46
2205	S32205	Duplex	36
2507	S32750	Superduplex	43
2707	S32707	Hyperduplex	49
3207	S33207	Hyperduplex	52

$$PRE_N = Cr + 3.3(Mo + 0.5\,W) + 16\,N$$

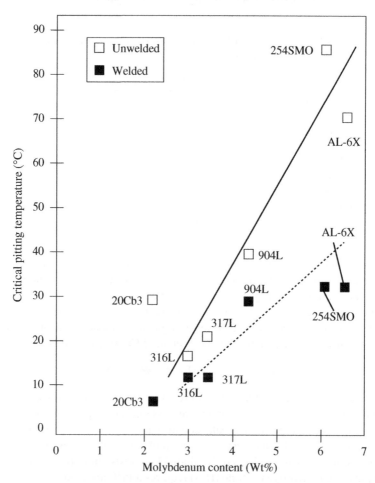

FIGURE 6.3 Critical pitting temperature (CPT) of commercial high-Mo austenitic stainless steels and their weld fusion zones. Open symbols are base metal values, and filled symbols are autogenous welds. The solid line represents the average base metal behavior, and the dotted line the weld metal behavior (Redrawn from Ref. [3]. © AWS).

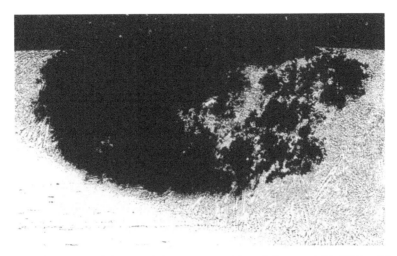

FIGURE 6.4   Severe pitting attack in superaustenitic stainless steel AL-6XN weld metal (© TWI).

A high CPT indicates good resistance to pitting. Note that the addition of Mo from 2 to over 6 wt% in austenitic stainless steels results in a significant increase in the pitting resistance of the base metal but that the weld metal response is not as large. This is primarily due to the partitioning of Mo during solidification. Since the partition coefficient (k) for Mo in stainless steel is on the order of 0.5, the core of the cells and dendrites will have Mo contents well below the average composition. For a 6% Mo alloy, the dendrite core would only achieve about 3 wt% Mo.

As a result of this, severe pitting corrosion attack can occur in the weld metal relative to the base metal, as shown in the cross section of the superaustenitic stainless steel AL-6XN weld in Figure 6.4. This situation can potentially be avoided by using a higher Mo filler metal to maintain the dendrite core composition above 5 wt%. It should be recognized, however, that the formation on an unmixed zone (UMZ) at the fusion boundary may negate this effect. The micrograph in Figure 6.5 shows such a UMZ in a weld on AL-6XN made using a high-Mo, Ni-base filler metal. Although the weld metal is now immune from pitting attack, the narrow UMZ is subject to local pitting.

Another alloying option to prevent pitting in stainless steels is to add tungsten (W) as a substitute for, or in addition to, Mo. Tungsten does not partition significantly during weld solidification, thereby preventing the reduction in pitting resistance at the dendrite core.

6.2.7   Intergranular Corrosion

Localized attack at, or adjacent to, grain boundaries with little or no attack of the grain interiors is appropriately called intergranular corrosion (IGC). This localized corrosion can be caused by impurities at the grain boundaries, an enrichment of an

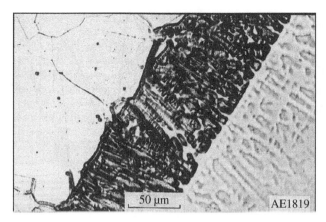

FIGURE 6.5 Unmixed zone (dark etching) that forms at AL-6XN fusion boundary when welded with a Ni-base filler metal (© AWS).

alloying element at the boundary, or, conversely, a depletion of an alloying element. Enrichment of aluminum alloy grain boundaries with iron can cause IGC in these materials. On the other hand, the depletion of chromium along grain boundaries in austenitic stainless steels results in IGC or intergranular attack (IGA). This phenomenon has been studied extensively in these alloys because of its potential for reducing the service life of components used in the power generation industry. A brief overview of IGC in austenitic stainless steels is included here.

IGC in 300-series austenitic stainless steels is of particular concern since it occurs in environments where the alloys would normally be expected to exhibit excellent corrosion resistance. The IGC in these alloys is particularly prevalent in the weld HAZ. Susceptibility to IGC results from the local depletion of chromium adjacent to the grain boundary due to chromium carbide (Cr-rich, $M_{23}C_6$) precipitation along the boundary. When the local chromium content drops below approximately 12 wt%, the region is no longer "stainless" and accelerated attack can occur. This phenomenon is also referred to as sensitization, since the material is made "sensitive" to IGA.

The schematic in Figure 6.6 represents the appearance of an austenitic stainless steel weld that has undergone IGA in the HAZ. On the surface of the weld exposed to the corrosive environment, there is often a linear area of attack roughly parallel to the fusion boundary. These are sometimes called "wagon tracks" because they are symmetric and parallel on either side of the weld. In cross section, severe attack (or weld "decay") can be observed along a "sensitized" band in the HAZ. Note that this band is at some distance from the fusion boundary. This is due to the fact that the carbide precipitation that leads to "sensitization" occurs in the temperature range from about 600 to 850°C (1110–1560°F). Above this temperature range, carbides go back into solution, and thus, the HAZ region adjacent to the fusion boundary that has been heated to higher temperatures is relatively free of carbides (assuming cooling rates are rapid enough to suppress carbide precipitation).

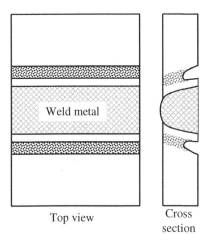

Top view   Cross
section

FIGURE 6.6   Illustration of intergranular attack in the HAZ of an austenitic stainless steel.

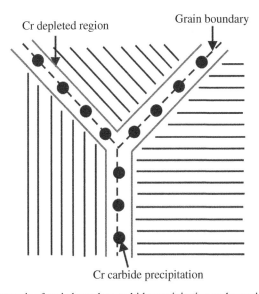

FIGURE 6.7   Schematic of grain boundary carbide precipitation and associated grain boundary chromium depletion.

In most austenitic stainless steels, Cr-rich, $M_{23}C_6$ carbides form preferentially along grain boundaries, as shown in the schematic in Figure 6.7. This precipitation requires short-range diffusion of Cr from the adjacent matrix and produces a Cr-depleted region surrounding the precipitate, as shown in Figure 6.8a. This reduces the local corrosion resistance of the microstructure and promotes rapid attack of the grain boundary region. This may be due, in part, to local galvanic attack where the Cr-depleted grain boundary is anodic to the surrounding matrix. The very

(a)

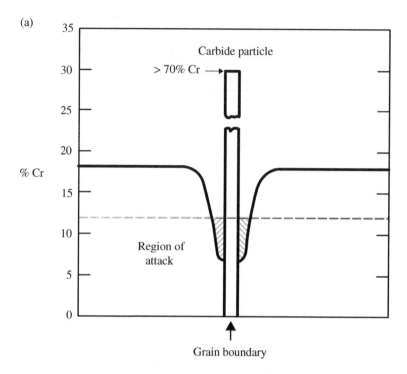

(b)

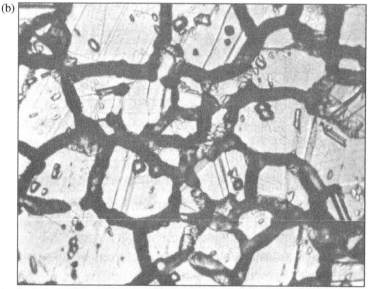

FIGURE 6.8 Intergranular corrosion. (a) Schematic of Cr depletion adjacent to the grain boundary carbide and (b) grain boundary attack in the HAZ of Type 304 (C = 0.06 wt%) (Courtesy of M.C. Juhas).

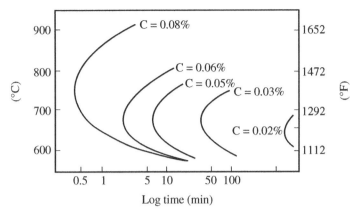

FIGURE 6.9 $M_{23}C_6$ time–temperature–precipitation curves for 18Cr–8Ni alloys with variable carbon content (From Ref. [4]. © McGraw-Hill).

small area of the grain boundary causes rapid attack, appearing as local "ditching" at the grain boundary, as shown in the metallographic section in Figure 6.8b. In extreme cases, the grains will actually drop out of the structure because of complete grain boundary attack and dissolution. This gives rise to the wagon track effect shown in Figure 6.6.

Carbon content has the most profound influence on susceptibility to IGC in austenitic stainless steels. The use of low-carbon ("L"-grade) alloys minimizes the risk of sensitization by slowing down the carbide precipitation reaction. The time–temperature–precipitation curves shown in Figure 6.9 demonstrate the effect of carbon content on the time to precipitation [4]. Note that with low carbon contents (C < 0.04 wt%), the nose of the curve is beyond 1 h, while for carbon levels from 0.06 to 0.08 wt%, the time for precipitation may be less than a minute. This difference demonstrates the benefit of the low-carbon austenitic stainless steel grades (so-called L grades such as 304L and 316L) for reducing or eliminating HAZ grain boundary sensitization during welding. The presence of residual stresses in the HAZ may also serve to accelerate the precipitation reaction.

In most cases, sensitization and subsequent IGC occur in the HAZ as a direct result of the weld thermal cycle. It should be noted, however, that the stress relief temperature range for most austenitic stainless steels overlaps the carbide precipitation range. Care must be taken not to sensitize the entire structure during PWHT. This is a particular concern with alloys containing more than 0.04 wt% C.

6.2.7.1 Preventing Sensitization It is possible to minimize or eliminate IGC in austenitic stainless steel welds by the following methods:

- Select base and filler metals with as low a carbon content as possible (L grades such as 304L and 316L).
- Use alloys that are "stabilized" by additions of niobium (Nb) and titanium (Ti). These elements are more potent carbide formers than chromium and, thus, tie up

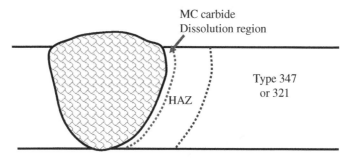

FIGURE 6.10 Location of knifeline attack that occurs in stabilized grades of austenitic stainless steels.

the carbon, minimizing the formation of Cr-rich, grain boundary carbides. They are also quite stable and resist dissolution during the weld thermal cycle.

- Use annealed material or anneal prior to welding to remove any prior cold work (cold work accelerates carbide precipitation).
- Use low weld heat input and low interpass temperature to increase weld cooling rates, thereby minimizing the time in the sensitization temperature range.
- In pipe welding, water cool the inside of the pipe after the root pass. This will help to eliminate sensitization of the ID resulting from subsequent passes.
- Solution heat treatment after welding. Heating the structure into the temperature range from 900 to 1100°C dissolves any carbides that may have formed along grain boundaries in the HAZ. The structure is then quenched (or rapidly cooled) from this temperature to prevent carbide precipitation during cooling. Note, however, that there are a number of practical considerations that tend to limit the usefulness of this latter approach. Distortion during quenching is a serious problem for many structures. Inability to quench complex pipe weldments is also a limiting factor.

6.2.7.2 Knifeline Attack IGC can also occur in certain situations in the stabilized grades, such as Types 347 and 321. This attack, shown schematically in Figure 6.10, may occur in a very narrow region just adjacent to the fusion boundary. It is called "knifeline" attack because the weld appears as if it was cut out with a knife. This type of attack occurs when the stabilizing carbides (NbC or TiC) dissolve at elevated temperatures in the region just adjacent to the fusion zone. Upon cooling, Cr-rich carbides will form faster than NbC or TiC, resulting in a narrow sensitized region. Farther from the fusion boundary, NbC and TiC do not dissolve and sensitization does not occur. Knifeline attack is associated with high heat input welds where the HAZ thermal cycle allows sufficient time for MC-type carbides to dissolve. This form of localized attack can usually be prevented by control of the welding procedure.

6.2.7.3 Low-Temperature Sensitization It has been observed that sensitization can actually occur after long exposures at low temperatures (<300°C) following an

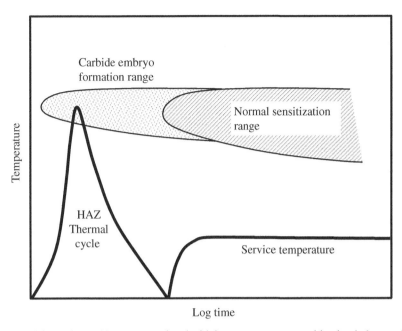

FIGURE 6.11   Thermal history associated with low-temperature sensitization in low-carbon austenitic stainless steels.

initial high-temperature thermal cycle, such as that experienced in the HAZ. This has come to be known as low-temperature sensitization (LTS) and has been a problem with stainless steel piping used in the power generation industry [5, 6]. LTS occurs because carbide "embryos" form in the HAZ during the original welding process and then grow to form carbide precipitates at low temperature. Figure 6.11 shows this phenomenon schematically. The long-term effect is the sensitization of the grain boundaries and the potential for IGC or intergranular stress corrosion cracking (IGSCC), even though the C curve for sensitization under normal conditions does not predict carbide precipitation during the thermal history of the weldment. This phenomenon has occurred in L-grade alloys, but does not seem to be a problem with stabilized grades such as Type 347.

6.2.8   Stress Corrosion Cracking

Stress corrosion cracking (SCC) refers to localized cracking resulting from the simultaneous presence of a tensile stress and a specific corrosive medium. Virtually all structural metals are susceptible to SCC given the appropriate combination of environment and stress, as indicated in Table 6.5. One of the earliest reports of SCC was "season cracking" of brass cartridge artillery shells. During periods of heavy rainfall or high humidity, cracks were often observed in brass cartridge cases at the point where the cartridge was crimped to the shell.

TABLE 6.5   Materials and environments leading to stress corrosion cracking

Alloy or alloy system	Environment
Aluminum alloys	NaCl solutions, seawater
Copper alloys	Ammonia vapors and solutions
Gold alloys	$FeCl_3$ solutions, acetic acid–salt solutions
Inconel	Caustic soda solutions
Lead	Lead acetate solutions
Magnesium alloys	Distilled water
Monel	Fused caustic soda, hydrofluoric acid
Nickel	Fused caustic soda
Carbon and low-alloy steels	Multiple
Stainless steel (austenitic)	Multiple, including seawater and H_2S
Titanium alloys	Fuming nitric acid, seawater, N_2O_4

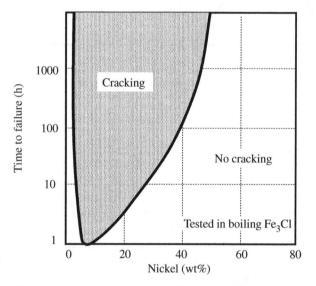

FIGURE 6.12   The Copson curve for predicting SCC susceptibility in stainless steels (Redrawn from Ref. [7]).

Stainless steels can also be susceptible to SCC, and these materials have been studied extensively because of their engineering importance. The presence of residual tensile stresses in the HAZ may accelerate corrosion attack and cracking along the sensitized grain boundaries. This is called IGSCC and may appear very similar to the IGC that was described in the previous section. Transgranular stress corrosion cracking (TGSCC) is also observed in austenitic stainless steels. This form of cracking is usually associated with Cl-bearing environments (seawater).

The Copson curve, shown in Figure 6.12, indicates the resistance of stainless steels to SCC in boiling magnesium chloride as a function of nickel content [7].

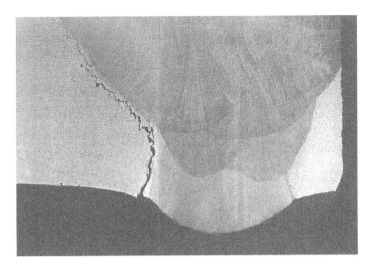

FIGURE 6.13 Intergranular SCC in the HAZ of Type 304 stainless steel (From Ref. [2]. © ASM).

The use of this aggressive environment is intended to accelerate the corrosion processes that would occur in other Cl-bearing environments (such as seawater). Note that the minimum of the resistance curve occurs in the range from 8 to 12% Ni. This is precisely the range within which many popular austenitic stainless steels alloys lie, such as Types 304 and 316.

As this figure indicates, SCC can be avoided by selecting alloys with either higher (>20%) or lower (<5%) nickel content. In the former case, the use of "super"austenitics or Ni-base alloys is common. In the low-Ni case, ferritic or duplex stainless steels are often selected. SCC has also been observed in caustic (high-pH) environments, such as in pulp and paper mills. It appears that the same rules apply in these environments as with Cl-bearing environments with respect to alloy selection to avoid caustic-induced SCC.

An example of IGSCC is shown in Figure 6.13 [2]. This is from a pipe weld in Type 304 stainless steel welded with Type 308 filler metal. Note that the cracking is specific to the region of the HAZ that has been sensitized during welding. An example of severe TGSCC in a Type 316 tubesheet after exposure to a caustic solution of sodium hydroxide in a pulp and paper mill is shown in Figure 6.14. This structure was exposed to the caustic solution for less than a year prior to failure. The residual stresses resulting from the weld in addition to imposed operating stresses led to the severe cracking seen in Figure 6.14. For this application, the Type 316 alloy was replaced with a duplex grade, Alloy 2205. This alloy has not exhibited any cracking after several years of service.

SCC is best avoided by proper alloy selection. The use of duplex and ferritic stainless steels in applications where austenitic grades would otherwise be selected can avoid SCC. Welding may exacerbate SCC in alloy systems that are otherwise resistant due to changes in microstructure and the presence of tensile residual stresses.

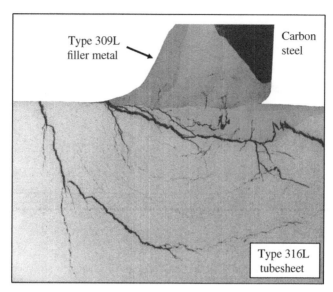

FIGURE 6.14   Transgranular SCC near the weld in Type 316L stainless steel after exposure to a caustic sodium hydroxide environment.

Sensitization can promote IGSCC in both austenitic (Fig. 6.13) and ferritic grades of stainless steel. Weld designs or conditions that generate high residual stress or create stress concentrations can also promote SCC. Postweld stress relief can sometimes be used to reduce these stresses and minimize susceptibility to SCC. But, as noted in Section 6.2.7.1, postweld stress relief needs to be done with care to avoid sensitization.

6.2.9   Microbiologically Induced Corrosion

MIC occurs in certain aqueous environments where aerobic or anaerobic bacteria literally attack the metal. MIC manifests itself as pitting, whereby the metal surface exhibits a small hole, or pit, and rapid attack occurs subsurface. In the past, MIC was probably often misinterpreted as normal pitting corrosion. However, it can be distinguished by the presence of "tubercules" of biological residue and corrosion products over the pit.

This form of corrosion has been observed in both freshwater and seawater and in many cases requires oxygenated water to support the metal dissolution reaction, as shown in Figure 6.15 [8]. In general, the presence of specific "metal-munching" bacteria, a warm-water (20–40°C) oxygenated environment, and specific materials are required to support MIC. This form of corrosion has been reported in a wide range of materials, including iron and steel, stainless steels, copper alloys, and aluminum alloys. Austenitic stainless steels seem to be particularly susceptible, and the presence of a two-phase austenite + ferrite microstructure, such as that present in many weld metals, seems to influence susceptibility. Studies have shown that MIC occurs preferentially in the ferrite phase.

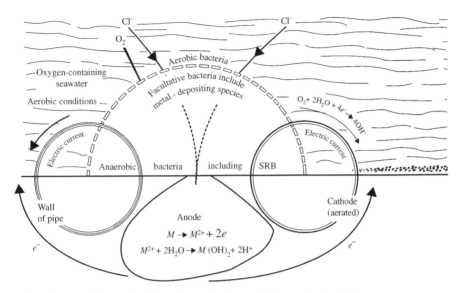

Reactions possible under tubercules created by metal depositing bacteria

FIGURE 6.15   Mechanism for microbiologically induced corrosion (MIC) (From Ref. [8]. © NACE).

FIGURE 6.16   MIC attack in Type 308 weld metal (Courtesy of Chris Hayes).

An example of MIC that occurred during the construction of a storage tank is shown in Figure 6.16. This is a shielded metal arc weld using E308-16 and joining two Type 304 plates in a butt weld configuration. The ferrite content of the weld metal was FN 10 (~10 vol%). During construction, water was allowed to accumulate in the bottom of the tank covering the weld shown in Figure 6.16. This created the environment for MIC, as evidenced by the severe attack of the weld metal.

6.3   CORROSION TESTING

Knowledge of corrosion behavior is an important part of alloy and filler metal selection for welded construction. While there is considerable corrosion data for most base metals, sufficient corrosion data to support the selection of welding processes and filler metals to prevent corrosion attack may not be available. In many situations, corrosion may represent the most significant factor relative to the service lifetime of a welded component. There are many examples of how rapid corrosion or corrosion-related cracking has led to premature, and sometimes catastrophic, failure.

Fortunately, there are many corrosion tests that have been developed over the years that can be used to quantify susceptibility to various the forms of corrosion. Many of these have been standardized by codes such as those published by the American Society for Testing and Materials (ASTM) and the National Association of Corrosion Engineers (NACE). Most of these test methods are designed for laboratory use and accelerate the attack that might be experienced in service. These include immersion tests, where samples are simply exposed to an aggressive environment and evaluated on a periodic basis, and electrochemical tests, where an electrical potential is applied to the material. ASTM Volume 03.02: Corrosion of Metals; Wear and Erosion describes many of these test techniques. Tests for IGC in stainless steels are described in ASTM A262. Some examples of commonly used corrosion tests that are applicable to testing and qualification of welds are briefly described here.

6.3.1   Atmospheric Corrosion Tests

The standard practice for conducting atmospheric corrosion tests is described in ASTM G50. For simple exposure tests, samples are mounted on a test rack and checked periodically. To determine, general corrosion rates, samples are simply weighed and weight loss is converted to corrosion rate in terms of mils/year (or mm/year). In other cases, the sample may be simply observed or photographed to determine the onset of staining, pitting, or other surface effects. Welded samples are often exposed in a U-bend configuration where the weld is bent in a longitudinal direction to provide equal straining in the weld metal and HAZ. The procedures for stress corrosion testing of welded samples for atmospheric exposure are described in ASTM G58.

6.3.2   Immersion Tests

In order to accelerate corrosion attack, samples are often immersed in an aggressive solution. These tests are widely used to evaluate susceptibility to pitting, crevice corrosion, and IGC. Many of these are described in ASTM standards. A partial list of such standards for determining susceptibility to IGC in stainless steels, nickel-base alloys, and aluminum alloys (5xxx series) is provided in Table 6.6. Procedures for sample preparation are very important since the surface condition will influence the degree of attack. These procedures are described in ASTM G31: Standard Practice for Laboratory Immersion Corrosion Testing of Metals.

TABLE 6.6   Immersion tests for evaluation of intergranular corrosion

Alloy	UNS number	Applicable ASTM standard	Solution	Immersion time (h)
Ferritic stainless steels				
Type 430	S43000			24
Type 446	S44600	A763-X	Ferric sulfate	72
26-1	S44625			120
Austenitic stainless steels				
Type 304/316	S30400/S31600	A262-A	Oxalic acid[a]	
		A262-B	Ferric sulfate	120
Type 321/347	S32100/	A262-C	Nitric acid	240
	S34700			
Nickel-base alloys				
Alloy 625	N06625	G28-A	Ferric sulfate	120
Alloy 690	N06690	A262-C	Nitric acid	240
Hastelloy C-4	N06455	G28-A	Ferric sulfate	24
Aluminum alloys				
5xxx alloys	A95005–95657	G67	Nitric acid	24

[a]Electrolytic test etched at $1\,A/cm^2$ for 1.5 min.

Most of these tests work quite well for welded samples. For base metal samples, weight loss is a good measure of corrosion rate. For samples containing welds, the attack may be localized and periodic removal of the sample from the solution is required to determine the location of the IGA. For example, a welded Type 304 stainless steel sample may undergo IGA in the HAZ well before any attack occurs in the base metal.

In some cases, it may be possible to use thermal simulation to produce samples with a HAZ microstructure. The sample shown in Figure 6.8b is a Type 304L stainless steel that was heated to 1300°C in a Gleeble™ thermomechanical simulator and then cooled at a specific rate to allow grain boundary carbide precipitation. This sample was then subjected to ASTM A262-A (oxalic acid) to produce the grain boundary attack shown.

There are also a number of immersion tests that are used to determine susceptibility to SCC. The sample types and configurations are described in ASTM G58. Two commonly used specimen types are the U-bend and C-ring configurations shown in Figure 6.17 [9]. Note that both of these use a bolt to apply the stress to the sample. This can sometimes create a problem with crevice corrosion unless precautions are taken to mask this area of the sample. Stainless steels and nickel-base alloys are often tested in boiling magnesium chloride to determine susceptibility to chloride SCC. This procedure is described in ASTM G36. Aluminum alloys of the 2XXX and 7XXX types are tested by immersion in a 3.5% NaCl solution. For the SCC immersion tests, the samples must be removed from the solution periodically to check for cracking.

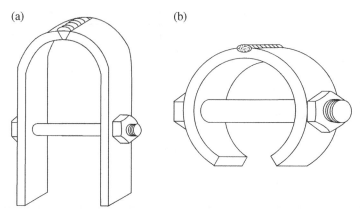

FIGURE 6.17   Welded specimen types for immersion corrosion tests. (a) U-bend and (b) C-ring (From Ref. [9]. © ASTM).

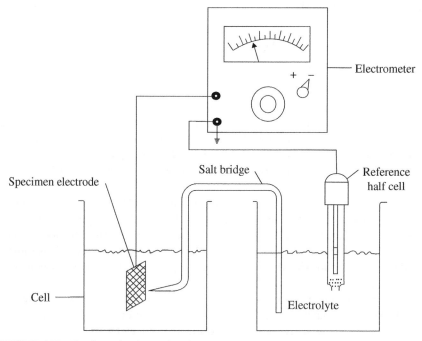

FIGURE 6.18   Configuration for performing electrochemical polarization tests (From Ref. [10]. © ASTM).

6.3.3   Electrochemical Tests

Electrochemical tests are commonly used to measure corrosion behavior in a wide range of metals. The basic procedure uses a potentiostat consisting of a reference electrode and a polarization cell as shown in Figure 6.18 [10]. The potential is

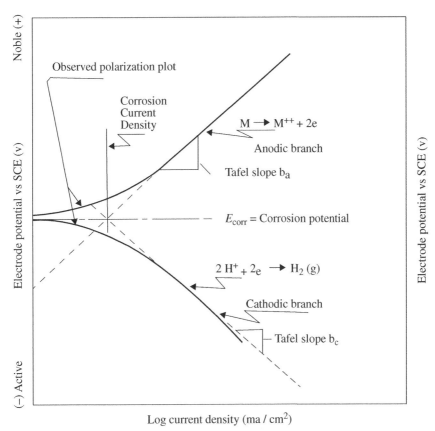

FIGURE 6.19 Tafel polarization plot from electrochemical testing (From Ref. [10]. © ASTM).

scanned from the active to passive region, and the current density in the sample is measured. An example of a polarization plot using this method is shown in Figure 6.19 [10]. Above the so-called open circuit potential of the material, the sample is anodic (active), and below this potential, it is cathodic. By measuring the slopes of these curves, a value of the corrosion current density at the open circuit potential can be determined. This value, taken from what is called the Tafel polarization plot, can be used to estimate the general corrosion behavior of a material. The procedure for estimating general corrosion rates is described in ASTM G105: Standard Practice for Calculation of Corrosion Rates and Related Information from Electrochemical Measurements.

Electrochemical methods can also be used to determine susceptibility to localized corrosion, such as pitting. An example of the potentiodynamic polarization response of a material exhibiting passive anodic behavior (such as a stainless steel) is shown in Figure 6.20 [10]. From this plot, a critical pitting potential can be determined that is related to the breakdown of the passive surface film on the sample. Another test

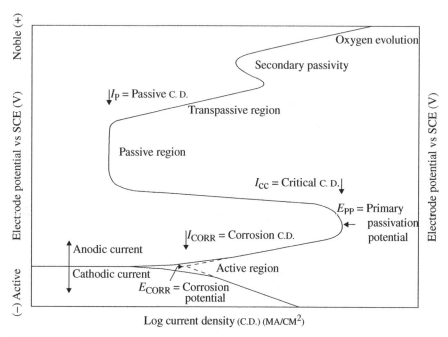

FIGURE 6.20 Hypothetical anodic and cathodic polarization plot for a material exhibiting passive anodic behavior (From Ref. [10]. © ASTM).

known as the electrochemical potentiostatic reactivation (EPR) test has been used to determine the degree of sensitization in stainless steels.

Electrochemical tests can be useful for determining the corrosion behavior of weld metals. Since the sample size is relatively small, all weld metal samples can be excised from weldments. It may also be possible to evaluate HAZ behavior, especially if a thermal simulator (such as the Gleeble) is used to generate samples of appropriate microstructure.

REFERENCES

[1] Fontana MG, Greene ND. *Corrosion Engineering*. New York: McGraw-Hill; 1967.

[2] ASM International. Handbook Committee. *ASM Handbook, Volume 13: Corrosion*. Materials Park, OH: ASM International; 1987.

[3] Garner A. Pitting corrosion of high alloy stainless steel weldments in oxidizing environments. Weld J 1983;62 (1):27–34.

[4] Peckner D, Bernstein IM. *Handbook of Stainless Steels*. New York: McGraw-Hill; 1977.

[5] Povich MJ. Low temperature sensitization of Type 304 Stainless Steel. Corrosion 1978;34(2):60–65.

[6] Povich MJ, Rao P. Low temperature sensitization of welded Type 304 Stainless Steel. Corrosion 1978;34(8):269–275.

[7] Copson HR. Effect of composition on stress corrosion cracking of some alloys containing nickel. In: Rhodin TN, editor. *Physical Metallurgy of Stress Corrosion Fracture.* New York: Interscience; 1959. p 247–272.

[8] Little B, Wagner P, Mansfeld F. Microbiologically influenced corrosion of metals and alloys. Int Mater Rev 1991;36 (6):253.

[9] ASTM G58-85. Standard practice for preparation of stress-corrosion test specimens for weldments.

[10] ASTM G3-89. Standard practice for conventions applicable to electrochemical measurements in corrosion testing.

7

FRACTURE AND FATIGUE

7.1 INTRODUCTION

Welded structures that are essentially free of defects may be susceptible to failure under certain environmental and loading conditions. In this chapter, the concepts of brittle fracture and fatigue are discussed in the context of welded construction. Welds may be particularly susceptible to these forms of failure because of changes in microstructure relative to the base metal and/or stress concentration effects that are associated with welds. The presence of preexisting fabrication defects will also influence the fracture and fatigue behavior of welded structures.

Incidences of catastrophic, brittle fracture in welded structures such as bridges, ships, and large storage tanks over the past 100 years have focused considerable attention on the concept of fracture mechanics and fracture-safe design. Brittle fracture is of great concern in structural members because it usually occurs without any prior plastic deformation and may proceed at speeds up to 7000 ft/s (2135 m/s) [1]. The Liberty ship failures during WWII are perhaps the most remarkable of these catastrophic brittle fractures. These ships were the first of an all-welded design, and many of the failures were eventually associated with stress concentrations at welds or with weld defects, some associated with hydrogen-induced cracking. Figure 7.1 shows a catastrophic failure of the USS Schenectady that was typical of these brittle fracture events [2]. Eventually, improved welding procedures and ship design resulted in a decrease in the number of brittle fracture failures, as shown in Figure 7.2 [3].

Welding Metallurgy and Weldability, First Edition. John C. Lippold.
© 2015 John Wiley & Sons, Inc. Published 2015 by John Wiley & Sons, Inc.

FIGURE 7.1   Brittle fracture at bulkhead welds in a Liberty ship (From Ref. [2]. © ASM).

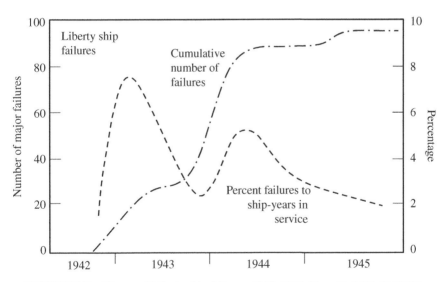

FIGURE 7.2   History of Liberty ship failures, 1942–1945 (From Ref. [3]. © AWS).

Fracture mechanics attempts to demonstrate the interrelation among materials, design, and fabrication in an effort to minimize the likelihood of catastrophic brittle fracture and to provide a means of predicting the "fitness for service" of a structure in a given service environment. This chapter provides a brief overview of the fundamental aspects of fracture mechanics as it relates to control of fracture and fatigue in welded structures. For a more comprehensive coverage of these topics, the reader is referred to the authoritative text by Rolfe and Barsom [1] and *ASM Handbook, Volume 19: Fatigue and Fracture* [2].

7.2 FRACTURE

Fracture mechanics describes approaches that can be used to predict fracture in materials assuming that there is a preexisting flaw in the material. These materials can exhibit brittle, ductile, or intermediate behavior. As shown in Figure 7.3, those materials (or structures) that are considered "brittle" exhibit linear-elastic behavior. This means that the region of plasticity that exists at the tip of an existing flaw under load is extremely small. As described later, material toughness in materials that exhibit linear-elastic behavior can be described using a stress intensity factor, K. As material ductility increases, plasticity at the crack tip increases and the material exhibits elastic-plastic behavior and other approaches must be used to describe toughness.

There are three primary factors that control the susceptibility of a structure to brittle fracture:

1. **Material toughness**. This property defines the ability of a material to carry a load or deform plastically in the presence of a crack, notch, or discontinuity.
2. **Crack size**. All brittle fractures must initiate from a crack or discontinuity of finite size. These discontinuities can vary from small defects such as weld

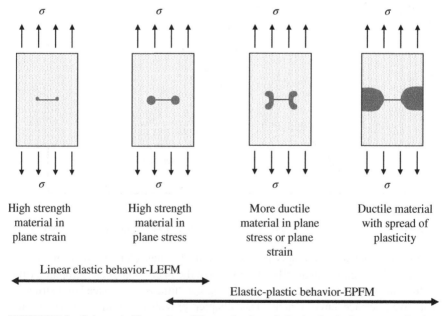

FIGURE 7.3   Schematic illustration of linear-elastic and elastic-plastic fracture mechanics.

undercut or crater cracks to large defects such as centerline solidification cracks or fatigue cracks. The critical crack (or flaw) size needed to cause brittle fracture is dependent on the material toughness.

3. **Stress level**. Tensile stresses are necessary for brittle fracture to occur. These stresses may be residual from the fabrication process, imposed in service, or, in general, a combination of both.

There are also many secondary factors that influence the susceptibility to brittle fracture including temperature, loading rate, stress concentration factors, etc. In general, these factors influence the three primary factors listed above to some degree.

Material toughness is a measure of the energy necessary to propagate a crack that has initiated at a discontinuity (i.e., crack or stress riser) in a structure. In many materials, particularly low- and medium-strength steels, this energy is a strong function of temperature. The general effect of temperature on fracture resistance (toughness) in a variety of structural materials is illustrated in Figure 7.4. Note the rapid decrease in fracture resistance of the low- and medium-strength steels below some "transition" temperature. For steels, this is usually termed the ductile-to-brittle transition temperature (DBTT). Such a dramatic transition was typical of the steels used in the Liberty ships. When the ambient temperature dropped below the transition temperature (such as in winter in waters of the North Atlantic), the resistance to fracture of these structures was extremely low.

Austenitic (fcc) materials, such as austenitic stainless steels and Ni-base alloys, do not show such a dramatic temperature dependence and exhibit good fracture

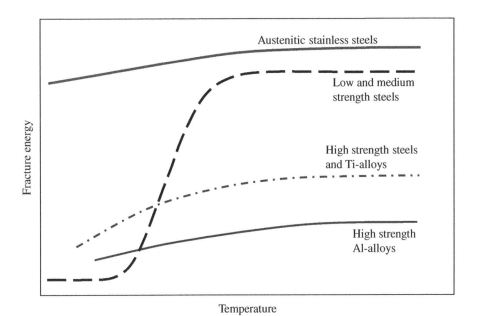

FIGURE 7.4   Illustration of toughness as a function of temperature for several materials.

toughness over a wide range of temperatures. The toughness of some austenitic stainless steels actually increases as the temperature decreases.

The stress intensity factor, K, was defined by Irwin to describe the magnitude of the elastic stress field surrounding a crack or structural discontinuity [4]. This factor is the basis for linear-elastic fracture mechanics (LEFM) and, thus, describes the stress intensity in materials that are essentially brittle, as, for example, in steels that are in service below the DBTT. The general expression for K is

$$K = \sigma \left(\pi a \right)^{1/2} Y \tag{7.1}$$

where σ is applied stress; a, crack length; and Y, dimensionless geometric factor.

A critical stress intensity factor can be defined depending on the loading rate and constraint, where:

- K_c is the critical stress intensity factor for static loading and plane stress conditions of variable constraint.
- K_{Ic} is the critical stress intensity factor for static loading and plane strain conditions of maximum constraint.
- K_{Id} is the critical stress intensity factor for dynamic loading (impact) and plane strain conditions of maximum constraint.

By knowing the critical value of K (K_c, K_{Ic}, K_{Id}) at failure for a given material of particular thickness and at a specific temperature and loading rate, the engineer can determine the magnitude (length) of cracks or discontinuities that can be tolerated in welds for a given design stress level. Conversely, the engineer can also recommend a design stress level knowing the size and orientation of flaws that would be expected in a given material using a designated welding process. From the standpoint of failure analysis, it is possible to use the critical stress intensity factor approach to determine if weld-induced flaws were responsible for failure or whether design deficiencies were at fault.

The stress intensity factor under plane strain conditions (low loading rates and linear-elastic behavior) is commonly used to estimate material toughness and applied in "fitness-for-service" analyses. Unfortunately, the determination of K_{Ic} is not straightforward, requiring specially prepared samples and careful data analysis. Conversely, the Charpy V-notch (CVN) sample and test technique are quite simple, and evaluation of the test results is straightforward, as described in Section 7.3. The plot shown in Figure 7.5 relates K_{Ic} and CVN toughness for impact toughness measured on the upper shelf of the toughness versus temperature plot. This linear relationship is described by the following equation:

$$\left[\frac{K_{Ic}}{\sigma_Y} \right]^2 = \frac{5}{\sigma_Y} \times \left[CVN - \frac{\sigma_Y}{20} \right] \tag{7.2}$$

where σ_Y is the material yield strength and CVN is the upper shelf impact toughness (or absorbed energy). Note that this relationship is only valid for high-strength steels.

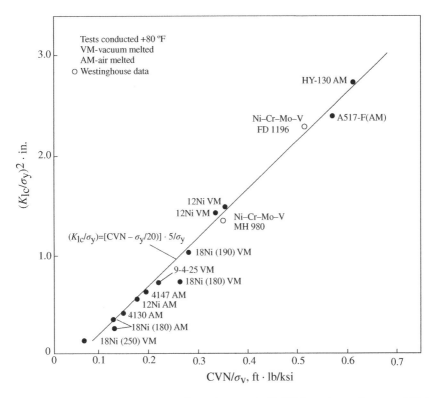

FIGURE 7.5   Relationship between K_{1C} and upper shelf CVN toughness for several high-strength steels. Note that the K_{1C} and CVN values have been normalized by the material yield strength (From Ref. [1]. © ASTM).

The general relationship between material toughness (K), applied stress (σ), and flaw size (a) is shown in Figure 7.6. Note that the curves representing K_c for a given material define many combinations of stress and flaw size that will cause failure (such as σ_f, a_f) in a material having a critical stress intensity, K_c, at a particular temperature. By reducing flaw size, stress level, or both (such as σ_o, a_o in Figure 7.6), failure will not occur. This can also be accomplished by selecting a material of higher toughness, as indicated by the dotted K_c curve.

7.3   QUANTIFYING FRACTURE TOUGHNESS

A number of standardized tests have been developed for measuring the toughness of a material. Many of these tests work well for assessing weld toughness. Certainly, the most widely used of these tests is the CVN test. The standard test sample is of dimensions $10\,\text{mm} \times 10\,\text{mm} \times 50\,\text{mm}$ in which a notch is machined, as shown in Figure 7.7 [5]. Note that subsize samples are also allowed as described in ASTM E2248-09. This sample is then fractured by impact and the energy absorbed by the impact is

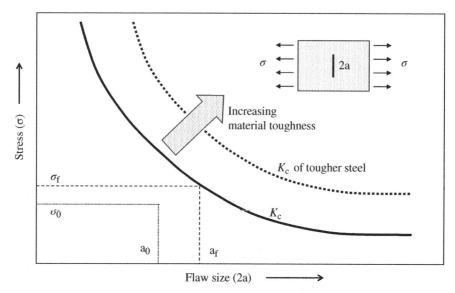

FIGURE 7.6 Relationship among applied stress, flaw size, and fracture toughness (K_c) (Adapted from Ref. [1]).

measured. For steels, impact tests can be performed over a range of temperatures in order to determine the upper shelf absorbed energy, the lower shelf absorbed energy, and the DBTT as illustrated in Figure 7.8. Some codes may specify a minimum impact toughness at a given temperature, such as 40 J minimum at –20°C (–5°F). Such a requirement requires fewer samples and can insure adequate toughness at the extreme of a service environment. The detailed procedure for conducting CVN impact tests is described in ASTM E23: Standard Test Methods for Notched Bar Impact Testing of Metallic Materials.

The CVN impact test is widely used to qualify welds. Samples can be prepared such that the notch resides in the weld metal or HAZ. It is also possible to simulate the HAZ in CVN blanks using a thermomechanical simulator such as the Gleeble™. These blanks are then machined into samples. Using this approach, the effect of peak temperature and cooling rate can be easily evaluated. For example, in steels, the transformation time, t_{8-5}, can be varied to determine the effect of preheat and heat input on the toughness of the HAZ. For duplex stainless steels, the transformation time, t_{12-8}, can be varied to determine the effect of grain size and ferrite–austenite balance on toughness.

Despite its versatility and widespread use, CVN data is only an indicator of the actual fracture toughness of the material, and the data cannot be used in fracture mechanics analyses. The reasons for this are (i) the sample is loaded by impact that is generally not representative of actual loading conditions, (ii) the machined notch is relatively blunt and results in plasticity at the notch tip except in brittle materials, and (iii) the small sample size does not allow for the plane strain conditions needed to generate valid K_{IC} values. Only in certain materials, such as high-strength steels

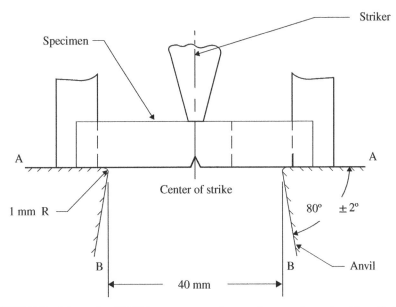

FIGURE 7.7   Schematic of CVN impact test specimen and test configuration (From Ref. [5]. © ASTM).

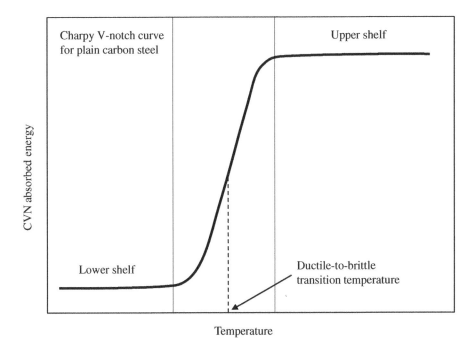

FIGURE 7.8   Representative CVN absorbed energy versus temperature curve for a plain-carbon steel.

(Fig. 7.5), can CVN toughness be correlated to K_{1C} fracture toughness. Other more complicated testing approaches have been developed to overcome the shortcomings of the CVN test.

In order to obtain a valid K_{1C} fracture toughness value for a material, a specimen must be used that maintains plane strain at the crack tip in order to insure linear-elastic behavior. Methods for determining K_{1C} are described in ASTM E399-09: Standard Method for Linear-Elastic Plane-Strain Fracture Toughness K_{1C} of Metallic Materials. A widely used sample type for these tests is known as the compact tension specimen (CTS). A standard CTS geometry is shown in Figure 7.9. The dimensions of the sample are dictated by the yield strength and expected fracture toughness of the material. In particular, the specimen thickness (Fig. 7.9b) must meet the following requirement: $B \geq 2.5 \ (K_{1C}/\sigma_{YS})^2$. For lower-strength ductile materials, the sample may have to be very thick (>50 mm) to achieve plane strain conditions. Thus, this test specimen is most applicable to higher-strength materials.

Prior to testing, a fatigue precrack is established at the tip of the notch. This sharp crack reduces the plasticity at the crack tip and insures the crack propagates under linear-elastic conditions. The specimen is then loaded in tension and a load versus displacement curve is generated. Based on this curve, a value of K_{1C} is obtained. The procedure for measuring fracture toughness using this method is described in ASTM E1820-09: Standard Test Method for Measurement of Fracture Toughness. The CTS can also be used to obtain elastic-plastic fracture toughness values (J_{1C}). The sample design may be slightly different, and the test is used to develop a *J–R* curve, which

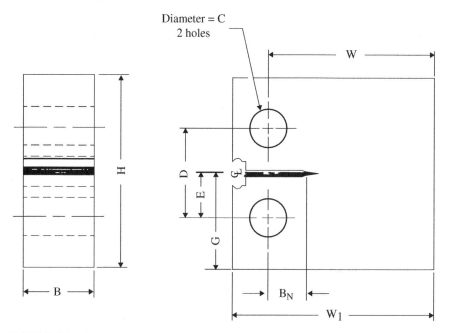

FIGURE 7.9 Compact tension specimen for determining K_{1C} fracture toughness (From ASTM E399-09. © ASTM).

measures the resistance (R) of the material to crack propagation. These procedures are also described in ASTM E1820-09.

Another popular test to determine fracture toughness is known as the crack tip opening displacement (CTOD) test. This test was originally developed at The Welding Institute and was published as a British Standard in 1979. It is described in ASTM E1290-08: Standard Test Method for Crack-Tip Opening Displacement (CTOD) Fracture Toughness Measurement. This test uses similar notched samples as those to determine elastic-plastic toughness, J_{1C}. The CTOD test has been found to be particularly useful for measuring fracture toughness in materials that transition from ductile to brittle behavior with decreasing temperature (such as steels). The CTOD test results are especially useful for determining the critical temperature at which the material shifts from linear-elastic to elastic-plastic fracture behavior. As such, this test has been widely used to evaluate welds in offshore platforms that operate at temperatures approaching 0°C (32°F).

Test specimens to determine valid K_{1C} and J_{1C} have geometric constraints that may make them difficult for evaluation of welded joints. Since the sample thickness needed to achieve plane strain conditions is normally a minimum of 25 mm, welded samples must consist of multipass welds. While evaluation of the weld metal is usually straightforward, developing valid data for the HAZ can be challenging. In some cases, a K-type joint configuration can be used to develop a HAZ that is normal to the plate surface and allows the notch to be located (and potentially propagate) within the HAZ. Even in this situation, there can be difficulties, since the notch may not be precisely located in the most critical region of the HAZ or the crack may wander from the HAZ into the adjacent weld metal or base metal.

Because of these issues and the expense of performing valid fracture toughness tests, the CVN test remains as the accepted method for qualifying many welds. Only in situations where fracture control is critical to a component's performance is fracture toughness testing warranted. This includes large offshore structures, thick-walled pressure vessels, ship structures, etc. In particular, ferritic materials are of special interest since they undergo a ductile-to-brittle transition with decreasing temperature that can lead to catastrophic failure if not properly managed.

7.4   FATIGUE

Fatigue is one of the most common failure mechanisms associated with welded construction. Even under ideal conditions, the weld almost always represents a location of stress concentration within the structure. By definition, fatigue is the formation of a crack (or cracks) under repeated application of loads that, taken individually, are insufficient to cause failure of the component, that is, not of sufficient magnitude to cause plastic yielding. All weld joints contain flaws or discontinuities in the form of geometric defects and/or metallurgical defects, such as cracks or "hard" spots. These defects are often located in regions of maximum stress concentration, and consequently, fatigue cracks may initiate shortly after the structure is put in service if the service conditions impose a periodic fluctuating stress.

A stress versus number of cycles or "*S–N* curve" is often used to describe the fatigue resistance of materials. These curves are generated by subjecting samples to multiple oscillating (sinusoidal) stress cycles until failure occurs. The *S–N* curve then represents the "fatigue life" of the material at a given peak oscillating stress level, as shown in Figure 7.10. For most nonferrous alloys, fatigue failure will occur at even low peak stresses. For ferrous alloys, however, there is usually a threshold stress, often called the fatigue limit, below which fatigue failure does not occur. It will be seen that this behavior is very important when using steels in structures where fatigue loading occurs.

The fatigue limit in steels is roughly proportional to the yield strength and usually occurs at a value of about 50% of yield strength. Thus, if peak oscillating stresses are maintained below $0.5\sigma_{YS}$, fatigue failure can be avoided. Note that this only applies to the situation where no cracks or other stress concentration points are present. If a small flaw or stress concentration (such as a weld) is present, the fatigue limit will be much lower.

The *S–N* curve actually consists of two components, a crack initiation phase and a propagation phase. The total fatigue life is thus the sum of these two components, as shown in Figure 7.11. At high stress levels, both the initiation and propagation components are quite short, while at low stress most of the fatigue life consists of the initiation phase. For a sample, or structure, that does not contain a crack or flaw that serves to concentrates stress, most of the fatigue life is spent in crack initiation. In structures that contain a flaw, there is no initiation phase, and the entire fatigue life is spent in crack propagation.

Most *S–N* curves, such as those represented in Figure 7.10, are obtained using smooth bar samples. As a result, *S–N* curves may not be very useful for predicting

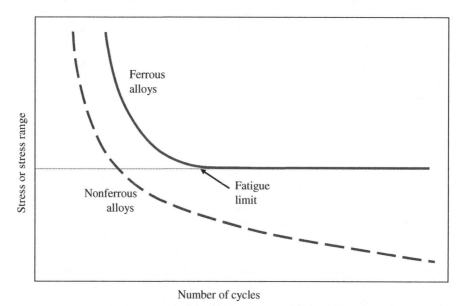

FIGURE 7.10   Typical *S–N* curves for ferrous and nonferrous alloys.

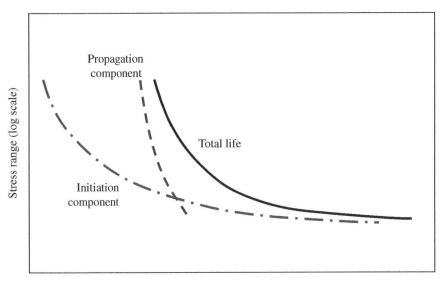

FIGURE 7.11 Initiation and propagation components of total fatigue life.

fatigue life in a structure that contains a crack or similar stress concentrator. When such a flaw is present, the local stress intensity can be quite high, even though the applied stress is low. In these situations, fatigue cracks may propagate at peak fluctuating stresses below the fatigue limit predicted by the *S–N* curve.

The presence of a weld in the structure can significantly alter the *S–N* performance of a material. As shown in Figure 7.12, the base metal (even with an internal defect) has a relatively high fatigue limit. When a welded connection is made to this base plate, the fatigue limit drops dramatically. This is because the weld tends to concentrate the stress in a specific location. This stress concentration effect decreases the number of cycles for crack initiation and severely reduces the fatigue strength of the structure.

Since preexisting flaws or stress concentrations can be expected to occur in all welds, the service life of many welded joints subjected to fatigue loading is dictated by the rate of crack propagation. In other words, crack initiation can be ignored and only crack propagation controls fatigue life. The fatigue life is then a function of the relative magnitude of the differential fluctuating stress ($\Delta\sigma$) and the rate of crack extension per fatigue cycle (*da/dN*). The effect of increasing the stress differential on the number of cycles to failure is illustrated in Figure 7.13.

Since fatigue failure can occur at stresses much lower than the elastic limit (yield stress) of the material, little or no plastic deformation may be associated with the fracture process. As a result, LEFM can often be used to describe fatigue crack growth in materials that are otherwise considered to be ductile.

Because LEFM can be used in most cases of fatigue crack growth, the stress intensity factor, K, which is proportional to the stresses near the crack tip (σ) and the crack

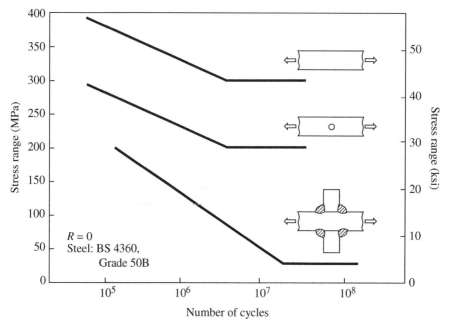

FIGURE 7.12   Effect of welded attachment on the fatigue behavior of a simple beam loaded in tension.

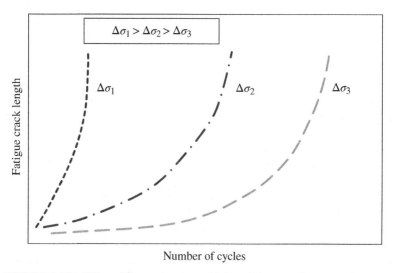

FIGURE 7.13   Effect of fluctuating stress ($\Delta\sigma$) on fatigue crack propagation rate.

length (a), provides a better method for predicting the rate of fatigue crack propagation (da/dN) for a given applied load. For certain conditions, the relationship between the range of applied stress intensity, (ΔK), and the crack growth rate can be expressed in the following way:

$$\frac{da}{dN} = C\left(\Delta K\right)^m \tag{7.3}$$

where C and m are constants depending upon the material and environment. This relationship is often called the Paris law for fatigue crack growth [6]. It predicts a linear relationship (log–log) between ΔK and the crack growth rate. In fact, this is only a special case, for over a large range of crack growth rates and stress intensity differentials (ΔK), the response curve shows three distinct regions, as shown in Figure 7.14.

In **Region 1**, a threshold value of fluctuating stress intensity, ΔK_{Th}, is required to initiate cracking. Microstructure, mean stress, and the environment have a major influence on crack growth in this region. For example, as the yield strength of the material increases, the stress intensity range to initiate cracking also increases. This behavior is shown in Figure 7.15 for steels ranging in yield strength from 50 to 200 ksi (345–1380 Mpa) [7]. This has important implications with respect to welded

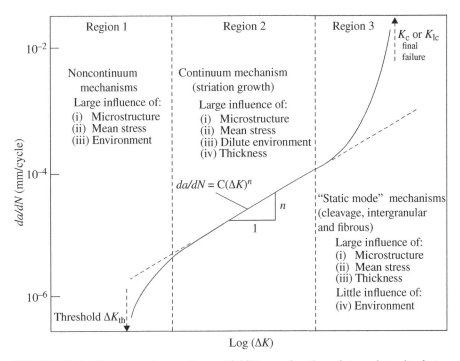

FIGURE 7.14 Fatigue crack growth rate (da/dN) as a function of stress intensity factor range (ΔK) (From Ref. [2]. © ASM).

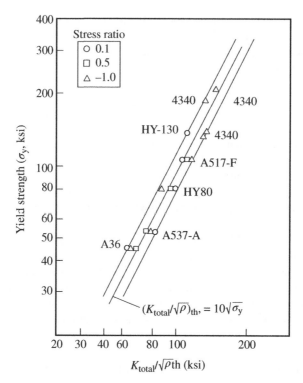

FIGURE 7.15 Dependence of fatigue crack initiation threshold on yield strength (From Ref. [7]. © ASTM).

joints, since the yield strength varies locally in the weld metal, HAZ, and base metal. In steels, fatigue crack initiation may be more difficult in the HAZ if the local strength (hardness) is higher than the surrounding weld metal and base metal.

The behavior in *Region 2* represents the "Paris law" behavior described by Equation 7.3. This region of the curve is usually representative of steady-state fatigue crack growth and can be used to predict the incremental distance of crack advance in a structure once a fatigue crack has initiated. Microstructure and environment have little effect on crack growth rate in this region. As will be shown later, crack growth rates among weld metals, HAZs, and base metals in steels show little variation within this steady-state crack growth regime. In *Region 3*, the rate of crack growth proceeds rapidly and approximates overload behavior. As with Region 1, there is a large influence of microstructure, mean stress, and material thickness.

As shown in Figure 7.16, there is little effect of steel type on the crack growth rate in the steady-state region. Note that martensitic and ferritic–pearlitic structural steels and austenitic stainless steels essentially overlay one another within a scatter band. Material type, strength level, and microstructure have a much more pronounced effect in Region 1 (initiation) and Region 3 (final failure) of this crack growth curve. Once the fatigue crack achieves steady-state growth, microstructure has little influence on the growth rate.

Stress intensity factor range (ΔK) (ksi $\sqrt{\text{in.}}$)

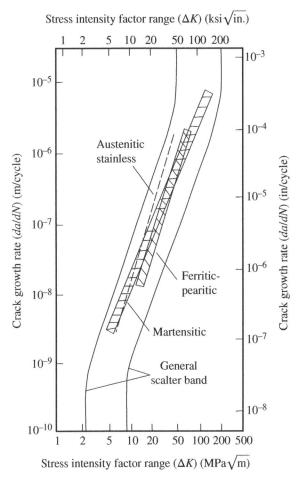

FIGURE 7.16 Steady-state crack growth rates for different steels (From Ref. [2]. © ASM).

Microstructure also has remarkably little influence on fatigue crack growth rate under plane strain conditions in welds. Figure 7.17 shows a compilation of data for different weld metals, a simulated HAZ, and a mild steel base metal. Similar behavior has been observed for other materials when comparing weld metals and base metals. For example, there is virtually no difference in fatigue crack growth rate between Type 304L stainless steel base metal and a Type 308 weld deposit containing a small volume fraction of ferrite.

The AWS D1.1: Structural Welding Code-Steel provides design guidelines for welded steel structures subjected to fatigue loading [8]. The curves shown in Figure 7.18 are for nonredundant structures and apply to a number of weld geometries. For a simple groove weld (butt weld between two members) in steel, the maximum stress range is approximately 15 ksi. For attachment welds, such as for a fillet-welded stiffener shown in Figure 7.18, the stress range drops below 10 ksi.

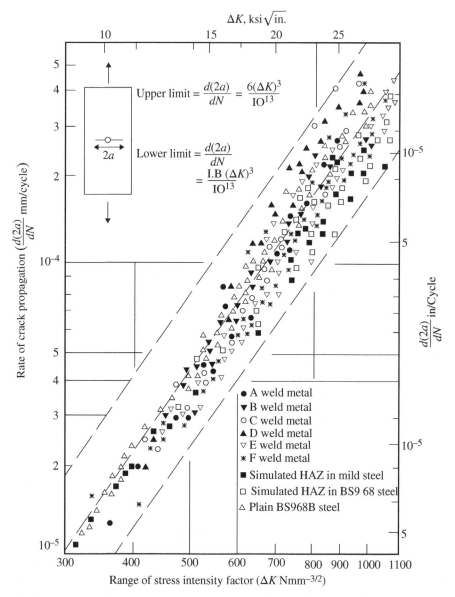

FIGURE 7.17 Fatigue crack growth rates in the weld metal, HAZ, and base metal of a 0.20% C structural steel (BS 968) (From Ref. [1]. © ASTM).

These curves provide design guidelines for specifying weld types and for avoiding fatigue failures in welded structures. They are widely used by structural and welding engineers for steel construction. Note that the fatigue limits are well below the yield strength of the steel and are designed to provide a safe stress range for structures that are subject to fatigue loading.

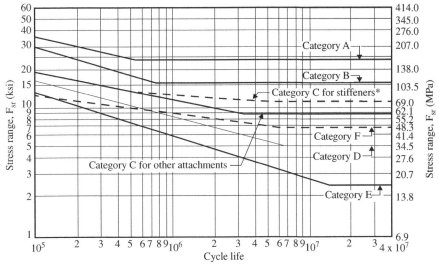

*Transverse stiffener welds on girder webs or flanges

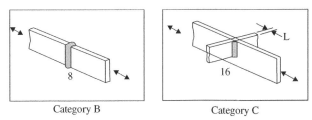

Category B   Category C

FIGURE 7.18 Example of fatigue limits for welded steel construction (From Ref. [8]. © AWS).

7.5 QUANTIFYING FATIGUE BEHAVIOR

There are two basic types of fatigue tests. The first type determines the total number of cycles for fatigue failure including both the initiation and propagation stages. For base metal behavior, the test specimens are usually of the "smooth bar" type where there are no preexisting notches present and the geometry and surface finish of the test sample are carefully controlled. Standard fatigue testing under constant amplitude axial tension is described in ASTM E466-07. This procedure can be used to develop *S–N* curves, such as those shown in Figure 7.12. Similar procedures can be used to test welded components, as illustrated in Figure 7.18. These tests will identify the total fatigue life and, for steels, the fatigue limit below which fatigue failure does not occur.

Although total fatigue life is important from a design standpoint, the data generated does not provide useful information regarding fatigue crack growth resistance. Such information is very important from a fracture-safe design point of view, where a flaw of some size is assumed and the rate of growth under an applied stress, or stress intensity, can be determined. For this reason, valid fracture mechanics samples are often used to determine crack growth rates, such as the CTS shown in Figure 7.9. The method for determining steady-state crack growth rates using this approach is described in ASTM E647-08. Similar to tests that determine the K_{IC} fracture toughness, fatigue crack growth rates are determined under plane strain conditions such that crack growth is predominantly elastic. The fatigue crack growth behavior shown in Figure 7.17 is an example of data that was gathered under plane strain (elastic) conditions. Similar samples can also be used to develop corrosion fatigue data.

7.6   IDENTIFYING FATIGUE CRACKING

Fatigue cracks may be difficult to distinguish visually from other types of weld defects (assuming there may have been preexisting fabrication defects). This is because fatigue cracking often initiates in regions of the weld where other types of defects may also be present. The fact that there is generally little, or no, plastic deformation accompanying fatigue crack growth results in extremely fine cracks, particularly during the initial stages of crack growth. As a result, fatigue cracks are often present as "hairline" defects that may escape detection. In addition, since fatigue cracks tend to form at points of stress concentration such as grooves, corners, or weld toes, all of which can act as traps for contaminants such as oil and dirt, these cracks are often disguised during the early stages of their formation.

Unlike an overload failure, there is normally little macroscopic deformation observable along the crack path since stresses are highly concentrated at the crack tip. As noted previously, steady-state fatigue crack propagation is not influenced by microstructural features, such as second phases, precipitates, or grain boundaries. The crack path tends to be very straight since propagation is incremental and proceeds in a uniform stress field in advance of the crack tip. On a microstructural scale, fatigue cracks may be transgranular, intergranular, or a combination of both. The specific path is a function of the material condition and the environment. For instance, in air, fatigue cracks in annealed austenitic stainless steels will be transgranular, but if the material is sensitized and tested in mildly corrosive environments, the crack path will be intergranular (corrosion fatigue).

In many cases, it is difficult to verify that a crack on the surface of a weld or near a weld is the result of fatigue merely from its appearance on that surface. For example, fatigue cracks and hydrogen cracks may initiate at similar locations. The proximity of the crack to a geometric discontinuity is usually a good indication that fatigue is involved, but positive identification usually requires examination of the fracture surface.

Fatigue cracks often exhibit distinct fracture surface features that distinguish them from other types of failure. Fatigue crack surfaces are frequently described as smooth and "silky" in appearance and often contain a number of distinguishing features (particularly in steels). These are shown in Figure 7.19.

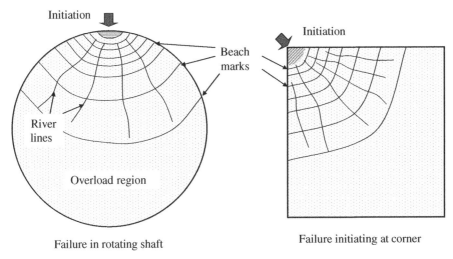

FIGURE 7.19   Macroscopic features observed on fatigue crack fracture surfaces.

7.6.1   Beach Marks

These are lines on the fracture surface that spread out in a "clamshell" pattern from the origin of the crack (the origin representing the hinge on the clamshell). These markings correspond to the location of the crack front during various stages of crack propagation and are usually readily evident to the observer upon inspection with the naked eye. They are called "beach marks" because the features resemble the ridges of sand that are left on the beach as the water retreats during low tide.

7.6.2   River Lines

These lines radiate from the origins of fatigue cracks and are formed as cracking occurs on slightly different planes that join by shearing through the remaining ligaments.

7.6.3   Fatigue Striations

These features, usually visible only at high magnification in the SEM, correspond to the incremental crack advance associated with each fatigue cycle. The distance between striations represents the value da/dN in the Paris law. As discussed later, fatigue striations are usually not observed in actual "field" failures.

Two examples of the macroscopic features associated with fatigue failures are shown in Figure 7.20 [9]. The arrows indicate the initiation points. The railhead failure (Fig. 7.20a) initiated on the top left corner of the rail (perhaps at a weld joining adjacent rails) and propagated down and to the right. The bands on the fracture surface represent the beach marks characteristic of fatigue failure. Below the beach mark region, failure occurred by overload. The fatigue failure in Figure 7.20b is a laboratory test sample in a medium-carbon, high-strength steel plate. In this case, the

(a)

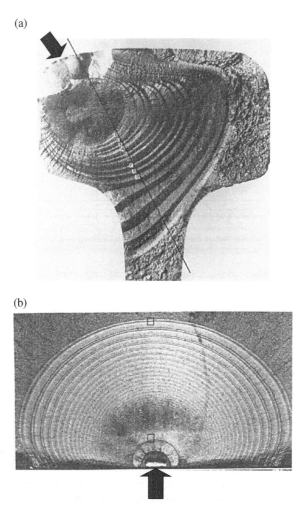

(b)

FIGURE 7.20 Examples of fatigue failures showing beach mark patterns. (a) AISI 1080 steel railhead and (b) D6AC steel plate with a premachined notch (From Ref. [9]. © ASM).

loading conditions are very well controlled, and uniform, almost concentric beach marks are generated.

At high magnification in an SEM, it is sometimes possible to observe fatigue striations. These features are usually associated with the steady-state stage of fatigue crack growth. The width of each striation represents the incremental advance of the fatigue crack per cycle (*da/dN*). Examples of fatigue striations in an aluminum alloy and Ni-base alloy tested under laboratory conditions are shown in Figure 7.21.

In actual field failures, it is unusual to observe fatigue striations. In steels, corrosion of the fracture surface can mask the striations. Also in cases where the oscillating load transitions from tension to compression, the fracture surface closes back down on itself and damages or destroys the striation pattern. Most fatigue failures are identified

(a)   (b)

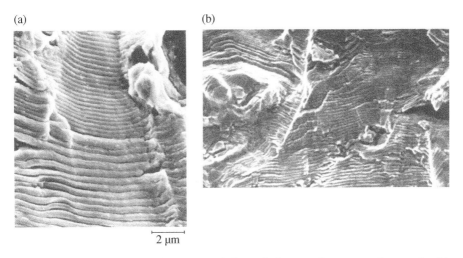

2 µm

FIGURE 7.21 Examples of fatigue striations during steady-state crack growth. (a) Aluminum alloy 7475-T7651 and (b) Ni-base alloy X-750 (From Ref. [9]. © ASM).

by macroscopic rather than microscopic features. The absence of fatigue striations on the fracture surface does not preclude fatigue as the failure mechanism.

Welds are often the preferred location for fatigue crack initiation because of both metallurgical and mechanical factors. A weld almost always produces a stress concentration point in the structure that can result in failure at relatively low loads and number of cycles to failure (Fig. 7.12). Preexisting defects such as solidification cracks or hydrogen-induced cracks also greatly reduce fatigue life since there is no initiation phase and crack propagation immediately enters the steady-state regime. Tensile residual stresses in and around the weld are typically additive to the applied loads and can result in even more rapid failure.

7.7 AVOIDING FATIGUE FAILURES

In general, the most effective method to avoid fatigue failure in welded structures is to use proper design guidelines that dictate material thickness, weld size, and maximum loads. For steels, the AWS D1.1: Structural Welding Code-Steel provides guidance on weld design and allowable loads for different types of welded joints and attachments, such as the example shown in Figure 7.18. In practice, it is often difficult to estimate the exact loads that a structure will be subjected to or to insure that the maximum allowable loads will not be exceeded. Periodic inspection of structures that undergo fatigue loading is a required practice in many industries. These inspections should become more frequently as the number of cycles (life of the structure) increases, since fatigue can progress quite rapidly once a fatigue crack has initiated. Since cracking is almost always initiated at the surface, either careful visual inspection or penetrant testing (after proper cleaning) is recommended.

REFERENCES

[1] Rolfe ST, Barsom JM. *Fracture and Fatigue Control in Structures*. Englewood Cliffs, NJ: Prentice-Hall; 1977.

[2] ASM International. Handbook Committee. *ASM Handbook, Volume 19: Fatigue and Fracture*; 1996. p 372.

[3] Bannerman DB, Young RT. Some improvements resulting from studies of weld ship failures. *Weld J* 1946;**25** (3):223–236.

[4] Irwin GW. Linear fracture mechanics, fracture transition, and fracture control. *Eng Frac Mech* 1968;**1** (2):241–247.

[5] ASTM E23. Standard test methods for notched bar impact testing of metallic materials, Volume 03.01. West Conshohocken, PA: American Society for Testing and Materials.

[6] Paris PC, Erdogan F. A critical analysis of crack propagation laws. *Trans ASME* 1963;**85** (3):528–534.

[7] Roberts R, Barsom JM, Fisher JW, Rolfe ST. *Fracture Mechanics for Bridge Design, FHWA-RD-78-69*. Washington, DC: Federal Highway Administration; 1977.

[8] *AWS D1.1-2010, Structural Welding Code—Steel*. Doral, FL: American Welding Society.

[9] ASM International. Handbook Committee. *ASM Handbook, Volume 12: Fractography*. 9th ed. Metals Park, OH: ASM International; 1987.

8

FAILURE ANALYSIS

8.1 INTRODUCTION

Engineers are frequently involved in failure investigations associated with welded components. It is a common assumption within the engineering community that when a failure occurs in a welded structure, the weld is either the primary culprit or somehow involved in the failure scenario. This is not always the case, of course, but it is important that the welding/material engineer is prepared to work in a structured manner in order to determine the cause of failures in welded structures. Failure analysis is an important part of a welding/material engineer's responsibility, since understanding the nature of failure can improve our ability to select materials and recommend designs that insure structural integrity. For example, understanding the basic nature of Liberty ship failures led to improvements in welding procedures and design that essentially eliminated brittle failure in these ships.

The primary goal of a failure investigation and the associated analysis is to determine the cause of the failure. Of perhaps equal importance is the ability of the investigator to recommend corrective action, which will avoid similar failures. The cause (or causes) of failures is occasionally readily evident and straightforward but normally involves a variety of factors, all of which contribute to a varying degree to the failure mechanism. It is the job of the failure investigator to weigh the contribution of these factors based on the evidence available and determine the most likely failure scenario. These investigations must be carefully conducted and reported,

Welding Metallurgy and Weldability, First Edition. John C. Lippold.
© 2015 John Wiley & Sons, Inc. Published 2015 by John Wiley & Sons, Inc.

since in some cases they may involve product liability, personal injury, or other situations where litigation may be involved.

In many cases, important information regarding the cause of failure can be gathered by examining the fracture surface. This evaluation, often called "fractography," has been greatly aided by the introduction of the scanning electron microscope (SEM) in the 1950s. This chapter will first discuss the importance and application of fractography and then provide a simple guide for conducting a failure investigation.

8.2 FRACTOGRAPHY

Fractography is an important tool that gives us insight into material properties and failure mechanisms. It has its roots in the Bronze Age with artisans and weapon makers observing fracture surfaces. Eventually, it evolved to the first application of optical microscopes and reproductions of fracture surfaces in the 1700s, to the first photographs of fractures in 1800s, and then to the development of modern electron fractography in the 1950s with the use of the transmission electron microscope (TEM) and the SEM to analyze fracture surfaces.

The development and use of the SEM for fracture analysis marked the beginning of modern fractography. The basic components of the SEM as well as the physical principles that govern the operation of an SEM will be described. For more detailed descriptions of analytical electron microscopy tools, the reader is referred to other textbooks on the subject [1, 2]. There are four principal fracture modes recognized in modern fractography: dimple rupture, cleavage, fatigue, and decohesive rupture. Regardless of fracture mode, fractures will propagate along a transgranular or intergranular (IG) path.

Typical fracture surfaces of hot cracks, solid-state cracks, and cold cracks will be presented and discussed. Fractography representative of solidification cracking, liquation cracking, ductility-dip cracking (DDC), strain-age cracking, reheat cracking, and hydrogen-induced cracking (HIC) is presented here.

8.2.1 History of Fractography

The term "fractography" was first coined in 1944 by Carl A. Zapffe after he successfully brought an optical microscope lens near enough to a jagged fracture surface to observe fracture details within individual grains. The *ASM Handbook on Fractography* states that "the purpose of fractography is to analyze the fracture features and to attempt to relate the topography of the fracture surface to the causes and/or basic mechanisms of fracture" [3].

Fractography is an important tool for the engineer. In conjunction with an understanding of the material microstructure and composition, the material processing, and the service environment, the engineer can gain an understanding of the causes and mechanisms of failures. This leads to a better understanding of material limitations and allows the engineer to take preventive measures against future failures.

Fracture surfaces have been analyzed since the beginning of the Bronze Age to draw correlations to variables in smelting and melting procedures to material properties and performance. In 1540, Vannoccio Biringuccio published in the *De La Pirotechnia* journal the first written description of fracture surfaces to estimate metal quality. His technique was used for quality assurance of both ferrous and nonferrous alloys. In 1627, there was the first recorded instance of determining grain sizes in metals.

In 1722, de Réaumur first reproduced fracture morphologies by meticulously producing hand-carved engravings. He classified seven different fracture types. The fracture types I–VII were characterized by varying amounts and sizes of "mirrorlike facets," "fibrous metal," and "woody appearance."

In the nineteenth century, there was a sharp decline in the use of fractography due to the rise of metallography. The metallurgist Floris Osmond was quoted as saying that there was "nothing either correct or useful" to be learned from studying fracture surfaces. Despite these setbacks, several significant advances were made. A paper by D.K. Tschernoff in 1879 first accurately illustrated the true shape of grains, and the first photographs of fracture surfaces were taken. B. Kirsh in 1889 described cup-and-cone tensile fracture and postulated the theory of fracture propagation in tensile specimens that is still in use today.

In the twentieth century, fractography experienced a comeback with the work of Zapffe with optical microscopes in the 1940s. Brittle fracture surfaces were observed at 1500–2000×, and cleavage facets and IG fatigue striations were reported. The greatest advances were made with the application of electron microscopes to the study of fracture surfaces. In the 1960s, theories about the micromechanism of ductile fracture, fatigue fracture evolution, and brittle fracture cleavage patterns were proposed.

Development of the TEM began in the 1920s with the discovery that magnetic fields could serve as lenses to focus electrons. In the 1950s, TEMs were first used to analyze fracture surfaces. However, due to the difficulty of sample preparation and interpretation of images, the TEM was soon replaced by the SEM as the tool of choice. Today, SEMs are used routinely to observe images of fractures at 20× to 50,000× magnification. Simple sample preparation, direct observation of specimens, large depth of focus, magnification capabilities over a large range, three-dimensional appearance of images, and ease of interpretation make the SEM a very useful tool for studying and understanding fracture surfaces.

8.2.2 The SEM

The SEM is made up of four systems: imaging, information, display, and vacuum systems. As shown in Figure 8.1, the imaging system is composed of the source and the electromagnetic lenses. The source is an electron gun that consists of a filament that generates electrons when heated to incandescence. The electrons are accelerated by a large potential difference between the filament (cathode) and the anode and directed through a series of condenser lenses that reduce the beam diameter from approximately 4000 to 10 nm. Scanning coils are used to move the beam along a

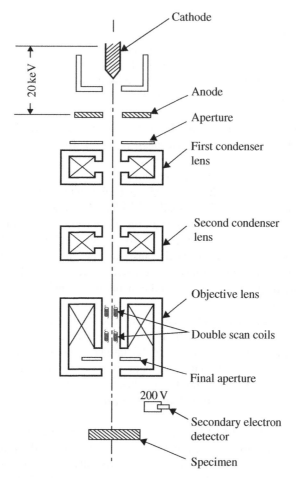

FIGURE 8.1   Schematic of a scanning electron microscope.

straight line over the width of the area known as the "raster." When the electron beam completes one line scan, it returns quickly to the line below the completed scan. By repeating this process, the entire raster area is scanned.

The information system includes the specimen, data signals, and detectors. Specimens must be grounded to prevent buildup of electrical charge. Sample preparation is quite simple compared to that needed for TEM specimens. For SEM sample preparation, the specimens need to be sized to fit in the vacuum chamber. Data signals are produced by elastic (electron nucleus) or inelastic (electron–electron) collisions. The elastic collisions produce backscattered electrons (BSE), which carry both topographical and compositional information. Inelastic collisions cause a release of secondary electrons (SE), X-rays, and heat photons. SE carry topographical data. Characteristic X-rays are used for chemical composition analysis.

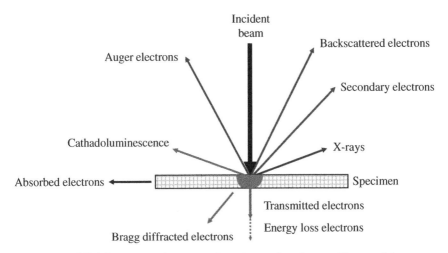

FIGURE 8.2 Electron beam–specimen interactions for metallic materials.

Analysis of metallic samples using electron microscopy techniques uses a number of "events" that occur when the electron beam interacts with the sample. These various interactions are shown in Figure 8.2. The signals most often used in fractographic analysis are described in the following.

SE are electrons in the specimen that are knocked out of their orbits by the incoming electrons and escape from the near surface. The SE signal is the most important in determining the surface topography and provides the image of the fracture surface.

BSE are incoming electrons that are scattered by the electronic structure of the specimen and end up escaping from the surface. The BSE signal provides good atomic number contrast and can provide some information on local composition differences, such as segregation in the weld metal or the presence of different phases.

X-rays are generated when an electron from an outer shell occupies an inner shell site. These inner shell electrons are displaced by the incoming electron beam and constitute the SE that are produced. An X-ray represents the photon of energy of specific wavelength that is given off when the lower energy site (inner shell) is occupied by the outer shell electron. Each element has specific X-ray wavelengths, and thus, these signals can be used to determine the composition of the specimen.

8.2.3 Fracture Modes

There are two basic paths that fractures can take: transgranular and IG. Some sources also refer to an "interphase" fracture path. Transgranular fractures pass through the grain interior. The term "intragranular" also refers to failure that occurs through grains. IG failure occurs along grain boundaries. Interphase fractures occur in materials that undergo a phase transformation during cooling. The fracture path propagates through the phase that offers the path of least resistance. The fracture path can be a combination of either transgranular, IG, or interphase. For example, many

Transgranular
(ductile rupture)

Intergranular

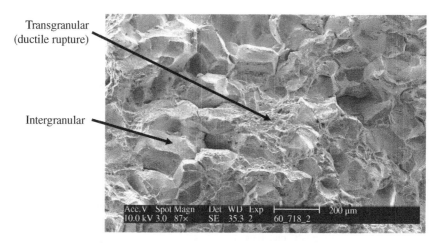

FIGURE 8.3   Mixed mode fracture behavior in Alloy 718 simulated HAZ.

fractures that initiate and propagate in an IG mode may eventually fail in a transgranular manner via overload.

There are four principle fracture modes: dimple rupture, cleavage, fatigue, and decohesive rupture. Regardless of the path that a fracture takes, fracture surfaces can exhibit any
single mode or combinations of fracture modes along the length of the entire crack.

In general, cracks propagate along the path of least resistance. The crack path is
determined by the material microstructure, which determines the material properties
at the grain boundaries and the grain interiors. Quite often, IG failure is associated
with a type of defect in welding. Characteristically, liquation cracking, DDC, strain-
age cracking, and reheat cracking are IG. In addition, coarse-grained microstructures
and hydrogen-embrittled materials can exhibit IG cracking.

The fractograph in Figure 8.3 shows a failure in Alloy 718 that is both transgranular and IG. In the transgranular region, the fracture passes through the grains and
exhibits a ductile mode of fracture. Individual grains cannot be distinguished on the
fracture surface. In the IG region, the grain faces are clearly visible and the features
of individual grains can be seen.

Dimple rupture is most often seen when overload is the primary cause of failure.
Dimple rupture occurs by a process called *microvoid coalescence* (MVC). As strain
increases in the microstructure, microvoids nucleate at local strain concentrations,
such as second phase particles, inclusions, grain boundaries, and dislocation pileups.
The microvoids grow and eventually coalesce to form a continuous network of "cup-
like" dimples. The fracture surface is composed of these peaks and valleys. Examples
of typical ductile dimple fracture in two steel samples are shown in Figure 8.4.

The size of dimples is affected by the number and distribution of nucleation sites.
Many nucleation sites result in many small dimples. For example, in weld metals that
contain many inclusions, the dimple rupture mode is associated with nucleation of
voids at these inclusions. Nucleation of microvoids at grain boundaries results in
IG dimple rupture.

(a)   (b)

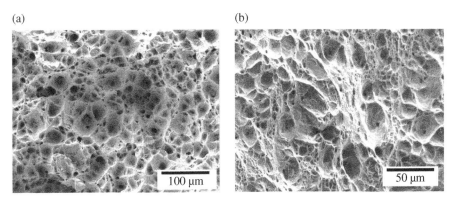

FIGURE 8.4 Examples of ductile rupture. (a) Elevated temperature (700°C) tensile failure of 2.25Cr–1Mo steel and (b) overload failure in HSLA steel (Courtesy of Xiuli Feng (a) and Xin Yue (b)).

Cleavage fracture is a low energy fracture that propagates along low index crystallographic planes known as "cleavage planes." They tend to propagate transgranularly. Theoretically, the fracture surface is perfectly smooth and featureless; however, because engineering materials are polycrystalline with grain boundaries, inclusions, dislocations, and second phases, distinct fracture features appear. These include cleavage steps, river patterns, feather markings (fan-shaped array), chevron (herringbone) patterns, and "tongues" (fracture deviation along twin orientations).

Cleavage fracture is often compared to rock candy fractures, and it is analogous to how minerals and gems cleave along certain crystallographic planes. When viewed immediately after fracture, cleavage surfaces usually appear shiny, since the cleavage planes reflect light. The old saying "the material crystallized and broke" refers to cleavage failures. Cleavage fractures are most common in ferritic steels, including C–Mn and HSLA steels, ferritic and duplex stainless steels, and high-strength steels. Cleavage does not generally occur in austenitic (fcc) steels. Two examples of cleavage fracture in stainless steel HAZs are shown in Figure 8.5.

The *fatigue fracture mode* is the result of repetitive or cyclic loading. Fatigue striations are often (but not always) the signature of fatigue cracking. As described in Chapter 7, fatigue cracking normally occurs in three stages: initiation, steady-state propagation, and overload.

Examples of fatigue fracture surfaces generated by laboratory testing are shown in Figure 8.6. These fractographs show finely spaced striations that are unique to fatigue cracks. Each striation is the result of the incremental crack advance associated with each loading cycle during steady-state crack growth. Striations are typical of fatigue crack fracture behavior in a variety of engineering materials.

Fatigue striations are not always evenly spaced and of equal size due to the variability of loading during actual service conditions. Because of this, it is difficult to exactly correlate the number of striations with the actual number of cycles, and often, estimates have to be used. Also, striations are not always observed on the fracture surface of fatigue cracks, particularly in actual field failures. This is because the

(a)

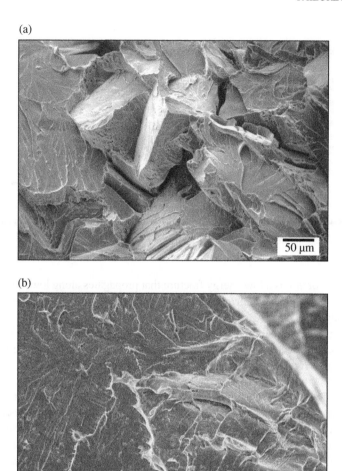

(b)

FIGURE 8.5   Examples of cleavage fracture mode. (a) Ferritic stainless steel HAZ and (b) duplex stainless steel HAZ.

cyclic loading may be tension/compression, where the compression stage causes crack closure and damage to the fracture surface. The examples shown in Figure 8.6 were from laboratory tests run in a tension/tension mode that avoids damage to the fracture surface by crack closure. Another reason for the absence of fatigue striations may be due to corrosion of the fracture surface by atmospheric exposure. This is particularly the case with steels where general corrosion of the fracture surface may be quite rapid. As noted in Chapter 7, the absence of fatigue striations on the fracture surface does not eliminate fatigue as a possible failure mechanism.

  Decohesive rupture is a fracture that exhibits little or no bulk plastic deformation and is not ductile rupture, cleavage, or fatigue. The most common form of decohesive rupture is *IG failure*, where crack initiation and propagation are along grain

(a)

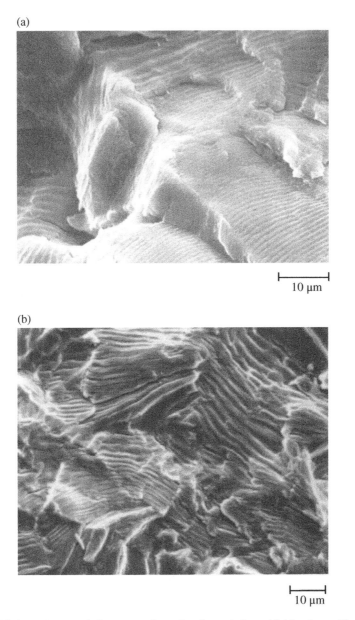

10 μm

(b)

10 μm

FIGURE 8.6 Fatigue crack fracture surfaces showing striations. (a) Aluminum Alloy 7050-T7651 and (b) commercially pure Ti (From Ref. [3]. © ASM).

boundaries. It is the result of either a reactive environment or a unique microstructure. It may occur as a result of impurities that reduce the cohesive strength along grain boundaries. Impurities that have a particularly damaging effect include hydrogen, oxygen, carbon, sulfur, phosphorus, antimony, arsenic, halides, gallium, mercury, cadmium, and tin.

The fracture process can be the result of a weakening of atomic bonds, reduction in surface energy for localized deformation, rupture of protective films, and anodic dissolution at active sites. IG stress corrosion cracking, HIC, liquid metal embrittlement, and HAZ and weld metal liquation cracking are classified as types of decohesive rupture due to the IG nature of fracture.

8.2.4 Fractography of Weld Failures

As described in the previous chapters, weld cracking can take various forms and is generally divided into three groups: (i) hot cracking that requires the presence of liquid films; (ii) subsolidus, elevated temperature (or "warm") cracking that occurs in the solid state; and (iii) cold cracking usually associated with hydrogen.

Hot cracking occurs due to the presence of liquid films and is always IG. There is usually evidence of these liquid films on hot crack fracture surfaces that helps distinguish hot cracking from other forms of weld cracking. As described in Chapter 3, hot cracking is divided into two different types: solidification cracking and liquation cracking. Solidification cracking occurs in the weld metal along solidification grain boundaries during solidification. Liquation cracking can occur in both the HAZ and in reheated weld metal.

Warm cracking occurs at elevated temperatures in the solid state, as described in Chapter 4. Types of warm cracking include DDC, strain-age cracking, and reheat (or postweld heat treatment (PWHT)) cracking. DDC is most often observed along migrated grain boundaries in the weld metal, normally in single-phase austenitic (fcc) materials. Strain-age cracking occurs in Ni-based superalloys during PWHT. It is IG. Reheat cracking occurs in low-alloy steels that contain Mo and V and in stabilized stainless steel alloys during PWHT. Reheat cracking is also IG.

Hydrogen cracking occurs at or near room temperature and is classified as *cold cracking*. Hydrogen cracking can be IG or transgranular. As discussed in Chapter 5, the fracture mode is a function of microstructure, hydrogen concentration, and stress intensity.

The following sections provide examples of fracture surfaces that are characteristic of these different forms of weld cracking.

8.2.4.1 Solidification Cracking Solidification cracking usually occurs along solidification grain boundaries (see Fig. 2.23). The fracture surface is characterized by smooth, rounded features indicative of the presence of liquid. The tips of solidification cells or dendrites that are exposed on the fracture surface give the surface an "eggcrate"-type appearance. Examples of solidification cracking in Ni-base alloy, duplex stainless steel, and aluminum alloy weld metals are provided in Figure 8.7.

In some alloys (Ni-base alloys and austenitic stainless steels), the crack surface morphology can change from dendritic to a flat, or shallow dendritic, appearance. This behavior was originally proposed by Matsuda as part of the modified generalized theory of solidification cracking [4] and is described in Section 3.2.1.4. According to Matsuda, this transition occurs due to a change in the nature of the solidification grain boundary

(a)

(b)

(c)

Fusion zone   HAZ

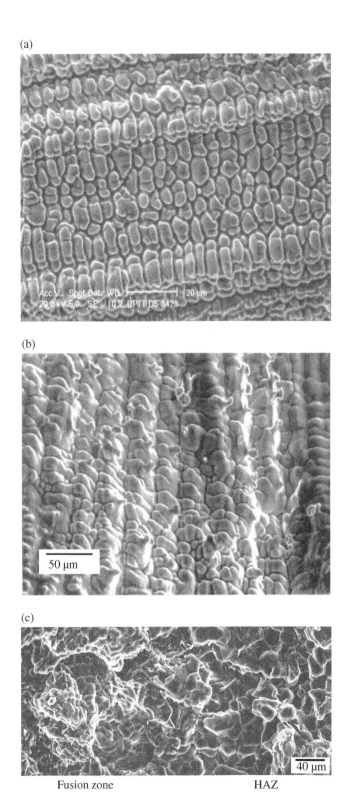

FIGURE 8.7 Solidification crack fracture surfaces. (a) Ni-base alloy, (b) duplex stainless steel, and (c) aluminum alloy (Courtesy of Dennis Harwig).

as the weld metal cools below the solidus. Another possibility is that the solidification crack with a dendritic morphology transitions to a ductility-dip crack with a flat fracture appearance. Since this transition is most prevalent in Ni-base alloys and fully austenitic stainless steels (both of which are susceptible to DDC), it is likely that such a transition in the type of cracking may explain the change in fracture morphology. In stainless steels that solidify as ferrite (FA solidification) and DDC is not observed, only a dendritic fracture morphology is observed. Alloys that do not exhibit this transition from dendritic to flat fracture are typically more resistant to solidification cracking.

Solidification crack fracture surfaces may also show evidence of eutectic reactions that occur during the terminal stages of solidification. This normally results in the decoration of the dendrite surface with a second phase. An example of this is shown in Figure 8.8 for a Nb-bearing Ni-base alloy. In this case, the γ/NbC

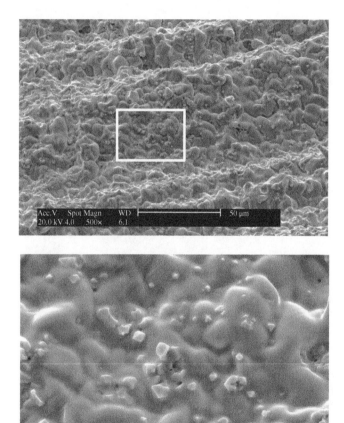

FIGURE 8.8   Solidification crack surface of a Nb-bearing Ni-base alloy showing NbC particles (Courtesy of Adam Hope).

eutectic reaction that occurs at the end of solidification results in the presence of NbC on the dendrite surface.

It is also important to note that the essentially fully dendritic nature of solidification crack fracture surfaces refutes the basic premise of the generalized theory of solidification cracking, which proposed a stage that includes solid–solid bridging. It is very unusual to see evidence of any solid-state fracture on a solidification crack fracture surface, suggesting that liquid films persist along solidification grain boundaries throughout the entire solidification range. This observation supports the strain theory of Pellini who proposed a segregate film stage that persists to the end of solidification [5] as described in Section 3.2.1.3.

The dendritic morphology of solidification cracks makes this form of cracking easy to identify and differentiate from other forms of weld metal cracking. For example, single-phase austenitic (fcc) weld metals may contain solidification cracks, weld metal liquation cracks, and ductility-dip cracks that can be difficult to distinguish based on optical metallography alone. The presence of a clean, dendritic fracture morphology, such as those shown in Figure 8.7, is a clear, distinguishing feature of solidification cracking.

8.2.4.2 *Liquation Cracking* Liquation cracking can occur in the HAZ or weld metal and results from local melting along grain boundaries. In the base metal HAZ, this melting occurs along grain boundaries in close proximity to the fusion boundary. In the weld metal, liquation during reheating of previously deposited weld metal can occur at either solidification grain boundaries or migrated grain boundaries. Examples of the fracture surface morphology of HAZ liquation cracking in a variety of alloy systems are shown in Figures 8.9–8.12.

Figure 8.9 shows a simulated HAZ of duplex stainless steel that was created in a thermomechanical simulator and has been pulled to failure at the nil ductility temperature (NDT) [6]. At this temperature, the steel is fully ferritic, and only very thin liquid

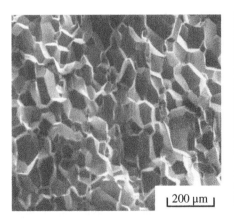

FIGURE 8.9 HAZ liquation crack surface in a duplex stainless steel (Ferralium 255) (From Ref. [6]. © AWS).

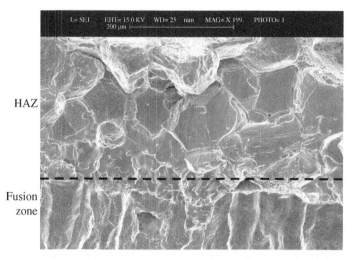

HAZ

Fusion
zone

FIGURE 8.10 HAZ and weld metal liquation cracking in a steel forging (Courtesy of Matt Johnson).

FIGURE 8.11 HAZ liquation crack surface in low-alloy steel HY-100 (From Ref. [7]. © AWS).

films are present along the grain boundaries. The HAZ liquation temperature range for this alloy (Ferralium 255) is very narrow, and this alloy is not susceptible to HAZ liquation cracking in normal practice.

Figure 8.10 shows an actual failure in a weld made in a steel forging. In this case, the liquation crack propagated into both the fusion zone and HAZ. The HAZ fracture surface is IG along prior austenite grain boundaries, while the fusion zone fracture surface shows columnar features typical of solidification growth.

Figure 8.11 is a simulated HAZ of HY-100, a low-alloy naval steel [7]. This fracture is also IG along prior austenite grain boundaries and shows only a thin coating of liquid film. This alloy is typically not susceptible to HAZ liquation cracking as evidenced by its narrow liquation temperature range (~60°C).

(a)

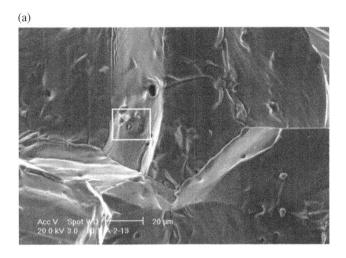

(b)

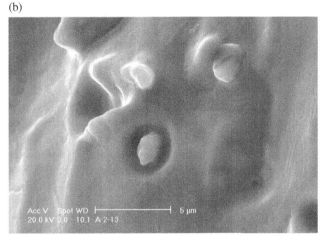

FIGURE 8.12 HAZ liquation crack fracture surface in a Ni-base superalloy (Waspaloy). (a) Intergranular fracture with liquid grain boundary liquid films and (b) inset from (a) showing constitutional liquation of TiC particles (From Ref. [8]. © AWS).

Figure 8.12 is an HAZ liquation crack in Waspaloy that has also been simulated in a thermomechanical simulator [8]. In contrast to Figure 8.9, this fracture surface reflects the presence of considerable liquid film that was present along the grain boundaries. Note that the grain faces are not as "clean" as those in Figures 8.9–8.11 indicating that considerable liquid film was present at the time of failure. At higher magnification, evidence of constitutional liquation can be seen on the fracture surface (Fig. 8.12b). These features represent TiC particles that have undergone constitutional liquation resulting in the particle appearing to sit in a depression that resulted from local melting at the interface with the matrix.

8.2.4.3 Ductility-Dip Cracking   DDC is most prevalent in austenitic (fcc) materials such as Ni-base alloys and austenitic stainless steels. In most cases, this form of cracking occurs in the weld metal but may also be present in the base metal HAZ. Ductility-dip cracks are normally macroscopically short and relatively straight. The crack surface appears flat at lower magnifications but can exhibit a variety of morphologies at higher magnifications. As described in Chapter 4, they occur along migrated grain boundaries in the weld metal and macroscopically can take on a columnar appearance. With optical metallography analysis alone, it may often be difficult to distinguish among solidification, weld metal liquation, and ductility-dip cracks. Fracture surface analysis may be the only way to confirm the nature of cracking.

DDC in a Ni-base alloy overlay (Filler Metal 52, ERNiCrFe-7) is shown in Figure 8.13. This is from a "boat sample" removed from a multipass overlay that contained indications when inspected using dye penetrant. The sample has been bent to open the crack in order to better observe the crack surface (Fig. 8.13a). Note that the fracture surface is macroscopically flat but at higher magnification exhibits some surface roughness. It is clearly distinct from a solidification crack surface morphology.

It has been shown that ductility-dip crack surface morphology is a function of temperature [9, 10]. At low temperatures in the DDC range (700–800°C for Ni-base alloys and stainless steels), the morphology is macroscopically flat and often with microscopic ductile dimples. This is often referred to as *ductile IG* fracture mode. An example of this morphology for a Type 310 stainless steel is shown in Figure 8.14a. This is from a strain-to-fracture sample tested at 750°C. At higher temperatures, the fracture surface exhibits more of a wavy pattern as shown in Figure 8.14b for Type 310 stainless steel at 950°C.

At higher temperatures, the waviness increases and there may be indications of slip lines on the fracture surface, as shown in Figure 8.15 for Type 310 stainless steel at 1100°C. At even higher temperatures within the ductility-dip temperature range (>1100°C), fracture surfaces once again become macroscopically flat with ductile dimples. This is due, in part, to the onset of recrystallization, which usually occurs at high strains at elevated temperatures.

It should be noted that the fracture appearance of ductility-dip cracks can vary greatly as a function of temperature and material. In weld metal, the failure path is always along migrated grain boundaries. The nature of these boundaries in terms of tortuosity, precipitation behavior, and crystallographic orientation can have a strong influence on fracture morphology.

8.2.4.4 Reheat Cracking   Reheat cracking is usually associated with low-alloy steels that contain Cr, Mo, and V and stabilized austenitic stainless steel alloys (such as Type 347). As described in Chapter 4, this form of cracking usually occurs during PWHT.

For reheat cracking to occur, there must be dissolution of carbides (Cr, Mo, V) during the "on-heating" thermal cycle. Upon reheating, intragranular precipitation of carbides and simultaneous relaxation of stresses occur. In low-alloy steels, there is strain localization at prior austenite grain boundaries, and failure occurs at or near the grain boundaries. The fracture surfaces appear clearly IG.

The IG fracture can be flat and generally featureless or may exhibit microductility (ductile IG). Low ductility IG failure during PWHT is thought to occur by two general

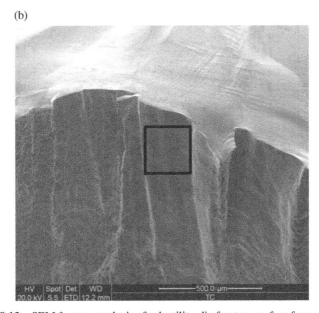

FIGURE 8.13 SEM fracture analysis of a ductility-dip fracture surface from a Ni-base Filler Metal 52 overlay. (a) Macro view, (b) higher magnification of region in (a), and (c) higher magnification of region in (b).

(c)

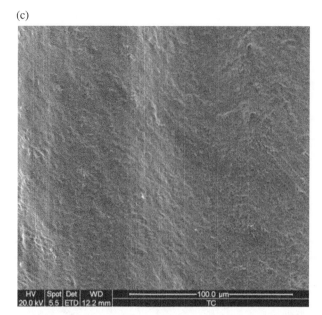

FIGURE 8.13   (Continued)

(a)

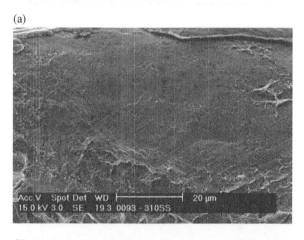

(b)

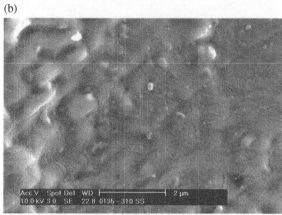

FIGURE 8.14   Ductility-dip crack surface in Type 310 stainless steel from strain-to-fracture test at (a) 750°C (1380 F) and (b) 950°C (1740 F) (Courtesy of Nathan Nissley).

(a)

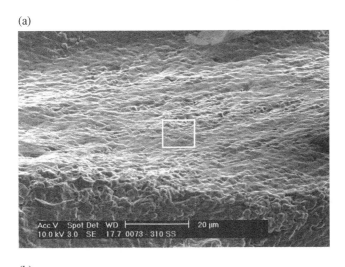

(b)

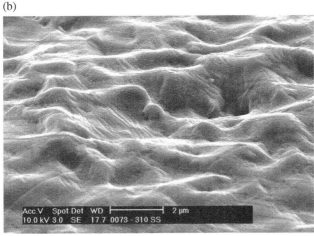

FIGURE 8.15 Ductility-dip crack surface in Type 310 stainless steel from strain-to-fracture testing at 1100°C (2010 F). (a) Low magnification and (b) higher magnification of region in (a) (Courtesy of Nathan Nissley).

mechanisms, segregation of impurity elements to prior austenite grain boundaries and intragranular precipitation strengthening with denuded (precipitate-free) zones near the grain boundaries. In the case of tramp element (P, S, Cu, Sn, As, Sb) segregation in the low-alloy steels, fracture surfaces are smooth and featureless. These elements (at trace levels) lower the cohesive strength across the grain boundaries, and failure occurs with little local ductility. When precipitation strengthening at grain interiors occurs and denuded zones form at grain boundaries, fracture tends to occur in a ductile IG mode. Temperature can also play a role, with failure at lower temperatures favoring a flat featureless morphology and higher temperatures tending toward ductile IG. Several examples of reheat cracking fracture morphology are provided in Figure 8.16.

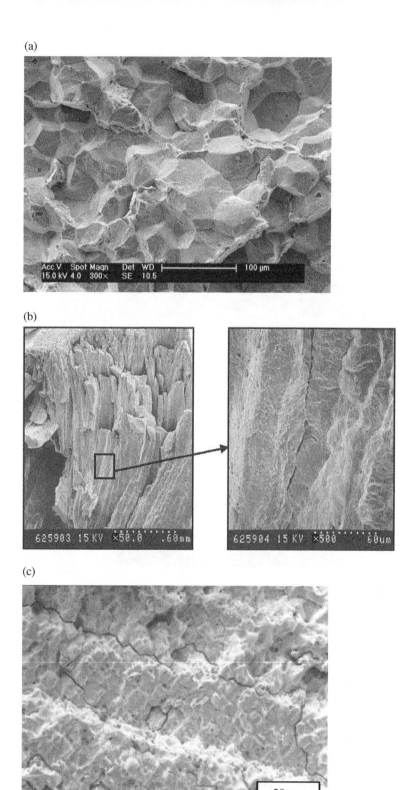

(a)

(b)

(c)

FIGURE 8.16   Reheat cracking fracture surfaces. (a) 2.25Cr–1Mo simulated HAZ, (b) INCO-A weld metal, and (c) Type 347 weld metal.

8.2.4.5 *Strain-Age Cracking* Strain-age cracking is associated with PWHT of Ni-base, precipitation-strengthened alloys (superalloys). As described in Chapter 4, this form of failure is usually associated with HAZ grain boundaries in very close proximity to the fusion boundary. Similar to reheat cracking, reformation of precipitates that dissolved during the welding process combined with stress relaxation leads to high strains at the grain boundary and subsequent cracking. The fracture surfaces are clearly IG and can be flat and featureless or exhibit microductility. Examples of strain-age crack surface morphology in Alloy 718 are provided in Figure 8.17. These fracture surfaces were produced using the strain-age cracking test described in Chapter 9 [11].

The fracture surface in both cases is macroscopically IG but microscopically shows either flat or ductile dimple features (ductile IG). The fracture surface in Figure 8.17b is

(a)

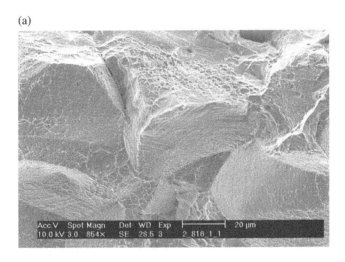

(b)

FIGURE 8.17 Strain-age cracking fracture morphology in Alloy 718. (a) Ductile intergranular and (b) mixed flat and ductile intergranular (Courtesy of Seth Norton).

particularly interesting since the adjacent grain faces exhibit either microductility or flat fracture. This suggests an effect of grain boundary orientation. It is also interesting that the flat grain faces show emergent slip lines associated with the fracture process.

8.2.4.6 Hydrogen-Induced Cracking   There is no single fracture morphology that is characteristic of HIC. Fractures can be IG with surfaces that are either flat or exhibit microductility. Fractures can also be transgranular with quasicleavage (QC) or MVC features. The effect of hydrogen concentration and stress intensity on fracture morphology was described by Beachem's model [12] as described in Chapter 5 and illustrated in Figures 5.5 and 5.6.

It is generally theorized that hydrogen migrates to and is trapped at dislocations, grain boundaries, micro-inclusions, and structural heterogeneities such as slag inclusions or martensite/ferrite interfaces. This trapped hydrogen coupled with residual and applied stresses eventually causes failure or decohesion. Hydrogen diffuses to the lattice just ahead of the crack tip and aids in whatever deformation process the matrix would normally allow. As HIC progresses, the fracture morphology can shift from one mode to another as the stress intensity, hydrogen concentration, and microstructure vary.

The three basic modes of HIC in steels are IG, QC, and MVC. All three of these modes can be observed on the fracture surface of an HSLA-100 implant test specimen, as shown in Figure 8.18 [13]. The implant test is described in Chapter 9.

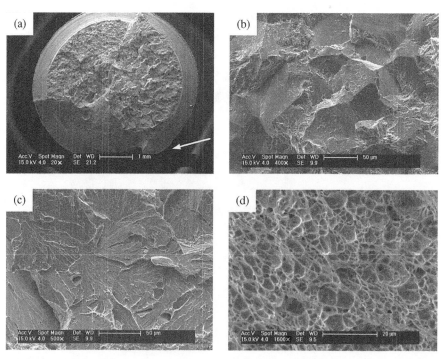

FIGURE 8.18   Fracture morphology of HSLA-100 implant specimen for HIC. (a) General fracture appearance (white arrow indicates the direction of crack growth), (b) intergranular, (c) quasicleavage, and (d) microvoid coalescence. Labels a, c, and d are observed in sequence along the crack propagation path (From Ref. [[13]. © AWS]).

Note that initial fracture in this specimen occurs in an IG mode and then quickly transitions to QC as the stress intensity increases. The QC mode then transitions to MVC (ductile dimple) and finally to ductile overload. Steels that have high resistance to HIC will exhibit only limited IG and QC failure modes. Other materials that are highly resistant to HIC (such as austenitic stainless steels) will only exhibit MVC, even at high hydrogen concentrations.

8.3   AN ENGINEER'S GUIDE TO FAILURE ANALYSIS

This section provides an overview of how a failure investigation should be conducted, including techniques and approaches to obtain the information needed to determine the cause of failure to a reasonable degree of engineering and scientific certainty. There are other more detailed texts that are available that provide more specifics. In particular, *ASM Handbook, Volume 11: Failure Analysis and Prevention* is a particularly good reference [14].

A failure analysis checklist is provided in Table 8.1 that should act as a guide when investigating either a fabrication or service failure. The sequence of investigative steps may vary depending on the nature of the failure, but in general, all of the items on this checklist should be considered during the course of the investigation. The following sections will provide additional insight regarding these items. This checklist is not meant to be all-inclusive. There may be other information that is required to support a failure investigation that is unique to specific types of failures.

TABLE 8.1   Failure investigation checklist

1.	Site visit
2.	Preliminary examination
3.	Interview personnel
4.	Collect background information
5.	Develop sectioning and testing protocol
6.	Remove sections for examination
7.	Conduct chemical composition analysis
8.	Macroscopic examination
9.	Samples for metallography/fractography
10.	Determine metallographic techniques
11.	Determine analytical techniques
12.	Mechanical testing
13.	Simulative testing
14.	NDE techniques
15.	Structural integrity assessment
16.	Consultation with other experts
17.	Final analysis and reporting

8.3.1 Site Visit

It is almost always helpful to visit the site of the failure, unless, of course, the failure has occurred within your own plant or in a situation with which you are totally familiar. Secondhand information, including photos taken by someone else and secondhand reporting, is usually not desirable since the details of the failure will be subject to someone else's interpretation or prejudice. If you cannot visit the failure site (perhaps the failure occurred on an offshore platform 200 ft below the ocean surface!), give explicit instructions as to the type of failure documentation you require and make sure you request sufficient photographic documentation.

The following is a partial list of things you might want to bring on the site visit:

- High-quality digital camera with a macro capability, a flash or strobe unit, and backup memory.
- Tape measure and scale markers (magnetic).
- Small handheld magnifying glass or a "loop."
- Flashlight and dental mirrors (for looking into tight places).
- Magnet (for material ID).
- Recording device or smartphone (this is often easier than trying to scribble notes that you won't be able to decipher later).
- Coveralls and boots (in the case of a field failure).
- Safety glasses and ear protection.
- Sample bags – "ziplock" bags work great.
- Identification tags and indelible marking pen.
- Fracture surface replication material.

Prior to the site visit, you should try to determine what materials and welding processes are involved and what the service environment was (where applicable). This will give you some "lead time" to formulate possible failure scenarios and help you decide on a "plan of attack" during your inspection. For example, if the material is an austenitic stainless steel operating in high-pressure steam, you may want to look for evidence of stress corrosion cracking.

Avoid conferring with other "experts" about the failure until you have completed your inspection. Your initial impressions and observations can often be biased by others whom you may consider well qualified. Remember, a good engineer always maintains his/her objectivity!

Spend plenty of time inspecting the failure and the surroundings. The mental (and actual) pictures that you obtain during this inspection may be invaluable later on when you're trying to decipher the evidence and don't have the key piece of information. The tendency is often to rush through the inspection at the expense of missing a valuable clue.

Photo documentation is extremely important. Photos should be taken from all angles and under different lighting situations. With digital photography, it is easy to check photo quality immediately and then retake a photo if it is not of acceptable quality. In many situations, you may get only one chance to photo document the failure, so you must insure that you get what you need. More is better than less—it is easier to delete photos later rather than lament over the one that was missed.

You should attempt to interview as many personnel as possible, especially those who were associated with the failure or knowledgeable of the system (component) that failed. This may include welders, shift foremen, welding engineers, design engineers, metallurgists, etc. Try to avoid secondhand information—you can usually discount any information you obtain from the plant manager, vice president, etc.—since their information is probably not even secondhand.

8.3.2   Collect Background Information

The information you may need will be very much subject to the type of failure you are investigating. The following is a list of items you should consider obtaining:

Material specifications. These would include any internal specifications used to order material or standard specifications that are referenced (ASTM, AWS, API, SAE, etc.).

Material qualification/testing reports. These may be in the form of in-house QA reports or qualification documents provided by the material supplier. They may include information on material composition, mechanical properties, inspection, etc.

Supplier of material. It may also be useful to determine how the material was processed (melting, thermomechanical processing, etc.) and the product form.

Fabrication records. This will include all in-house processing of the material prior to welding. Items may include forging, pressing, heat treatment, cleaning, plating, machining, etc.

Welding records. Obtain all documentation concerning the welding qualification and procedure. These may be in the form of procedure qualification records (PQRs) and weld procedure specifications (WPSs).

Postweld operations. Following welding, determine if the component was heat treated, cleaned, machined, etc.

Inspection records. Determine if any visual or NDT inspections were performed and obtain documentation. This may also include proof test results or destructive evaluations on random parts.

Service history. When investigating service failures, a knowledge of the service history is essential in determining the nature of failure. Unfortunately, in most instances, this information is usually not readily available or relatively sketchy. In collecting service histories, pay special attention to environmental details, loading conditions, abnormal or accidental overloads, variations in temperature, temperature extremes, startup and shutdown conditions, etc.

Abnormal conditions. Determine if there was anything abnormal about the design of the failed component or the conditions under which it was operating. Check the service and repair record of similar components. Try to determine if the failure is an isolated incident or generic in nature.

8.3.3   Sample Removal and Testing Protocol

It will often be necessary to develop a protocol for sample removal and evaluation. This will almost always be the case when other parties are involved and there is a possibility of litigation. Under the latter circumstances, the protocol will have to be agreed upon by the various parties and then strictly followed.

From your earlier visual examination of the failure, you have identified certain areas either on the fracture surface or adjacent to the fracture that may provide valuable information as to the cause of the failure. These areas may represent potential failure initiation sites or regions, which are representative of the metallurgical condition of the material at the time of failure. If available, it is often useful to examine similar samples (exemplars) that have not failed since this may provide some insight into the cause of failure.

8.3.4   Sample Removal, Cleaning, and Storage

You should develop an identification scheme that is both simple and logical and, most importantly, avoids any chance of mistaken sample identity after removal. You should take extreme precautions and be very explicit when specifying specimen removal techniques. Make sure that valuable evidence is not destroyed during removal. Many failure investigations have been clouded by the inadvertent destruction of a fracture surface during specimen removal. Also, pay particular attention to specimen removal methods. You should opt for "cold" removal methods, which do not alter the microstructure. Oxyacetylene or plasma cutting techniques can reheat critical areas of the microstructure, resulting in phase transformations or other metallurgical transformations that can disguise or destroy valuable evidence. In situations where these techniques are necessary, precautions should be used to prevent overheating of the region of interest.

In many cases, the fracture surface or region of interest will be contaminated by exposure to the environment, mishandling, or other factors. This will often require that the area of interest be cleaned to remove the contamination. In the case of dirt, grease, and other surface contaminants, this can usually be removed with solvents, soap and water, or mild cleaning solutions.

In the case of oxidation (particularly "rust" on steels), there are some special techniques that can be used to remove the oxidation without damaging (attacking) the underlying surface. One such technique that is very effective in removing oxidation from steels without damage to the underlying metal uses a reagent solution consisting of concentrated hydrochloric acid (HCl) and 10 g/l of 1,3-n-butyl-2 thiourea (known as DBT). This stock solution is then diluted 50/50 with distilled water prior to use [15]. The sample is then immersed in this solution under ultrasonic agitation for a period of seconds to minutes until the oxidation is removed.

8.3.5   Chemical Analysis

Often, chemical composition information will not be available for the materials in question. Many materials are purchased to a given material specification, but the material test records are not available. In cases where multiple heats of the same material have been used, it may be difficult to determine the exact composition of the material of interest. The composition of the materials involved in the failure can often provide valuable insight into the nature of the failure. The level of impurity elements (such as S, P, O, N) may be particularly critical when considering the weldability of the material or the performance of the welded structure in the intended service environment.

Various analytical techniques can be used to determine the chemical composition of materials. Generally, some trade-offs among expense, accuracy, and reliability are required when using these techniques. Some common analytical techniques that are useful in material identification and analysis are listed below:

Wet chemical analysis. The most accurate and most expensive of the analysis techniques, this technique involves dissolution of the material in a chemical medium (usually an acid). Reaction products of the desired species are then carefully weighed and compared with the weight of the original sample to obtain a weight percent. See *ASTM Standard E50, Volume 3.05: Apparatus, Reagents, and Safety Considerations for Chemical Analysis of Metals, Ores, and Related Materials* for more information.

Emission spectroscopy. This is probably the most common method in use today for obtaining bulk chemical analysis. The method is relatively inexpensive ($50-100/ sample) and can be reasonably accurate if the analysis system is calibrated with proper standards (always ask about calibration). In addition, relatively small samples can be analyzed, which is often helpful if only small pieces of material can be obtained from near the failure site or within the weld. The sensitivity of this technique to low atomic weight elements (lighter than aluminum) is variable. Carbon, nitrogen, sulfur, phosphorus, and boron contents should be determined by other methods. See ASTM practices E350-E354 (Volume 3.05) for details of this and other techniques.

Combustion technique—thermal conductivity method. This technique is most popularly known as the "LECO method" after the manufacturer of the equipment. This method has been used to accurately measure carbon, oxygen, nitrogen, and sulfur. Hydrogen may also be determined by the LECO method, although the results are semiquantitative at best. Again, proper standards must be used to calibrate the equipment. Multiple analyses are recommended. See *ASTM E1019 (Vol. 3.05): Determination of Carbon, Sulfur, Nitrogen, and Oxygen in Steel, Iron, Nickel and Cobalt Alloys by Various Combustion and Fusion Techniques*.

Infrared and ultraviolet spectroscopy. These techniques can be used for identifying organic materials associated with the failure, such as paint, oil, grease, rubber, and plastic. In particular, a technique known as Fourier transform infrared (FTIR) spectroscopy is widely used for this type of analysis. ASTM E1252 (Vol. 3.06) describes the general methods for infrared spectroscopy.

Other techniques. Many other techniques are available for analyzing specific elements, compounds, corrosion products, etc. An analytical chemist should be consulted for help in analyzing unusual materials, or if highly precise measurements are required.

8.3.6 Macroscopic Analysis

Once you have selected and removed the samples you feel are pertinent to the failure, make a very thorough macroscopic examination of the samples. This will include both visual inspection and optical microscopy up to 100× magnification. A binocular microscope with good depth of field resolution is essential for this examination.

Again, document the samples photographically. You want to have a permanent record of their appearance before you start sectioning for more in-depth analysis. Among the things you are looking for in this examination are the following:

- Fracture surface appearance, for example, dull, shiny, facetted, striations, porosity, etc.
- Presence of oxidation or corrosion products
- Macroscopic yielding and shear lips
- Secondary cracking
- Weld surface features such as undercut or arc strikes

8.3.7 Selection of Samples for Microscopic Analysis

Your macroscopic examination should give you a good idea of the samples necessary for further examination at higher magnifications. This would include samples for:

- Fracture surface analysis—SEM or other surface analysis such as Auger spectroscopy
- Metallography—optical or electron microscopy
- Microanalytical evaluation—SEM or electron probe microanalysis (EPMA)

Again, use a simple identification system to label the samples that will be sectioned. Use both a sketch and photographic documentation to preserve identity and traceability. This type of scheme is very useful for later explaining the sample location and selection strategy to others and for inclusion in your report.

8.3.8 Selection of Analytical Techniques

There are many analytical techniques available to support a failure investigation. These include techniques for bulk sample analysis and those that provide topographic, metallurgical, or chemical information at very high magnifications.

X-ray diffraction (XRD) of bulk samples can provide information regarding the phases or precipitates that are present in the material. This technique is not limited to metals and can be used to evaluate the structure of materials that are crystalline in nature. XRD patterns are generated by characteristic wavelength XRD from a crystal lattice.

The SEM is a powerful and often used tool in failure analysis since it can provide both topographic and chemical composition information. Also, sample preparation is quite straightforward. However, for samples that are contaminated, special cleaning techniques may be required to prevent "charging" of the sample by the electron beam.

EPMA provides much more accurate composition analysis on the microscopic scale relative to the SEM. This technique is useful for evaluating segregation patterns in welds or determining chemical composition at specific locations. The resolution is on the order of 1–2 μm. Samples for EPMA are generally mounted in a conductive medium and prepared metallographically since composition information is most easily obtained off a flat surface.

Analysis of metallic samples using electron microscopy techniques uses a number of "events" that occur when the electron beam interacts with the sample. These various interactions have been described in Section 8.2.2. In summary, the *SE* signal is used primarily to determine surface topography and provide the image of the fracture surface. Metallographic samples that have been etched can also be examined using the SE image to provide better resolution than is possible in an optical microscope. The *BSE* signal provides good atomic number contrast and can be used to estimate local composition differences. Many modern SEMs allow mixing of the SE and BSE signals to provide both topographical and composition information. Finally, *X-rays* that are generated by the electron beam interaction with the specimen can be used to provide semiquantitative composition analysis. Although not as precise as EPMA, the information gathered can be useful in identifying different phases and compounds. The resolution is limited to 1–2 μm. This technique is not accurate for analysis of light elements (C, O, N) unless the instrument is equipped with a special detector.

8.3.9   Mechanical Testing

Mechanical tests are often useful for determining the "fitness" of the material in its service environment. Testing can also be performed for fabrication failures to determine if the material meets specifications. These tests may include:

- Macro- or microhardness
- Tensile
- Fracture toughness
- Fatigue, creep fatigue, and corrosion fatigue
- Stress rupture
- Weldability tests

Many of these tests have been standardized, and procedures are available in *ASTM Volume 3.01: Metals Mechanical Testing; Elevated and Low Temperature Tests: Metallography*. Recommendations on possible weldability tests are provided in Chapter 9.

8.3.10   Simulative Testing

In some cases, it may be helpful to try to simulate the conditions of a failure in a controlled laboratory test. This type of testing is designed to replicate the loading conditions, environment, and material variables, which may have contributed to the failure. Simulative tests are generally expensive and usually only approximate the service or fabrication conditions that led to the failure. In the case of service failures, it is often difficult to determine the actual operating conditions (loading spectrum, temperature history, environment, etc.). Use caution when interpreting data from these tests.

8.3.11   Nondestructive Evaluation Techniques

Nondestructive evaluation (NDE) techniques are often useful for determining the extent of cracking in a component that may not be obvious through visual inspection. These techniques can also be used when deciding sample selection or sectioning plans. NDE is also very useful for inspecting similar components or regions adjacent to the main fracture.

One of the most common of these techniques is known as *penetrant inspection*. It involves cleaning the piece of interest, applying a dye that will penetrate into cracks, and then applying a developer that will reveal where any cracks or defects are present on the surface. This technique is only useful for cracks that are on the surface of a component, such as a fatigue crack. For bulk analysis, X-ray radiography, ultrasonic-based, or eddy current techniques are required.

8.3.12   Structural Integrity Assessment

The principles of fracture mechanics can occasionally be used to evaluate the structural integrity of the material in the context of its service or fabrication environment. Various numerical analysis methods, such as finite element and finite difference analysis, can also be used to estimate the stress distribution in the vicinity of a crack or discontinuity. These methods, except in very simple systems, are approximate at best, so use caution when interpreting the results.

8.3.13   Consultation with Experts

If the failure involves areas outside your expertise (corrosion, fracture mechanics, witchcraft, etc.), consult an expert. He/she may have considerable experience with similar failures and can often be invaluable in helping to determine the cause. The expense ($1500-2500/day) may seem exorbitant, but the consultant may save you considerable time and expense on some of the previous steps. These consultants may have access to information that is key to the failure investigation. There are many consulting engineering firms that specialize in structural integrity assessment and failure analysis. Many university professors also engage in external consulting. Remember that your ultimate goal is to accurately determine the cause of failure, so be open to seeking help from experts.

8.3.14   Final Reporting

You are now ready to assimilate all the information that has been gathered into a <u>concise</u> report. Reporting formats vary, but the following sections would normally be included in a failure analysis report:

Summary (or executive summary). This should be a very concise description of what was done and the major findings.

Background. Provide a history of the fabrication procedures and/or service behavior of the component in question. This section must provide enough information to support the findings that you will report. In some cases, detailed information can be included in Appendices to reduce the length of this section.

Methodology. This section is very important, since it is necessary to precisely describe how samples were selected, the sectioning schemes used, methodology for sample identification, and analytical techniques used.

Significant results. While the investigation may have generated many results (micrographs, fractographs, mechanical tests, chemical analyses, etc.), it is best to report these as concisely as possible. The use of tables is a good way to summarize data. Information that is more peripheral, or that you want to include for completeness, can be provided in an Appendix.

Discussion of results. Discussion may be included in the previous section, but in some cases, it may be useful to have this as a separate section.

Conclusions. These are short concise statements that capture the major points of the investigation. It is usually best to number these, since it makes reference to specific conclusions much easier.

Recommendations. Based on the results of the investigation, this section can be used to recommend remedial action, changes in design or material selection or address other corrective issues.

Additional work (if analysis is not complete). If the conclusions are incomplete or the cause of the failure is still not clear, additional analysis or testing may be recommended to complete the analysis.

References. Provide any references that you relied on in your report. These may include previous reports, other expert opinions, handbook data, journal articles, or other sources of technical information.

Appendices. The use of appendices is a good way to be inclusive in your reporting while reducing the "clutter" in the body of the report. For example, it may be important to report the results of mechanical testing that provides data from each sample tested. These details can be summarized in the body of the report using a table with supporting information in an Appendix. Most failure investigations result in hundreds (or thousands) of photographs. These are best included in an Appendix.

8.3.15 Expert Testimony in Support of Litigation

Engineers often become involved in litigation involving product liability, patent infringement, personal injury/death, and other situations that must be resolved in a court of law. The *plaintiff* in the case is the person/entity who files the case or complaint, and the *defendant* is the party who must defend against the complaint. There are many law firms that specialize in product liability and personal injury cases.

As an engineer, you may be requested/required to participate in these cases as an expert witness in order to provide testimony and an opinion regarding the details of the case. In cases where a failure has occurred, you may be requested to conduct an investigation to determine the "root cause" of the failure.

If the case proceeds to trial, you will normally be asked to provide an expert opinion in the form of a written document and will then be "deposed" by the lawyers from the other side. Depositions represent sworn testimony that allows the opposing lawyers to question the expert regarding his/her opinion prior to trial. These can be quite intensive, and the expert is expected to be well prepared to defend his/her opinion based on sound scientific and engineering principles. If the case goes to trial, you will be expected to provide the same opinion/testimony in the presence of a jury.

For testimony at a jury trial, remember that most jurors have little or no technical background and will have a difficult time comprehending complex engineering principles. The effective expert knows how to simplify engineering concepts such that the jury (and the judge) can grasp the important points of your testimony.

REFERENCES

[1] Joy DC, Romig AD Jr, Goldstein JI, editors. *Principles of Analytical Electron Microscopy*. New York: Plenum Press; 1986.

[2] Goldstein JI, Newbury DE, Echlin P, Joy DC, Romig AD Jr, Lyman CE, Fiori C, Lifshin E. *Scanning Electron Microscopy and X-Ray Microanalysis*. New York: Plenum Press; 1992.

[3] ASM International. Handbook Committee. *ASM Handbook, Volume 12: Fractography*. 9th ed. Metals Park, OH: ASM International; 1987.

[4] Matsuda F. Solidification crack susceptibility of weld metal. In: David SA, Vitek JM, editors. *Recent Trends in Welding Science and Technology*. Metals Park, OH: ASM International; 1990. p 127–136.

[5] Apblett WR, Pellini WS. Factors which influence weld hot cracking. Weld J 1954;33 (2):83s–90s.

[6] Nelson DE, Baeslack III WA, Lippold JC. An investigation of weld hot cracking in duplex stainless steels. Weld J 1987;66 (8):241s–250s.

[7] Caron JL, Babu SS, Lippold JC. The weld heat-affected zone liquation cracking susceptibility of naval steels. Weld J 2013;92 (4):110s–123s.

[8] Qian M, Lippold JC. Effect of multiple postweld heat treatment cycles on the weldability of Waspaloy. Weld J 2002;81 (11):233s–238s.

[9] Nissley NE. Intermediate temperature embrittlement in Ni-base weld metals [PhD Dissertation]. Columbus, OH: The Ohio State University; 2006.

[10] Collins MG, Ramirez A, Lippold JC. An investigation of ductility-dip cracking in Ni-base filler metals, Part 3. Weld J 2004;83 (2):39s–49s.

[11] Norton SJ, Lippold JC. Development of a gleeble-based test for postweld heat treatment cracking susceptibility, *Trends in Welding Research VI*. Columbus, OH: ASM International; 2003. p 609–614. Proceedings of the 6th International Conference, April 15–19, 2002, Callaway Gardens, GA. ASM International Metal Park, OH.

[12] Beachem C. A new model for hydrogen-assisted cracking (hydrogen "embrittlement"). Metall Trans 1972;3:441–455.

[13] Yue X, Feng X, Lippold JC. Quantifying heat-affected zone hydrogen cracking in high strength naval steels. Weld J 2013;92 (9):265s–273s.

[14] ASM International. Handbook Committee. *ASM Handbook, Volume 11: Failure Analysis and Prevention*. 9th ed. Metals Park, OH: ASM International; 1986.

[15] Kayafas I. Technical note: corrosion product removal from steel fracture surfaces for metallographic examination. Corrosion 1980;36 (8):443–445.

9

WELDABILITY TESTING

9.1 INTRODUCTION

Weldability testing is widely used to assess the fabricability of materials by various welding processes and to determine the service performance of welded construction. Although there are hundreds of weldability tests that have been used to assess material performance, very few of these tests have been standardized. For mechanical testing of welded joints, there are a number of standardized tests that can be used. Some of these have been described in Chapter 7 and include tensile, fatigue, and fracture toughness tests. The hardness test, while not actually a test that quantifies mechanical performance, can be a useful indicator of strength and ductility.

Many of the weldability tests that have been developed over the years are specific to certain types of weld cracking. For example, the widely used Varestraint test was developed to assess solidification and liquation cracking susceptibility. There have been a number of tests that have been developed for other forms of cracking, including reheat, ductility dip, strain age, relaxation, and hydrogen-induced cracking (HIC). Only a few of these have been standardized. ISO 17641 and 17642 provide procedures for hot cracking and cold cracking tests, respectively. These standards include the Varestraint and PVR tests for solidification cracking and the controlled thermal severity (CTS) and Y-groove (Tekken) tests for HIC. A summary of the weldability tests that are covered by the ISO standards is provided in Table 9.1. It should be noted that, in some cases, the specific details regarding the test procedure are not included in these standards.

Welding Metallurgy and Weldability, First Edition. John C. Lippold.
© 2015 John Wiley & Sons, Inc. Published 2015 by John Wiley & Sons, Inc.

TABLE 9.1   Summary of weldability tests covered under ISO standards 17641 and 17642

ISO standard	Test type	Test technique
17641-2	Hot cracking, self-restraint	T-joint
		Weld metal tensile
		Longitudinal bend
17641-3	Hot cracking, externally loaded	Hot tensile
		Varestraint
		Flat tensile (PVR)
17642-2	Cold cracking, self-restraint	Controlled thermal severity (CTS)
		Tekken (Y-groove)
		Lehigh (U-groove)
17642-3	Cold cracking, externally loaded	Implant

This chapter will describe a number of weldability tests that have been used in the Welding and Joining Metallurgy Laboratory at Ohio State University (OSU). Most of these tests are simulative in nature, as opposed to "self-restraint" tests. The simulative test attempts to evaluate the behavior of the material under controlled conditions that allows the influence of important variables (temperature, time, stress, strain, etc.) related to the cracking mechanism to be determined. It is often argued that the self-restraint tests are more appropriate for assessing weldability since the actual weld geometry can be used with these tests. The results tend to be of the "go or no-go" type—the test weldment cracks or it does not. While these tests can be useful for procedure development, they often do not provide the type of metallurgical information needed to make appropriate material selection for a given application. The danger always exists that if the self-restraint test is not more severe than the actual application, test results may not predict performance.

9.2   TYPES OF WELDABILITY TEST TECHNIQUES

Weldability test techniques can be broadly grouped into four categories: mechanical, nondestructive, service performance, and specialty. With the exception of the inspection (nondestructive) techniques, these tests are destructive in nature, requiring sectioning of welds and special sample preparation. The destructive evaluation techniques are obviously not applicable in a manufacturing or quality control environment but should be implemented at the material selection and procedure qualification stage.

Specialty tests to evaluate susceptibility to various metallurgically related defects, such as solidification cracking or hydrogen cracking, should be conducted in the alloy development or selection stage. Failure to address issues related to susceptibility to cracking during fabrication can result in considerable cost and development time penalties.

The simulative tests described in the following sections involve the use of tension or bending to simulate high restraint levels in welds. These tests operate under strain control, strain rate control, or stress control. Most are capable of providing some measure of cracking susceptibility that can be used to compare different materials and that may provide important metallurgical information.

Hot ductility tests simply measure the strength and ductility of materials at elevated temperatures. Since weld cracking is often associated with insufficient ductility relative to the imposed strain at high temperature, an understanding of the material's hot ductility will provide insight into the potential for cracking.

9.3   THE VARESTRAINT TEST

The variable restraint, or Varestraint, test was developed in the 1960s by Savage and Lundin at Rensselaer Polytechnic Institute (RPI) [1]. It was devised as a simple, augmented strain-type test that would isolate the metallurgical variables that cause hot cracking. Since its introduction, numerous modifications have been made to the original test, and a number of variants are used worldwide. The test itself has never been fully standardized, and a variety of test techniques are in use today, although ISO 17641 does provide some guidance on the use of this test. In the 1990s, Lin and Lippold at OSU developed test techniques that allow the magnitude of the crack-susceptible region (CSR) to be determined using Varestraint testing [2].

There are three basic types of Varestraint tests, as shown in Figure 9.1. The original test was a longitudinal type, whereby bending is applied along the length of the weld that is being made on the sample. This type of test produces cracking in both the fusion zone and adjacent HAZ. Since in some cases it is desirable to separate HAZ liquation cracking from solidification cracking, the spot and transverse tests were developed. The transverse test applies bending strain across the weld and generally restricts cracking to the fusion zone (there is little or no strain applied to the HAZ). The spot Varestraint test uses a small spot weld to develop a susceptible HAZ microstructure. The sample is then bent when the weld is still molten in order to isolate cracking in the HAZ.

The gas tungsten arc (GTA) welding process is most commonly used for these tests, although other fusion welding processes can also be used. Welds are usually autogenous, although filler metal deposits can be tested using special samples. Bending is normally very rapid (2–10 in./s), but some slow bending versions of the test have been developed.

The augmented strain is controlled by the use of interchangeable die blocks, where the strain (ε) in the outer fibers of sample is given by

$$\varepsilon = \frac{\mathbf{t}}{2\mathbf{R} + \mathbf{t}} \tag{9.1}$$

where $\mathbf{t}$ is the sample thickness and $\mathbf{R}$ is the radius of the die block.

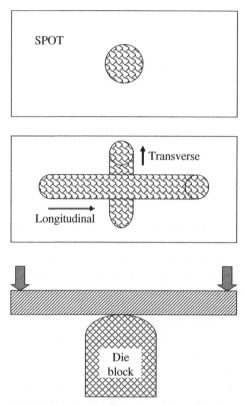

FIGURE 9.1   Different types of Varestraint tests.

A number of metrics have been developed to quantify cracking susceptibility using the Varestraint test. Most of these require the measurement of crack lengths on the surface of the as-tested sample using a low-power (20–50×) microscope. Total crack length is the summation of all crack lengths, while maximum crack length (MCL) is simply the length of the longest crack observed on the sample surface. More recently, the concept of *maximum crack distance* (MCD) has been used to allow more accurate determination of the cracking temperature range.

The strain to initiate cracking, defined as the *threshold strain*, has also been used to quantify cracking susceptibility. The *saturated strain* represents the level of strain above which MCD does not change with additional applied strain. It is important to test at saturated strain when determining the maximum extent of the CSR, as described in the following sections.

9.3.1   Technique for Quantifying Weld Solidification Cracking

This section describes a technique developed by Lin and Lippold [2] for quantifying weld solidification cracking susceptibility using the Transvarestraint test. A schematic illustration of the test is shown in Figure 9.2. Above the saturated strain level, the MCD

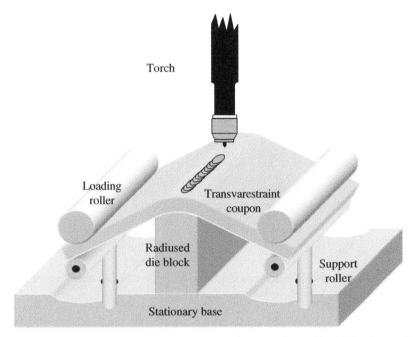

FIGURE 9.2   Schematic of the Transvarestraint test for evaluating weld solidification cracking susceptibility.

does not increase with increasing strain. This indicates that the solidification crack has propagated the full length of the CSR. Typical cracking as it appears at the trailing edge of the weld pool is shown in Figure 9.3 for an austenitic stainless steel sample tested at 5% strain. By testing over a range of augmented strain, an MCD versus strain plot such as that shown in Figure 9.4 can be generated. In this manner, the threshold strain for cracking to occur and the saturated strain above which the MCD does not increase can be identified. The typical strain range over which samples are tested is 0.5–7%.

Most stainless steels and Ni-base alloys exhibit saturated strain levels between 3 and 7%. Threshold strain levels are generally in the range from 0.5 to 2.0%, depending on the alloy and its solidification behavior. Although the threshold strain may, in fact, be an important criterion for judging susceptibility to weld solidification cracking, the MCD above saturated strain is much easier to determine and provides a measure of the solidification cracking temperature range (SCTR).

In order to determine SCTR, the cooling rate through the solidification temperature range is measured by plunging a thermocouple into the weld pool. The time over which cracking occurs is approximated by the MCD above saturated strain divided by the solidification velocity. Using this approach, SCTR can be calculated using the following relationship, where V represents the welding velocity:

$$\text{SCTR} = \text{cooling rate} \times \frac{\text{MCD}}{V} \tag{9.2}$$

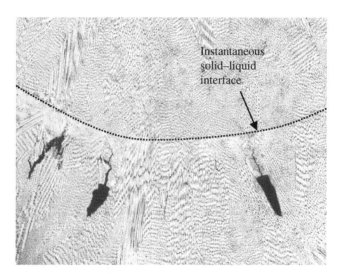

FIGURE 9.3   Weld solidification cracking in a Type 310 stainless steel Transvarestraint specimen. Plan view of sample surface.

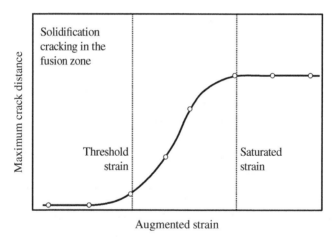

FIGURE 9.4   Maximum crack distance in the fusion zone versus applied strain during Transvarestraint testing.

The concept for determining SCTR using this approach is shown in Figure 9.5. By using a temperature rather than a crack length as a measure of cracking susceptibility, the influence of welding variables (heat input, travel speed, etc.) can be eliminated. SCTR then represents a metallurgically significant, material-specific measure of weld solidification cracking susceptibility.

SCTR values for a number of austenitic stainless steels and Ni-base alloys are provided in Table 9.2. Alloys that solidify as primary ferrite (such as some heats of Types 304 and 316L stainless steel) have low SCTR values, typically less than 50°C. Austenitic

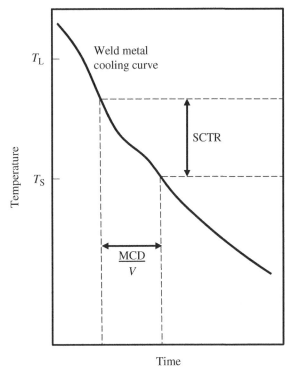

FIGURE 9.5 Method for determining the solidification cracking temperature range (SCTR) using the cooling rate through the solidification temperature range and maximum crack distance (MCD) at saturated strain.

TABLE 9.2 **Solidification cracking temperature range (SCTR) values for several stainless steels and Ni-base alloys obtained using the Transvarestraint test**

Alloy	Primary solidification mode	SCTR (°C)
Austenitic stainless steels		
Type 304L	BCC—ferrite	31
Type 316L	BCC—ferrite	49
Type 310	FCC—austenite	139
A-286 (PH type)	FCC—austenite	418
Nickel-base alloys		
Alloy C-22	FCC—austenite	50
Alloy 617	FCC—austenite	58
Alloy 230W	FCC—austenite	95
Alloy Hast W	FCC—austenite	145
Alloy Hast X	FCC—austenite	190

TABLE 9.3   Recommended variables and variable ranges for Transvarestraint testing of stainless steels and Ni-base alloys[a]

Variable	Range
Minimum specimen length	3.5 in. (89 mm)
Minimum specimen width	3.0 in. (76 mm)
Welding process	Gas tungsten arc welding
Arc length	0.05–0.15 in. (1.25–3.8 mm)
Maximum voltage changes	±1–1.5 V
Current	160–190 A
Travel speed	4–6 in./min (1.7–2.5 mm/s)
Augmented strain	0.5–7%
Ram travel speed	6–10 in./s (152–254 mm/s)

[a]From Ref. [3].

stainless steels that solidify as austenite and Ni-base alloys exhibit SCTR values from 50 to over 200°C. The SCTR data allows a straightforward comparison of cracking suscep- tibility. These values may also allow alloy selection based on restraint conditions. For example, in high-restraint situations, SCTR values below 100°C may be required to prevent cracking, while for low-restraint weldments, 150–200°C may be sufficient.

Finton [3] used a statistical approach to evaluate the variables associated with transverse Varestraint testing. This study used both austenitic stainless steels (Types 304 and 310) and Ni-base alloys (Alloys 625 and 690) to determine the statistical impor- tance of different variables and to establish variable ranges in which testing should be conducted to give reproducible results. Based on this study, Finton recommended the variable ranges in Table 9.3 for use with stainless steels and Ni-base alloys. It should be noted that this testing was conducted using 0.25 in. (6.35 mm) thick material. The weld- ing parameter range was designed to produce a weld bead that was approximately 0.375 in. (10 mm) wide on the plate surface. The ram travel speed (bending rate) is set relatively high in order to establish a fixed temperature gradient at the solid–liquid inter- face at the instant of bending. Actual test results showed that there was little difference in MCD over the bending rate range of 2–10 in./s for the alloys tested.

9.3.2   Technique for Quantifying HAZ Liquation Cracking

Susceptibility to HAZ liquation cracking can be quantified using both the Varestraint test and the hot ductility test, as described by Lin et al. [4]. The Varestraint test tech- nique for HAZ liquation cracking differs from that described previously for weld solidification cracking, since it uses a stationary spot weld to generate a stable HAZ thermal gradient and microstructure. The technique developed by Lin et al. used both an "on-heating" and "on-cooling" approach to quantify HAZ liquation cracking. A schematic of the test is shown in Figure 9.6.

The on-heating test is conducted by initiating the GTA spot weld, ramping to the desired current level, and then maintaining this current level until the weld pool size stabilizes and the desired temperature gradient in the HAZ is achieved. For the

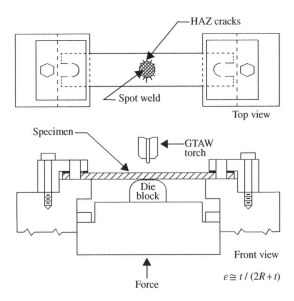

FIGURE 9.6 Schematic illustration of the spot Varestraint test (From Ref. [4]. © AWS).

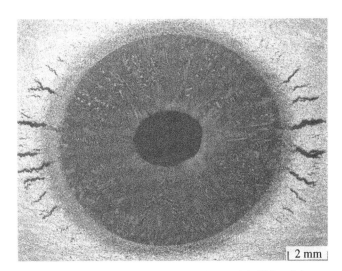

FIGURE 9.7 Spot Varestraint test sample (plan view) of A-286 stainless steel tested at 5% strain.

austenitic stainless steels evaluated by Lin *et al.* [4], a spot weld surface diameter of approximately 12 mm was achieved after a weld time at constant current of 35 s. The arc is then extinguished and the load immediately applied to force the sample to conform to the die block. By using no delay time between extinction of the arc and application of load, HAZ liquation cracks initiate at the fusion boundary and propagate back into the HAZ along liquated grain boundaries (Fig. 9.7).

(a)

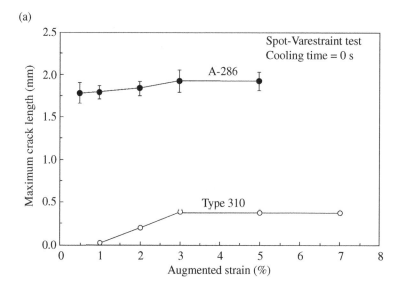

(b)

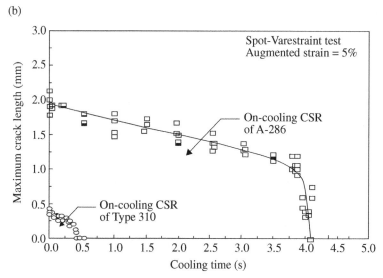

FIGURE 9.8 On-heating (a) and on-cooling (b) HAZ liquation cracking in Type 310 and A-286 as measured by the spot Varestraint test (From Ref. [4]. © AWS).

By plotting MCL versus strain, a "saturated strain" can be determined, which defines the strain above which the MCL does not change. Examples of on-heating MCL versus strain plots for Type 310 and A-286 stainless steels are shown in Figure 9.8a. Note that a threshold strain for cracking cannot be identified for A-286 and that both materials achieve saturated strain at 3%.

For the on-cooling test, the same procedure as described earlier is used, but after the arc is extinguished, there is a delay before the sample is bent. By controlling the delay, or cooling, time, the weld is allowed to solidify, and the temperature in the

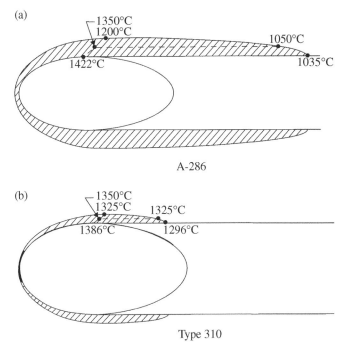

FIGURE 9.9 The thermal crack-susceptible region (CSR) as determined from the spot Varestraint test for (a) A-286 and (b) Type 310 (From Ref. [4]. © AWS).

partially melted zone (PMZ) drops until eventually the liquid films along grain boundaries are completely solidified. By plotting MCL versus cooling time, the time required for liquid films in the PMZ to solidify can be determined. This is shown in Figure 9.8b. Note that over 4 s is required before cracking disappears in A-286, indicating that the grain boundary liquid films persist to quite low temperatures.

By measuring the temperature gradient in the HAZ using implanted thermocouples, it is possible to determine the thermal CSR surrounding the weld within which HAZ liquation cracking is possible. As described by Lin *et al.* [4], this is done by converting the MCL at saturated strain for both on-heating and on-cooling spot Varestraint tests to a temperature by multiplying the MCL by the temperature gradient in the HAZ. Using this approach, it is possible to describe the region around a moving weld pool that is susceptible to HAZ liquation cracking.

Plots of the thermal CSR for A-286 and Type 310 are shown in Figure 9.9. Note that the width of the CSR at the periphery of the weld is 222°C for A-286 and only 61°C for Type 310. It is also interesting that, based on the on-cooling data, the PMZ in A-286 does not fully solidify until the temperature reaches 1035°C (1895°F), while for Type 310, it is 1296°C (2365°F). This technique provides a quantifiable method to determine the precise temperature ranges within which cracking occurs and allows differences in HAZ liquation cracking susceptibility among materials to be readily measured. Although this procedure was developed using austenitic stainless steels, it works equally well with Ni-base alloys.

9.4   THE CAST PIN TEAR TEST

The cast pin tear test (CPTT) was originally developed by Hull [5] to evaluate the solidification cracking susceptibility of a wide range of stainless steel compositions and allowed the development of equivalency relationships for predicting ferrite content and solidification cracking susceptibility of these steels. Small samples of the desired composition were levitation melted and then poured into molds and allowed to solidify. By controlling the length and width of the molds, the restraint in the solidified "pins" could be varied, and a relationship between pin length and onset of cracking was used to determine cracking susceptibility. A modified version of the CPTT was recently developed at OSU and has been used to evaluate the weld solidification cracking susceptibility of a number of steels, Ni-base alloys, and dissimilar combinations of these materials.

The original motivation for the development of the modified CPTT was to devise a self-restraint test that could be used to test alloys that are highly susceptible to weld solidification cracking. With proper mold design, it has now been shown that this test can be used for many conventional materials that have moderate susceptibility to solidification cracking. This test was also developed as a method to facilitate alloy development. Since the CPTT requires very little material, cracking susceptibility can be determined with as little as 500 g of material. The test can also be used to study dilution effects by melting desired mixtures of dissimilar materials.

The original CPTT developed at OSU uses a water-cooled copper hearth and an arc welding torch to melt and contain small charges of material. Recently, an induction melting coil has been added to the system that allows levitation and induction melting of a small button (10–20 g) made using the arc melting system. A schematic of this system is shown in Figure 9.10. This system has the added advantage of temperature monitoring of the molten charge temperature. The apparatus can be programmed such that casting can be initiated when the sample reaches a preprogrammed temperature, usually on the order of 100°C above the liquidus for most materials. All melting and casting is conducted in an inert atmosphere.

Once the molten charge achieves the desired temperature, it is dropped into copper molds ranging in length from 0.5 to 2.0 in. (12.5–50 mm). The mold contains a "foot" feature at the bottom, which anchors the pin, and solidification proceeds upward to the head of the pin. When solidification restraint becomes high enough, small circumferential cracks form on the outside surface of the pin, and with increased pin length, complete separation due to solidification cracking can occur. Selection of different copper-based mold materials to adjust the thermal conductivity of the mold allows some control over the cooling rate. In general, the cooling rates achievable in the CPT tests are equivalent to those experienced during actual arc welding. A detailed description of this methodology is provided elsewhere [6, 7].

A typical cast pin is shown in Figure 9.11. Standard sample diameter is 0.375 in. (9.5 mm) between the foot and head of the pin. Since the pin solidifies from the bottom up, solidification cracks tend to form just below the head of the pin, as shown in Figure 9.11. Cracking data from this test is plotted as percent circumferential cracking versus pin length, as illustrated by the schematic in Figure 9.12. Materials that

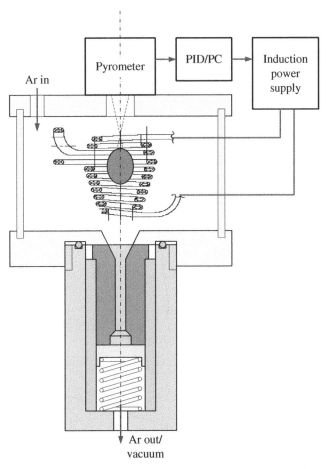

FIGURE 9.10   The OSU cast pin tear test (CPTT) arrangement using levitation melting.

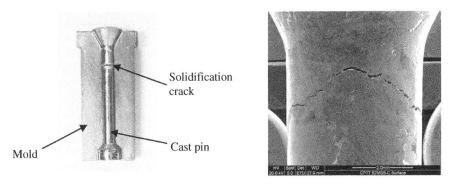

FIGURE 9.11   Example of cast pin with cracking near top of the pin.

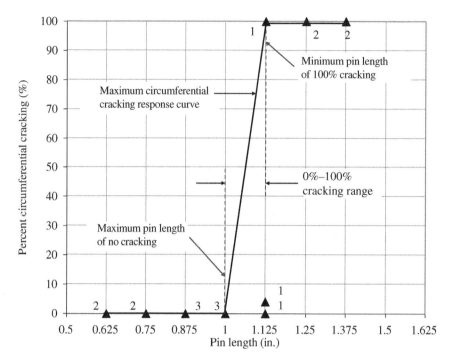

FIGURE 9.12   Evaluation method for the cast pin tear test (From Ref. [7]. © Springer).

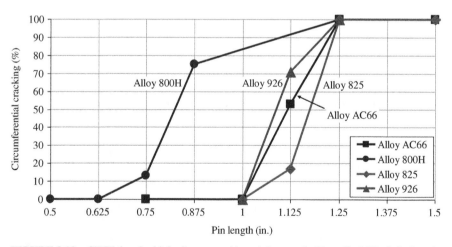

FIGURE 9.13   CPTT data for high-alloy austenitic stainless steels (From Ref. [7]. © Springer).

transition to 100% circumferential cracking at the shortest pin lengths are most susceptible to solidification cracking. Results for several high-alloy austenitic stainless
steels are shown in Figure 9.13 [7].

As noted previously, a major advantage of the modified CPTT is that very little
material is required to determine cracking susceptibility. The cast pins range in

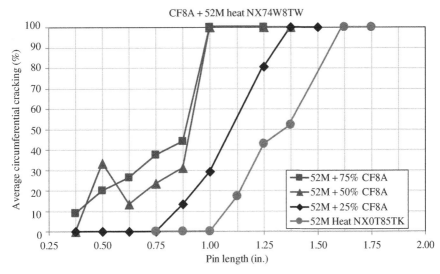

FIGURE 9.14 CPTT results showing effect of dilution of Ni-base Filler Metal 52M by austenitic stainless steel casting CF8A (From Ref. [7]. © Springer).

weight from approximately 10 to 16 g (0.5–2.0 in. pin). Because of this, an entire evaluation program can be conducted with less than 500 g of material. This is a tremendous advantage over such tests as the Varestraint and PVR tests, which requires experimental materials to be processed to plate or sheet form in order to be tested.

The CPT apparatus can also be used to determine the effect of dilution on cracking susceptibility. The CPTT results in Figure 9.14 show the effect of dilution of Ni-base Filler Metal 52M (ERNiCrFe-7A) by an austenitic stainless steel cast alloy (CF8A). As has been experienced in actual practice, high levels of dilution (>30%) by the stainless steel can lead to solidification cracking during overlay of 52M on the stainless steel.

9.5 THE HOT DUCTILITY TEST

Elevated temperature ("hot") ductility can provide some indication of a material's weldability, since cracking is usually associated with an exhaustion of available ductility. Most hot ductility testing involves both heating (or "on-heating") and cooling (or "on-cooling") tests. In order to properly simulate the heating and cooling rates associated with welding, special equipment has been developed to rapidly heat and cool small laboratory samples. The most widely used test machine for accomplishing this was developed by Savage and Ferguson at RPI in the 1950s [8]. The inventors called this machine the "Gleeble™," a namesake whose origin is a closely guarded secret of RPI graduates. The machine is now produced commercially by Dynamic Systems, Inc. in Poestenkill, NY [9].

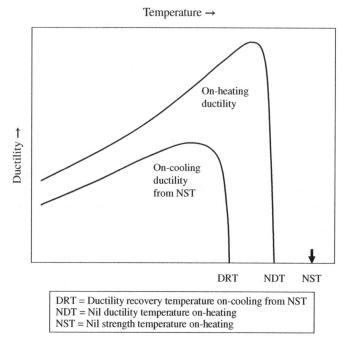

FIGURE 9.15   Schematic illustration of a hot ductility signature showing on-heating and on-cooling ductility curves and the location of the nil ductility temperature (NDT), nil strength temperature (NST), and ductility recovery temperature (DRT).

The Gleeble uses resistance heating (I^2R) to heat small samples under precise temperature control and cools them by conduction through water-cooled copper grips. Quench systems are also available for accelerated cooling. Heating rates up to 10,000°C/s are possible. The Gleeble also is capable of mechanically testing samples at any point along the programmed thermal cycle.

Hot ductility testing allows a ductility "signature" to be developed for a material. This signature will exhibit several distinct features, as shown schematically in Figure 9.15. When ductility is measured on heating, most materials will exhibit an increase in ductility with increasing temperature followed by a fairly abrupt drop. This drop is associated with the onset of melting. The temperature at which ductility drops to zero is termed the nil ductility temperature (NDT). At the NDT, the material still has measurable strength. Additional tests are conducted to determine the point at which the strength is zero, termed the nil strength temperature (NST).

To determine the on-cooling ductility curve, samples are heated to a temperature between NDT and NST, cooled to a preprogrammed temperature, and tested. The point at which measurable ductility is observed (~5%) is termed the ductility recovery temperature (DRT). At the DRT, liquid formed upon heating to the NST has solidified to the point that the sample exhibits measurable ductility when pulled in tension.

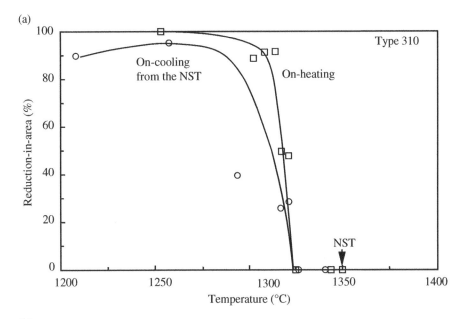

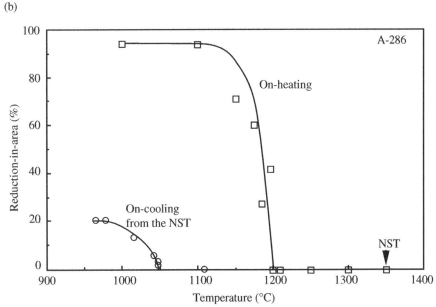

FIGURE 9.16 Hot ductility test results for (a) Type 310 and (b) A-286 austenitic stainless steels (From Ref. [4]. © AWS).

Examples of hot ductility test results for stainless steel alloys Type 310 and A-286 from the work of Lin *et al.* [4] are shown in Figure 9.16. These are the same materials that were described previously for the spot Varestraint test. Note that the on-heating and on-cooling ductility curves for Type 310 are nearly identical and

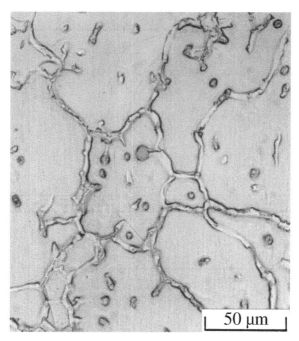

FIGURE 9.17   A-286 after heating to the nil strength temperature (1350°C) (From Ref. [4]. ©AWS).

that the NDT and DRT are essentially equivalent. The NST is only 25°C above the NDT. This indicates that grain boundary liquid films that form between the NDT and NST on-heating solidify on cooling at approximately 1325°C and ductility quickly recovers. Because of the narrow temperature range over which liquid films are present in the HAZ/PMZ, this material would have good resistance to HAZ liquation cracking. In contrast, the on-heating and on-cooling ductility curves are quite different for A-286. The NDT for A-286 is approximately 1200°C, while the NST is 1350°C. After heating to the NST, on-cooling ductility does not recover until approximately 1050°C. Thus, grain boundary liquid films are present over a 300°C temperature range below the NST. This produces a very wide HAZ/PMZ region and renders the material susceptible to HAZ liquation cracking. A photomicrograph of an A-286 sample heated to the NST temperature is shown in Figure 9.17. Note the complete coverage of the grain boundaries by liquid films.

Similar to the spot Varestraint test, the hot ductility results (NDT, NST, and DRT from the hot ductility curves) can be used to define the CSR in the HAZ. In advance of the weld, liquation will begin when the material is heated above the NDT. The difference between the NDT and the liquidus temperature at the fusion boundary then defines the width of the on-heating liquation temperature range. For convenience, the NST is used to approximate the liquidus temperature.

In the cooling portion of the HAZ, liquid films in regions heated between NDT and NST will solidify. The value of DRT varies as a function of the peak temperature and is the lowest at the point closest to the fusion boundary. The difference between NST and DRT will then represent the maximum temperature range over which liquid films are present in the HAZ. This value can then be used to define the liquation cracking temperature range (LCTR). The LCTR represents a measure of liquation cracking susceptibility, similar to SCTR for the weld metal. From Figure 9.16, LCTR values for Type 310 SS and A-286 are 25 and 300°C, respectively.

Unfortunately, there is no standardized procedure for generating hot ductility curves. ISO 17641-3 describes the hot ductility test and the on-heating and on-cooling behavior but does not provide specifics in terms of how the curves are generated. The following summarizes the procedure that is currently used in the Welding and Joining Metallurgy Laboratory at OSU for generating hot ductility data:

1. Standard specimens are 0.25 in. (6.35 mm) in diameter and 4.0 in. (100 mm) long with threaded ends.

2. Specimen free span (Gleeble jaw spacing) is 1.0 in. (25 mm). Tests are normally run in a vacuum chamber with argon shielding. This protects the sample from excessive oxidation so that fracture surface analysis can be conducted if needed. The chamber is evacuated with the roughing pump (10^{-2} torr), backfilled with argon, and then repeated.

3. NST is determined by heating at linear rate of 200°F/s (111°C/s) under a static load of 10 kg until the sample fails. This test is conducted at least twice and the average value computed.

4. On-heating tests are conducted by heating to the desired peak temperature at a rate of 200°F/s (111°C/s) and then pulling the samples to failure at a rate of 2 in./s (50 mm/s). The sample is held at the test temperature for 1 s prior to testing.

5. On-cooling tests are performed after heating the sample to the NST at 200°F/s (111°C/s) and then cooling to the desired temperature at 50°C/s. On-cooling samples were also pulled to failure at a rate of 2 in./s. Again, a 1 s hold at the test temperature is used.

It should be noted that step 5 may need to be altered for some materials, since some alloys undergo extensive liquation at the NST. This can be a particular problem with some Ni-base alloys where extensive liquation can lead to cracking during cooling prior to reaching the desired on-cooling test temperature. In these cases, a peak temperature midway between the NST and NDT should be selected. Peak temperatures for on-cooling tests that are at or below NDT are not appropriate because insufficient grain boundary liquation occurs and the microstructure will not be representative of the PMZ that forms in many alloys.

The hot ductility test can also provide information on solid-state ductility losses, such as those that result in ductility-dip cracking (DDC). For stainless steels and Ni-base alloys, ductility dips may be observed in the temperature range from 800 to 1100°C (1470–2010°F). Although the hot ductility test may be an indicator of potential DDC susceptibility, the effects can be quite subtle. The strain-to-fracture (STF) test, described in the next section, was developed to provide better information regarding DDC susceptibility.

Other investigators have used different testing conditions, particularly heating and cooling rates and stroke rates during testing. Again, there is no standardized procedure for hot ductility testing, and the effects of the testing variables have not been comprehensively evaluated. The procedures recommended here represent "best practices" determined at OSU.

9.6   THE STRAIN-TO-FRACTURE TEST

In order to better quantify susceptibility to DDC, the STF test was developed in 2002 by Nissley and Lippold at OSU [10, 11]. The STF test employs a "dogbone" tensile sample with a GTA spot weld applied in the center of the gage section. The spot weld is made under controlled solidification conditions using welding current downslope control. This results in an essentially radial array of grain boundaries within the spot weld. A schematic illustration of a STF sample is shown in Figure 9.18.

Samples are then tested in a Gleeble thermomechanical simulator at different temperatures and strains. For stainless steels and Ni-base alloys, temperature and strain ranges are typically from 650 to 1200°C (1200–2190°F) and 0 to 20%, respectively. After testing at a specific temperature–strain combination, the sample is examined under a binocular microscope at 30× to determine if cracking has occurred and any cracks observed at this magnification are simply counted.

Using this data, a temperature versus strain envelope is developed that defines the regime within which DDC may occur. Both a threshold strain for cracking (ε_{min}) and ductility-dip temperature range (DTR) can be extracted from these

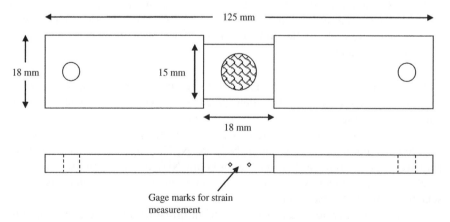

FIGURE 9.18   Schematic of the strain-to-fracture test sample with GTA spot weld in the gage section.

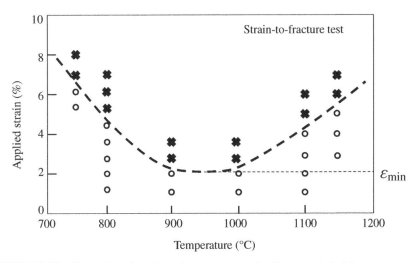

FIGURE 9.19 Illustration of strain-to-fracture test results. Open symbols (o) represent samples with no cracks, and crossed symbols (×) represent samples where cracks are observed.

curves. A schematic STF curve across the entire DDC temperature range is shown in Figure 9.19. Note that a range of strains and temperatures must be investigated to develop the entire curve.

Since the introduction of this test in 2002, the method for assessing susceptibility to DDC has evolved in order to better differentiate materials. Many of the austenitic (fcc) materials exhibit ε_{min} values in the range from 2 to 4% and have similar DTRs—normally in the range from 700 to 1150°C. However, when evaluating the increase in the number of cracks as a function of increasing strain above the threshold strain, a significant difference in behavior has been observed, particularly in Ni-base filler metals. It has also been noted that the minimum in the strain versus temperature curve for most materials occurs at approximately 950°C (1740°F). For this reason, a new measure for assessing susceptibility to DDC using the STF test has been used, which simply tests the material over a range of strain at a single temperature (950°C). Using this approach, both a threshold strain for cracking and the rate of increase in cracking with applied strain are identified. This behavior is shown schematically in Figure 9.20 (same as Fig. 4.12). Materials with the highest threshold strain to initiate cracking and/or the shallowest slope for number of cracks versus applied strain represent the most DDC-resistant materials. This test technique and the methods to interpret results continue to be optimized as an effective technique for assessing susceptibility to DDC.

9.7 REHEAT CRACKING TEST

Reheat cracking is generally encountered during postweld heat treatment (PWHT) of steels and Ni-base superalloys and includes stress relief cracking (steels) and strain-age cracking (SAC) (Ni-base alloys) as described in Chapter 4. Numerous tests have

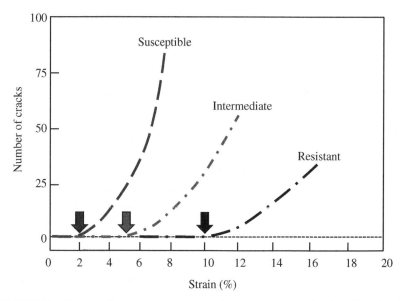

FIGURE 9.20 Strain-to-fracture test results at a single temperature showing DDC threshold and increase in cracking with applied strain. The arrows represent the threshold strain for cracking (ε_{min}) for different materials.

been developed to quantify susceptibility to this general form of cracking, including both self-restraint and simulative tests. Some of these were briefly described in Sections 4.3.6 and 4.4.3.

The approach described here was developed by Norton and Lippold [12] at OSU in an attempt to more closely simulate the actual conditions experienced by the HAZ during welding. The Gleeble is used to impose a simulated thermal cycle on the specimen. After reaching the peak temperature, the sample is cooled to room temperature with the application of a small amount of strain in order to achieve a net residual stress in the sample at room temperature. The amount of strain (stroke) applied during cooling is carefully controlled in order to generate stress levels that are slightly below the yield strength of the material at room temperature. The Gleeble jaws are then locked in place and the sample reheated to a desired PWHT temperature and held for a period of up to 4 h. Stress (load) on the sample is continuously monitored. A schematic of the thermal and mechanical cycles and resultant stress (load) response of a sample susceptible to reheat cracking is shown in Figure 9.21.

As illustrated, considerable stress relaxation occurs upon heating to the PWHT temperature, and then the stress in the sample begins to increase with time at the PWHT temperature (see Figure 4.35 as an example). This is the result of a precipitation reaction that is characteristic of materials that are susceptible to reheat cracking. If the stress level is high enough, the sample will fail and the test ends. In many cases, however, the sample does not fail. In the early stages of test development, it was found that tremendous (and unacceptable) scatter is characteristic of this testing

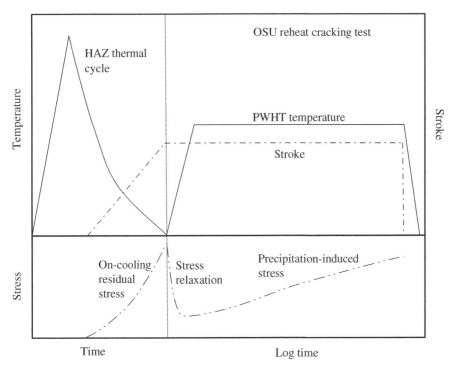

FIGURE 9.21 OSU reheat cracking test.

approach as with others that use similar "hold until failure" techniques. For example, under identical conditions, one sample may fail in an hour, while another may not fail after 10 h.

Rather than use a failure time criterion for quantifying reheat cracking susceptibility, Norton developed a procedure such that after variable hold times (1–4 h) at different temperatures, the sample was pulled to failure at the test temperature and the ductility measured. The hot ductility following PWHT was used to develop a multivariate polynomial for calculating the ductility as a function of PWHT temperature and time. The collected data for susceptible materials exhibited a parabolic curve, so the model was chosen to be a second-order polynomial. Using this approach, ductility versus temperature curves can be developed as a function of PWHT time. An example of ductility versus temperature curves for Waspaloy and Alloy 718 determined using this method is shown in Figure 9.22 (same as Fig. 4.36). The coefficients of determination (R^2) for the Waspaloy and Alloy 718 surface plot polynomials are 0.92 and 0.91, respectively. The regression models show good fit to the measured data over the range of tested times and temperatures both by their high coefficients of determination and the ability to predict the ductility of samples. The curves in Figure 9.22 are consistent with actual experience in welding these alloys. Alloy 718 is generally quite resistant to SAC, while Waspaloy is considered moderately susceptible (see Figs. 4.32 and 4.34).

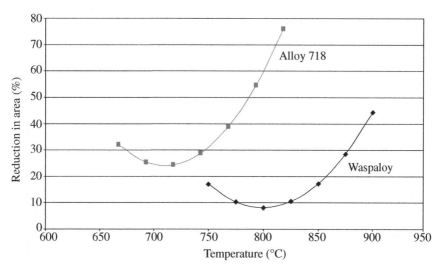

FIGURE 9.22 Results of reheat cracking test for Ni-base alloys 718 and Waspaloy (From Ref. [12]. © ASM).

It should be noted that this test is not limited to the evaluation of Ni-base superalloys but can also be applied to other materials that are susceptible to PWHT cracking, such as Cr–Mo–V steels and stainless steels (Type 347). This test can also potentially be used to evaluate the susceptibility to relaxation cracking, which occurs after extended service exposure at elevated temperature.

As noted previously, one issue with this test is that the sample does not always fail. With the newest generation of Gleeble thermomechanical simulators, it is possible to switch between stroke control (controlling the motion of the jaws) and load control. Using this capability, it is possible to apply a controlled load to the sample as it is heated to the PWHT temperature. Ideally, the load (stress) is maintained at a level below the yield stress of the material during the heating cycle. When the desired temperature is reached, the stroke control is reestablished and the jaws are fixed in place. This potentially allows the effect of restraint level to be evaluated as a function of temperature. Such an approach is currently under investigation in the Welding and Joining Metallurgy Laboratory at OSU.

9.8 IMPLANT TEST FOR HAZ HYDROGEN-INDUCED CRACKING

The original implant test was developed by Henri Granjon at the French Welding Institute (Institut de Soudure) in the 1960s. It was devised as a simple laboratory test to evaluate the susceptibility of the base metal HAZ to HIC. There have been many modifications to this test since it was originally introduced, but perhaps the most significant was the use of a helical notch rather than a fixed notch location to insure that equal stress concentration existed throughout the entire HAZ. This allows failure to occur in the most susceptible region of the HAZ. The test is described in ISO 17642-3.

(a)   (b)

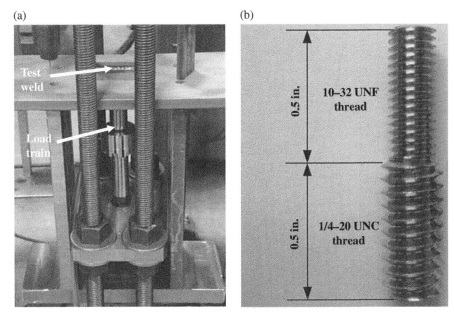

FIGURE 9.23 OSU implant test setup. (a) Loading fixture and (b) sample (Courtesy of Xin Yue).

The sample geometry and test configuration used at OSU are shown in Figure 9.23. The sample is preplaced into a hole drilled in plain-carbon steel plate, and then a weld is deposited over the sample to create a HAZ microstructure in the implant test sample. After cooling to room temperature, a constant load is applied using a simple hydraulic mechanical system with a feedback loop to maintain the desired load. The test ends when the sample fails or after 24 h if there is no failure. By using different loads, a relationship between load (stress) and failure time is established, as shown in Figure 9.24 (same as Fig. 5.23).

The implant test can be used to investigate the effect of consumable type, hydrogen level, preheat, heat input, and other variables on HAZ HIC. It is an excellent weldability test technique to use in any alloy development study, since the amount of material is minimal. Implant test samples weigh only a few grams. More details regarding the application of this test technique can be found elsewhere [13, 14].

9.9 GAPPED BEAD-ON-PLATE TEST FOR WELD METAL HIC

Most of the test procedures developed for HIC have focused on cracking in the HAZ. With the recent introduction of low-carbon, high-strength steels, susceptibility to HAZ HIC has decreased and the potential for weld metal HIC has increased. One of the few tests designed specifically to study weld metal HIC is the gapped bead-on-plate (G-BOP) test introduced by McParlan and Graville [15, 16]. This test was evaluated by Marianetti at OSU, who studied a number of test variables and recommended a test procedure for the G-BOP test [17].

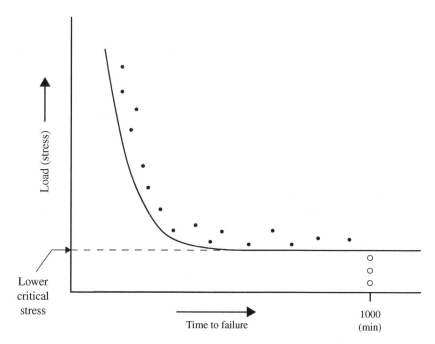

Note: Solid dots = failure.
       Open dots = no Failure.

FIGURE 9.24 Typical load (stress) vs. time-to-failure relationship for the implant test.

The G-BOP test assembly consists of two large steel test plates with a narrow gap (slot) machined at the intersection. The large plate thickness is necessary to prevent distortion during welding, thereby insuring high restraint at the root of the weld made over the gap between the plates. Fortunately, since this is a weld metal test, the plates can be low-cost, plain-carbon steel. A buttering technique can be used to eliminate the effect of base metal dilution. A schematic of the G-BOP test sample is shown in Figure 9.25 (same as Fig. 5.20) [18].

As with other weldability tests, there is considerable variation in the manner in which the test is performed. Important variables include the method of clamping the specimens, clamping force, incubation time after welding (time allowed for HIC to occur), and the technique for heat tinting following incubation time to reveal the extent of HIC. Marianetti systematically evaluated each of these variables in order to understand their influence [17]. Based on his results, he made the following recommendations regarding the test technique:

1. The test block geometry shown in Figure 9.25 is recommended. Variation in gap width from 0.375 to 1.625 mm was found to have little effect on the test results. A gap width 0.75 mm provides reproducible results.

2. The method of clamping can significantly affect the results, especially when preheat is applied to the clamped test blocks. A more uniform clamping force can be

applied and maintained by using a 0.5 in. (13 mm) bolt for clamping (as shown in Fig. 9.25). A torque of 110 ft-lbs is used to apply the clamping force.

3. An incubation time (hold time after welding) of 24 h is recommended. It was determined that most of the cracking occurs within the first hour after welding.

4. Heat tinting to reveal the crack surface resulting from testing should be performed while the test blocks are still clamped together. Releasing the clamping force prior to heat tinting can lead to some crack extension due to the "hinging" effect of the test blocks.

5. Heat tinting is accomplished using an oxyacetylene torch and heating the test area to a dull red color (~500°C). The test blocks are then separated, and the fraction of cracked and uncracked weld cross section is measured. An example is shown in Figure 9.26.

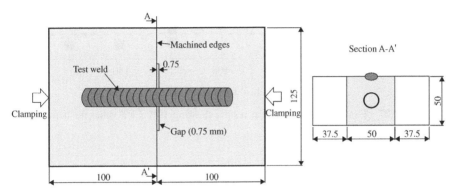

FIGURE 9.25   G-BOP test sample configuration. The circle in Section A-A′ represents the location of the bolt recommended by Marianetti [17] for loading the sample (Modified from Ref. [18]. © Springer).

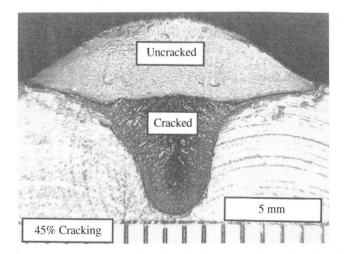

FIGURE 9.26   Weld cross section after heat tinting to reveal area fraction of hydrogen cracking (From Ref. [17]).

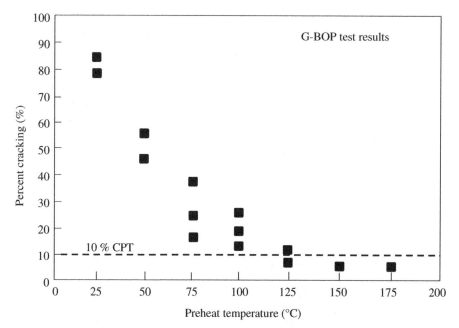

FIGURE 9.27 Example of G-BOP test results showing 10% CPT value for determining minimum preheat. (From Ref. [17]).

Percent cracking is then plotted versus preheat temperature to determine the effect of preheat, as shown in the schematic in Figure 9.27. A value of 10% crack preheat temperature (10% CPT) or the preheat temperature where cracking is 10% is then used as a recommended <u>minimum</u> preheat to prevent weld metal HIC. Based on the results shown in Figure 9.27, a minimum preheat temperature of 150°C would be recommended.

9.10 OTHER WELDABILITY TESTS

There are many other weldability tests that have not been described in this chapter. A comprehensive review of tests for hydrogen cracking has been published by Kannengiesser and Boellinghaus [18]. Other test techniques for hot cracking and solid-state cracking have been referenced in Chapters 3 and 4, respectively. Again, most of these tests have not been standardized, and test procedures may vary widely among different laboratories. While weldability testing can provide valuable information regarding the fabrication and service performance of materials, caution should be used in applying these results and/or making comparisons to data that has been published in the open literature.

REFERENCES

[1] Savage WF, Lundin CD. The Varestraint test. Weld J 1965;44 (10):433s–442s.

[2] Lippold JC, Lin W. Weldability of commercial Al–Cu–Li alloys. Proceedings of ICAA5. Aluminum Alloys—their Physical and Mechanical Properties. In: Driver JH, *et al.*, editors. Zurich-Uetikon: Transtec Publications; 1996. p 1685–1690.

[3] Finton T. Standardization of the Transvarestraint test [MS Thesis]. Columbus, OH: Ohio State University; 2003.

[4] Lin W, Lippold JC, Baeslack WA III. An investigation of heat-affected zone liquation cracking, Part 1: A methodology for quantification. Weld J 1993;71 (4):135s–153s.

[5] Hull FC. Cast-pin tear test for susceptibility to hot cracking. Weld J 1959;38 (4):176s–181s.

[6] Alexandrov BT, Nissley NE, Lippold JC. Evaluation of weld solidification cracking in Ni-base superalloys using the cast pin tear test. 2nd International Workshop on Hot Cracking; May 2008; Berlin. Berlin/Heidelberg: Springer-Verlag. p 193–214.

[7] Alexandrov BT, Lippold JC. Use of the cast pin tear test to study solidification cracking. Weld World 2013;57 (5):635–648.

[8] Nippes EF, Savage WF. An investigation of the hot ductility of high-temperature alloys. Weld J 1955;34 (4):183s–196s.

[9] Poestenkill NY. Dynamic Systems Inc. Available at www.gleeble.com. Accessed July 5, 2014.

[10] Nissley NE, Lippold JC. Ductility-dip cracking susceptibility of austenitic alloys. Proceedings of the 6th International Conference on Trends in Weld Research, April 15–19, 2002 Callaway Gardens, GA. Materials Park, OH: ASM International; 2003. p 64–69.

[11] Nissley NE, Lippold JC. Development of the strain-to-fracture test for evaluating ductility-dip cracking in austenitic alloys. Weld J 2003;82 (12):355s–364s.

[12] Norton SJ, Lippold JC. Development of a Gleeble-based test for postweld heat treatment cracking susceptibility. Proceedings of the 6th International Conference on Trends in Welding Research, April 15–19, 2002 Callaway Gardens, GA. Materials Park, OH: ASM International; 2003. p 609–614.

[13] Yue X, Lippold JC. Evaluation of heat-affected zone hydrogen-induced cracking in Navy steels. Weld J 2013;92 (1):20s–28s.

[14] Yue X, Feng X, Lippold JC. Effect of diffusible hydrogen level on heat-affected zone hydrogen-induced cracking of high-strength steels. Weld World 2013;58 (1):101–109.

[15] McParlan M, Graville BA. Hydrogen cracking in weld metals. Weld J 1976;55 (4):95s–102s.

[16] McParlan M, Graville BA. Development of the G-BOP test for weld metal cracking. IIW Document IX-922-75; International Institute of Welding, Paris, France, 1975.

[17] Marianetti C. The development of the G-BOP test and the assessment of weld metal hydrogen cracking [MS Thesis]. Columbus, OH: Ohio State University; 1998.

[18] Kannengiesser T, Boellinghaus T. Cold cracking tests: an overview of present technologies and applications. Weld World 2013;57 (1):3–37.

APPENDIX A

COMPOSITION OF SELECTED STEELS

Welding Metallurgy and Weldability, First Edition. John C. Lippold.
© 2015 John Wiley & Sons, Inc. Published 2015 by John Wiley & Sons, Inc.

Base metals

Type	Grade	Composition (wt%) (single values are maxima)								
		C	Mn	Si	Cr	Ni	Mo	V	Cu	Others
A36	—	0.25	0.8–0.12	0.15–0.4	—	—	—	—	—	—
A508	2	0.27	0.5–1.0	0.4	0.25–0.45	0.5–1.0	0.55–0.70	0.05	0.20	—
A508	22	0.11–0.15	0.3–0.6	0.35	2.0–2.5	0.25	0.9–1.1	0.02	0.25	—
A514	F	0.1–0.2	0.6–1.0	0.15–0.4	0.4–0.65	0.7–1.0	0.4–0.6	0.03–0.08	0.15–0.5	B: 0.0005–0.006
A516	65	0.24	0.85–1.2	0.15–0.4	—	—	—	—	—	—
A533	B	0.25	1.15–1.5	0.15–0.40	—	0.4–0.7	0.45–0.6	—	—	—
A543	B	0.20	0.4	0.15–0.4	1.0–1.9	2.25–4.0	0.2–0.65	—	—	—
A588	A	0.19	0.8–1.25	0.3–0.65	0.4–0.65	0.4	—	0.02–0.1	—	—
A710	A	0.07	0.4–0.7	0.4	0.6–0.9	0.7–1.0	0.15–0.25	—	1.0–1.3	Nb: 0.02 min
A945[a]	65	0.10	1.1–1.65	0.1–0.4	0.2	0.5–1.0	0.08	0.1	0.35	Ti: 0.007–0.02
HY-80	—	0.12–0.18	0.1–0.4	0.15–0.35	1.0–1.8	2.0–3.25	0.2–0.6	0.03	0.25	Ti: 0.02
HY-100	—	0.12–0.20	0.1–0.4	0.15–0.35	1.0–1.8	2.0–3.50	0.2–0.6	0.03	0.25	Ti: 0.02
HSLA-100	—	0.06	0.3–0.6	0.4	0.45–0.75	1.5–2.0	0.3–0.55	—	1.0–1.3	Nb: 0.02–0.06
HP9-4-20	—	0.17–0.23	0.2–0.4	0.2	0.65–0.85	8.5–9.5	0.9–1.1	0.06–0.12	—	Co: 4.25–4.75

[a]Also known as HSLA-65.

APPENDIX B

NOMINAL COMPOSITION OF STAINLESS STEELS

Welding Metallurgy and Weldability, First Edition. John C. Lippold.
© 2015 John Wiley & Sons, Inc. Published 2015 by John Wiley & Sons, Inc.

| Type | UNS no. | Microstructure | Nominal composition (wt%) | | | | | | | | | | | |
			C	Mn	Si	Cr	Ni	Mo	Nb	Cu	N	Al	Ti	Others
Cast alloys														
CA6N	Casting	Martensitic	0.04	0.25	0.50	11.50	7.00	—	—	—	—	—	—	—
CA6NM	Casting	Martensitic	0.04	0.50	0.50	12.8	4.0	0.70	—	—	—	—	—	—
CA15	Casting	Martensitic	0.10	0.50	0.75	12.75	—	—	—	—	—	—	—	—
CA15M	Casting	Martensitic	0.10	0.50	0.30	12.75	—	0.60	—	—	—	—	—	—
CD3MCuN	Casting	Duplex	0.02	0.60	0.55	25.30	6.10	3.35	—	1.65	0.28	—	—	—
CD3MN	Casting	Duplex	0.02	0.75	0.50	22.25	5.50	3.00	—	—	0.20	—	—	—
CD4MCu	Casting	Duplex	0.03	0.50	0.50	25.50	5.20	2.00	—	3.00	—	—	—	—
CD4MCuN	Casting	Duplex	0.03	0.50	0.50	25.50	5.20	2.00	—	3.00	0.18	—	—	—
CD6MN	Casting	Duplex	0.04	0.50	0.50	25.50	5.00	2.10	—	—	0.20	—	—	—
CE3MN	Casting	Duplex	0.02	0.75	0.50	25.00	7.00	4.50	—	—	0.20	—	—	—
CE8MN	Casting	Duplex	0.04	0.50	0.75	24.00	9.50	3.80	—	—	0.20	—	—	—
CF3/CF3A	Casting	Austenitic	0.02	0.75	1.00	19.00	10.00	—	—	—	—	—	—	—
CF3M	Casting	Austenitic	0.02	0.75	0.75	19.00	11.00	2.50	—	—	—	—	—	—
CF3MN	Casting	Austenitic	0.02	0.75	0.75	19.50	11.00	2.50	—	—	0.15	—	—	—
CF8/CF8A	Casting	Austenitic	0.04	0.75	1.00	19.50	9.50	—	—	—	—	—	—	—
CF8C	Casting	Austenitic	0.04	0.75	1.00	19.50	10.50	—	0.60	—	—	—	—	—
CF8M	Casting	Austenitic	0.04	0.75	1.00	19.50	10.50	2.50	—	—	—	—	—	—
CF10	Casting	Austenitic	0.07	0.75	1.00	19.5	9.5	—	—	—	—	—	—	—
CF10M	Casting	Austenitic	0.07	0.7	5	1.00	19.5	10.5	2.50	—	—	—	—	—
CF20	Casting	Austenitic	0.10	0.75	1.00	19.50	9.50	—	—	—	—	—	—	—
HC	Casting	Ferritic	0.2	0.5	1.0	28.0	—	—	—	—	—	—	—	—
HD	Casting	Duplex/austenitic	0.2	0.8	1.0	28.0	5.5	—	—	—	—	—	—	—
HK30	Casting	Austenitic	0.30	1.0	1.0	26.0	20.0	—	—	—	—	—	—	—
HK/HK40	Casting	Austenitic	0.40	1.0	1.0	26.0	20.0	—	—	—	—	—	—	—
HP	Casting	Austenitic	0.55	1.0	1.2	26.0	35.0	—	—	—	—	—	—	—

(Continued)

Wrought alloys

Type	UNS no.	Microstructure	C	Mn	Si	Cr	Ni	Mo	Nb	Cu	N	Al	Ti	Others
						Nominal composition (wt%)								
13-8Mo PH	S13800	Martensitic PH	0.03	0.10	0.05	12.75	8.00	2.25	—	—	—	1.12	—	—
15-5 PH	S15500	Martensitic PH	0.04	0.50	0.50	14.75	4.50	—	0.30	3.50	—	—	—	—
17-4 PH	S17400	Martensitic PH	0.0	0.50	0.50	16.25	4.00	—	0.30	4.00	—	—	—	—
17-7 PH	S17700	Semiaustenitic PH	0.05	0.50	0.50	17.00	7.00	—	—	—	—	1.00	—	—
201	S20100	Austenitic	0.08	6.50	0.50	17.0	4.50	—	—	—	0.20	—	—	—
201L	S20103	Austenitic	0.02	6.5	0.40	17.0	4.50	—	—	—	0.20	—	—	—
201LN	S20153	Austenitic	0.02	7.0	0.40	16.8	4.5	—	—	—	0.20	—	—	—
Gall-Tough™	S20161	Austenitic	0.08	5.0	3.5	16.5	5.0	—	—	—	0.14	—	—	—
—	S20162	Austenitic	0.08	6.0	3.5	18.8	8.0	1.50	—	—	0.15	—	—	—
202	S20200	Austenitic	0.08	8.8	0.50	18.00	5.00	—	—	—	0.20	—	—	—
204	S20400	Austenitic	0.02	8.0	0.50	16.0	2.25	—	—	—	0.22	—	—	—
Nitronic 30	S20400	Austenitic	0.02	8.0	0	16.0	1.6	—	—	—	—	0.23	—	—
205	S20500	Austenitic	0.18	14.8	0.50	17.2	1.4	—	—	—	0.36	—	—	—
Nitronic 50	S20910	Austenitic	0.04	5.00	0.40	22.00	12.50	2.25	0.20	—	0.30	—	—	V: 0.20
Nitronic 60	S21800	Austenitic	0.05	8.00	4.00	17.00	8.50	—	—	—	0.13	—	—	—
Nitronic 40	S21900	Austenitic	0.04	9.00	0.50	20.2	6.50	—	—	—	0.28	—	—	—
Nitronic 33	S24000	Austenitic	0.04	13.00	0.40	18.00	3.00	—	—	—	0.30	—	—	—
Nitronic 32	S28200	Austenitic	0.08	18.00	0.50	18.00	—	1.00	—	1.00	0.50	—	—	—
301L	S30103	Austenitic	0.02	1.00	0.50	17.00	7.00	—	—	—	0.10	—	—	—
301LN	S30153	Austenitic	0.02	1.00	0.50	17.00	7.00	—	—	—	0.15	—	—	—
302	S30200	Austenitic	0.08	1.00	0.40	18.00	9.00	—	—	—	—	—	—	—
303	S30300	Austenitic	0.08	1.00	0.50	18.00	9.00	—	—	—	—	—	—	S: 0.20
303Se	S30323	Austenitic	0.08	1.00	0.50	18.00	9.00	—	—	—	—	—	—	Se: 0.2, S: 0.13
304	S30400	Austenitic	0.04	1.00	0.40	19.00	9.25	—	—	—	—	—	—	—
304L	S30403	Austenitic	0.02	1.00	0.40	19.00	10.00	—	—	—	—	—	—	—

Grade	UNS No.	Type	C	Mn	Si	Cr	Ni	Mo		Cu	N	Ti	Other
304H	S30409	Austenitic	0.07	1.00	0.40	19.00	9.25	—	—	—	—	—	—
304N	S30451	Austenitic	0.04	1.00	0.40	19.0	9.2	—	—	—	0.13	—	—
304LN	S30453	Austenitic	0.02	1.00	0.40	19.00	10.00	—	—	—	0.13	—	—
—	S30454	Austenitic	0.02	1.00	0.50	19.0	9.5	—	—	—	0.23	—	—
308	S30800	Austenitic	0.04	1.00	0.50	20.00	11.00	—	—	—	—	—	—
309	S30900	Austenitic	0.10	1.00	0.40	23.00	13.50	—	—	—	—	—	—
309S	S30908	Austenitic	0.04	1.00	0.40	23.00	13.50	—	—	—	—	—	—
309H	S30909	Austenitic	0.07	1.00	0.50	23.00	13.50	—	—	—	—	—	—
301	S30100	Austenitic	0.08	1.00	0.75	17.00	7.00	—	0.05	—	—	—	—
310	S31000	Austenitic	0.15	1.00	0.75	25.00	20.50	—	—	—	—	—	—
310S	S31008	Austenitic	0.04	1.00	0.40	25.00	20.50	—	—	—	—	—	—
310H	S31009	Austenitic	0.07	1.00	0.40	25.00	20.50	—	—	—	—	—	—
254SMo	S31254	Austenitic	0.01	0.50	0.3	20.00	18.00	6.25	—	0.75	0.20	—	—
27-7Mo	S31277	Austenitic	0.01	1.50	0.40	21.8	27.0	7.2	—	1.0	0.35	—	—
DP-3™	S31260	Duplex	0.02	0.50	0.50	25.0	6.5	3.0	—	0.50	0.20	—	W: 0.30
314	S31400	Austenitic	0.15	1.00	2.25	24.50	20.50	—	—	—	—	—	—
316	S31600	Austenitic	0.04	1.00	0.40	17.00	12.00	2.20	—	—	—	—	—
316L	S31603	Austenitic	0.02	1.00	0.40	17.00	12.00	2.20	—	—	—	—	—
316H	S31609	Austenitic	0.07	1.00	0.40	17.00	12.00	2.20	—	—	—	—	—
316Ti	S31635	Austenitic	0.04	1.00	0.40	17.00	12.00	2.20	—	—	—	0.50	—
316N	S31651	Austenitic	0.04	1.00	0.40	17.00	12.00	2.20	—	—	0.13	—	—
316LN	S31653	Austenitic	0.02	1.00	0.40	17.00	12.00	2.20	—	—	0.13	—	—
317	S31700	Austenitic	0.04	1.00	0.40	19.00	13.00	3.30	—	—	—	—	—
317L	S31703	Austenitic	0.02	1.00	0.40	19.00	13.00	3.30	—	—	—	—	—
317LN	S31753	Austenitic	0.02	1.00	0.40	19.00	13.00	3.30	—	—	0.16	—	—
2205 (former)	S31803	Duplex	0.02	1.00	0.50	22.0	5.5	3.0	—	—	0.14	—	—
321	S32100	Austenitic	0.04	1.00	0.40	18.00	10.50	—	—	—	—	0.40	—
321H	S32109	Austenitic	0.07	1.00	0.40	18.00	10.50	—	—	—	—	0.50	—
2205	S32205	Duplex	0.02	1.00	0.50	22.5	5.5	3.2	—	—	0.17	—	—

(Continued)

Type	UNS no.	Microstructure	Nominal composition (wt%)											
			C	Mn	Si	Cr	Ni	Mo	Nb	Cu	N	Al	Ti	Others
2304	S32304	Duplex	0.02	1.25	0.50	23.00	4.2	0.30	—	0.32	0.12	—	—	—
	S32520	Duplex	0.02	0.75	0.40	25.0	6.8	3.5	—	1.25	0.28	—	—	—
255	S32550	Duplex	0.02	0.75	0.50	25.5	5.5	3.4	—	2.00	0.18	—	—	—
654SMo™	S32654	Austenitic	0.01	03.0	0.25	24.5	22.0	7.5	—	0.45	0.50	—	—	—
2507	S32750	Duplex	0.02	0.60	0.40	25.00	7.00	4.00	—	—	0.28	—	—	—
Zeron 100	S32760	Duplex	0.02	0.50	0.50	25.0	7.0	3.5	—	0.75	0.25	—	—	W: 0.75
329	S32900	Duplex	0.04	0.50	0.40	25.5	3.5	1.50	—	—	—	—	—	—
	S32906	Duplex	0.02	1.0	0.25	29.0	6.6	2.05	—	—	0.35	—	—	—
7 Mo Plus™	S32950	Duplex	0.02	1.00	0.30	27.5	4.4	1.75	—	—	0.25	—	—	—
330	S33000	Austenitic	0.05	1.00	1.20	18.50	35.50	—	—	—	—	—	—	—
	S33228	Austenitic	0.06	0.50	0.20	27.0	32.0	—	—	—	—	—	—	Ce: 0.08
347	S34700	Austenitic	0.04	1.00	0.40	18.00	11.00	—	0.8	—	—	—	—	—
347H	S34709	Austenitic	0.07	1.00	0.40	18.00	11.00	—	0.60	—	—	—	—	—
403	S40300	Martensitic	0.08	0.50	0.25	12.25	—	—	0.80	—	—	—	—	—
405	S40500	Ferritic	0.04	0.50	0.50	13.00	—	—	—	—	—	0.20	—	—
409 (former)	S40900	Ferritic	0.05	0.50	0.50	11.2	—	—	—	—	—	—	0.40	—
409	S40910	Ferritic	0.02	0.50	0.50	11.2	—	—	0.10	—	—	—	0.30	—
409	S40920	Ferritic	0.02	0.50	0.50	11.2	—	—	0.05	—	—	—	0.35	—
409	S40930	Ferritic	0.02	0.50	0.50	11.2	—	—	0.25	—	—	—	0.25	—
	S40940	Ferritic	0.03	0.50	0.50	11.1	—	—	0.40	—	—	—	—	—
	S40945	Ferritic	0.02	0.50	0.50	11.2	—	—	0.29	—	—	—	0.12	—
	S40975	Ferritic	0.02	0.50	0.50	11.2	0.75	—	—	—	—	—	0.50	—
	S40976	Ferritic	0.02	0.50	0.50	11.1	0.88	—	0.40	—	—	—	—	—
	S40977	Ferritic/martensitic	0.02	0.50	0.50	11.5	0.60	—	—	—	—	—	—	—
410	S41000	Martensitic	0.11	0.50	0.50	12.50	—	—	—	—	—	—	—	—
	S41003	Ferritic/martensitic	0.02	0.75	0.50	11.5	—	—	—	—	—	—	—	—
410S	S41008	Martensitic	0.04	0.50	0.50	12.5	—	—	—	—	—	—	—	—

414	S41400	Martensitic	0.08	0.50	0.50	12.50	2.00	—	—	—	—	—	—
	S41425	Martensitic	0.03	0.75	0.25	13.5	5.5	1.75	—	0.09	—	—	—
410NiMo	S41500	Martensitic	0.03	0.75	0.30	12.75	4.50	0.75	—	—	—	—	—
416	S41600	Martensitic	0.08	0.60	0.50	13.00	—	—	—	—	—	—	S: 0.20
	S41603	Ferritic/martensitic	0.04	0.60	0.50	13.0	—	—	—	—	—	—	S: 0.20
420	S42000	Martensitic	0.20	0.50	0.50	13.00	—	—	—	—	—	—	—
	S42010	Martensitic	0.22	0.50	0.50	14.2	0.60	0.62	—	—	—	—	—
422	S42200	Martensitic	0.30	1.15	0.25	11.50	0.75	1.10	—	—	—	—	V: 0.25; W: 1.1
429	S42900	Ferritic	0.06	0.50	0.50	15.00	—	—	—	—	—	—	—
430	S43000	Ferritic	0.06	0.50	0.50	17.00	—	—	—	—	—	—	—
439	S43035	Ferritic	0.03	0.50	0.50	18.00	—	—	—	—	0.60	—	—
431	S43100	Martensitic	0.10	0.50	0.50	16.00	2.00	—	—	—	—	—	—
434	S43400	Ferritic	0.06	0.50	0.50	17.00	—	1.00	—	—	—	—	—
436	S43600	Ferritic	0.06	0.50	0.50	17.00	—	1.00	0.50	—	—	—	—
	S43932	Ferritic	0.02	0.50	0.50	18.0	—	—	0.50	—	—	—	—
	S43940	Ferritic	0.02	0.50	0.50	18.0	—	—	0.50	—	0.35	—	—
440A	S44002	Martensitic	0.70	0.50	0.50	17.00	—	—	—	—	—	—	—
440B	S44003	Martensitic	0.85	0.50	0.50	17.00	—	—	—	—	—	—	—
440C	S44004	Martensitic	1.10	0.50	0.50	17.00	—	—	—	—	—	—	—
442	S44200	Ferritic	0.10	0.50	0.50	20.5	—	—	—	—	—	—	—
444	S44400	Ferritic	0.01	0.50	0.50	18.50	—	2.10	0.20	—	0.30	—	—
	S44500	Ferritic	0.01	0.50	0.50	20.0	—	—	0.40	—	1.10	—	—
446	S44600	Ferritic	0.10	0.75	0.50	25.00	—	—	—	—	—	—	—
	S44625	Ferritic	0.005	0.20	0.02	26.00	—	1.00	0.10	—	—	—	C + N < 0.025
26-1(XM-27)	S44627	Ferritic	0.002	0.05	0.20	26.0	—	1.0	—	—	—	—	C + N < 0.025
29-4	S44700	Ferritic	0.005	0.20	0.10	29.00	—	4.0	0.20	—	—	—	—
	S44735	Ferritic	0.015	0.50	0.50	29.0	—	3.9	—	—	0.40	—	—
29-4-2	S44800	Ferritic	0.005	0.20	0.10	29.00	2.25	4.0	—	—	—	—	—
660 (A286)	S66286	Austenitic PH	0.04	1.00	0.50	14.75	25.50	1.25	—	—	2.10	0.20	V: 0.30; B: 0.005

(Continued)

Type	UNS no.	Microstructure	Nominal composition (wt%)											
			C	Mn	Si	Cr	Ni	Mo	Nb	Cu	N	Al	Ti	Others
JBK-75	—	Austenitic PH	0.02	0.000	0.000	15.0	30.0	1.25	—	—	0.000	0.25	2.15	V: 0.25; B: 0.001
AL-6XN	N08367	Austenitic	0.02	1.00	0.50	21.00	24.50	6.50	—	—	0.22	—	—	—
800	N08800	Austenitic	0.05	0.75	0.50	21.0	32.5	—	—	—	—	0.38	0.38	—
800H	N08810	Austenitic	0.07	0.75	0.50	21.0	32.5	—	—	—	—	0.38	0.38	—
	N08811	Austenitic	0.08	0.75	0.50	21.0	32.5	—	—	—	—	0.38	0.38	—
904L	N08904	Austenitic	0.01	1.00	0.50	21.0	25.50	4.50	—	1.50	—	—	—	—
	N08926	Austenitic	0.01	1.00	0.25	20.0	25.0	6.50	—	1.0	0.20	—	—	—

2.1 FILLER METALS FOR STAINLESS STEELS

AWS classification	UNS no.	Composition (wt%) (single values are maxima)							
		C	Cr	Ni	Mo	Mn	Si	Cu	Others
Martensitic alloys									
E410-XX	W41010	0.12	11.0–13.5	0.7	0.75	1.0	0.90	—	—
ER410	S41080	0.12	11.5–13.5	0.6	0.75	0.6	0.5	—	—
E410TX-X	W41031	0.12	11.0–13.5	0.60	0.5	0.6	1.0	—	—
E410NiMo-XX	W41016	0.06	11.0–12.5	4.0–5.0	0.4–0.7	1.0	0.90	—	—
ER410NiMo	S41086	0.06	11.0–12.5	4.0–5.0	0.4–0.7	0.6	0.5	—	—
E410NiMoTX-X	W41036	0.06	11.0–12.5	4.0–5.0	0.4–0.7	1.0	1.0	—	—
ER420	S42080	0.25–0.40	12.0–14.0	0.6	0.75	0.6	0.5	—	—
Ferritic alloys									
E409Nb-XX	—	0.03	11.0–14.0	0.6	0.75	1.0	0.9	0.75	Nb: 0.5–1.5
ER409	S40900	0.08	10.5–13.5	0.6	0.50	0.8	0.8	0.75	Ti: 10xC-1.5
ER409Cb	S40940	0.08	10.5–13.5	0.6	0.50	0.8	1.0	0.75	Nb: 10xC-0.75

E409TX-X	W41031	0.10	10.5–13.5	0.6	0.50	0.8	1.0	0.5	Ti: 10xC–1.5
E430-XX	W43010	0.10	15.0–18.0	0.6	0.50	1.0	0.9	0.75	—
E430Nb-XX	—	0.10	15.0–18.0	0.6	0.75	1.0	0.9	0.75	Nb: 0.50–1.50
ER430	S43080	0.10	15.5–17.0	0.6	0.75	0.6	0.5	0.75	—
ER446LMo	S44687	0.015	25.0–27.5	Note	0.75–1.5	0.4	0.4	Note	Ni+Cu=0.5 max
Austenitic alloys									
ER219	S21980	0.06	19.0–21.5	5.5–7.0	0.75	8.0–10.0	1.00	—	N: 0.10–0.30
ER308	S30880	0.08	19.5–22.0	9.0–11.0	0.75	1.0–2.5	0.30–0.65	—	—
E308	W30810	0.08	18.0–21.0	9.0–11.0	0.75	0.5–2.5	1.0	—	—
ER308H	S30880	0.04–0.08	19.5–22.0	9.0–11.0	0.75	1.0–2.5	0.30–0.65	—	—
ER308L	S30883	0.03	19.5–22.0	9.0–11.0	0.75	1.0–2.5	0.30–0.65	—	—
E308L	W30813	0.04	18.0–21.0	9.0–11.0	0.75	0.5–2.5	1.0	—	—
ER308Si	S30881	0.08	19.5–22.0	9.0–11.0	0.75	1.0–2.5	0.65–1.00	—	—
ER308LSi	S30888	0.03	19.5–22.0	9.0–11.0	0.75	1.0–2.5	0.65–1.00	—	—
ER309	S30980	0.12	23.0–25.0	12.0–14.0	0.75	1.0–2.5	0.30–0.65	—	—
E309	W30910	0.15	22.0–25.0	12.0–14.0	0.75	0.5–2.5	1.0	—	—
ER309L	S30983	0.03	23.0–25.0	12.0–14.0	0.75	1.0–2.5	0.30–0.65	—	—
E309L	W30917	0.04	22.0–25.0	12.0–14.0	0.75	0.5–2.5	1.0	—	—
ER309Si	S30981	0.12	23.0–25.0	12.0–14.0	0.75	1.0–2.5	0.30–0.65	—	—
ER309LSi	S30988	0.03	23.0–25.0	12.0–14.0	0.75	1.0–2.5	0.30–0.65	—	—
ER310	S31080	0.08–0.15	25.0–28.0	20.0–22.5	0.75	1.0–2.5	0.30–0.65	—	—
E310	W31010	0.08–0.20	25.0–28.0	20.0–22.5	0.75	1.0–2.5	0.75	—	—
ER316	S31680	0.08	18.0–20.0	11.0–14.0	2.0–3.0	1.0–2.5	0.30–0.65	—	—
E316	W31610	0.08	17.0–20.0	11.0–14.0	2.0–3.0	0.5–2.5	1.0	—	—
ER316H	S31680	0.04–0.08	18.0–20.0	11.0–14.0	2.0–3.0	1.0–2.5	0.30–0.65	—	—
ER316L	S31683	0.03	18.0–20.0	11.0–14.0	2.0–3.0	1.0–2.5	0.30–0.65	—	—
ER316Si	S31681	0.08	18.0–20.0	11.0–14.0	2.0–3.0	1.0–2.5	0.65–1.00	—	—
ER316LSi	S31688	0.03	18.0–20.0	11.0–14.0	2.0–3.0	1.0–2.5	0.65–1.00	—	—

(Continued)

AWS classification	UNS no.	Composition (wt%) (single values are maxima)							
		C	Cr	Ni	Mo	Mn	Si	Cu	Others
ER317	S31780	0.08	18.5–20.5	13.0–15.0	3.0–4.0	1.0–2.5	0.30–0.65	—	—
E317	W31710	0.08	18.0–21.0	12.0–14.0	3.0–4.0	0.5–2.5	1.0	—	—
ER317L	S31783	0.03	18.5–20.5	13.0–15.0	3.0–4.0	1.0–2.5	0.30–0.65	—	—
ER330	N08331	0.18–0.25	15.0–17.0	34.0–37.0	0.75	1.0–2.5	0.30–0.65	—	—
ERE347	S34780	0.08	19.0–21.5	9.0–11.0	0.75	1.0–2.5	0.30–0.65	—	Nb: 10xC–1.0
E347	W34710	0.08	18.0–21.0	9.0–11.0	0.75	0.5–2.5	1.0	—	Nb: 8xC–1.00
ER347Si	S34788	0.08	19.0–21.5	9.0–11.0	0.75	1.0–2.5	0.65–1.00	—	Nb: 10xC–1.0

Duplex alloys

AWS classification	UNS no.	Composition (wt%) (single values are maxima)								
		C	Cr	Ni	Mo	Mn	Si	Cu	N	Others
E2209-XX	A5.4	0.04	21.5–23.5	8.5–10.5	2.5–3.5	0.5–2.0	1.00	0.75	0.08–0.20	—
ER2209	A5.9	0.03	21.5–23.5	7.5–9.5	2.5–3.5	0.50–2.0	0.90	0.75	0.08–0.20	—
E2209TX-X	A5.22	0.04	21.0–24.0	7.5–10.0	2.5–4.0	0.5–2.0	1.0	0.5	0.08–0.20	—
E2552-XX	A5.4	0.04	24.0–27.0	4.0–6.0	1.5–2.5	1.0	1.00	2.5–3.5	0.08–0.22	—
E2553-XX	A5.4	0.06	24.0–27.0	6.5–8.5	2.9–3.9	0.5–1.5	1.00	1.5–2.5	0.10–0.25	—
E2553TX-X	A5.22	0.04	24.0–27.0	8.5–10.5	2.9–3.9	0.5–1.5	0.75	1.5–2.5	0.10–0.20	—
ER2553	A5.9	0.04	24.0–27.0	4.5–6.5	2.9–3.9	1.5	1.0	1.5–2.5	0.10–0.25	—
E2593-XX	A5.4	0.04	24.7–27.0	8.5–11.0	2.9–3.9	0.5–2.5	1.00	1.5–3.0	0.08–0.25	—
E2594-XX	A5.4	0.04	24.0–27.0	8.0–10.5	3.5–4.5	0.5–2.0	1.00	0.75	0.20–0.30	—

APPENDIX C

COMPOSITION OF NICKEL-BASE ALLOYS

Welding Metallurgy and Weldability, First Edition. John C. Lippold.
© 2015 John Wiley & Sons, Inc. Published 2015 by John Wiley & Sons, Inc.

Composition (wt%) (single values are maxima unless noted)

Alloy	UNS no.	C	Cr	Fe	Mn	Ni	Mo	Ti	Al	Si	Others
Nickel											
200	N02200	0.15	—	0.40	0.35	99.0 min	—	—	—	0.35	Cu: 0.25
201	N02201	0.02	—	0.40	0.35	99.0 min	—	—	—	0.35	Cu: 0.25
205	N02205	0.15	—	0.20	0.35	99.0 min	—	—	—	0.15	Cu: 0.15
233	N02233	0.15	—	0.10	0.30	99.0 min	—	—	—	0.10	Cu: 0.10
253	N02253	0.02	—	0.05	0.003	99.9 min	—	—	—	0.005	Cu: 0.10
270	N02270	0.02	—	0.005	0.001	99.97 min	—	—	—	0.001	Cu: 0.001
Nickel–copper alloys											
400	N04400	0.30	—	2.50	2.00	63.0–70.0	—	—	—	0.50	Cu bal.
401	N04401	0.10	—	0.75	2.25	40.0–45.0	—	—	—	0.25	Cu bal.
404	N04404	0.15	—	0.50	0.10	52.0–57.0	—	—	0.05	0.10	Cu bal.
405	N04405	0.30	—	2.50	2.00	63.0–70.0	—	—	—	0.50	Cu bal.
Nickel–chromium, nickel–chromium–iron, and nickel–chromium–molybdenum alloys											
600	N06600	0.15	14–17	6–10	1.0	72.0 min	—	—	—	0.5	—
601	N06601	0.1	21–25	Bal.	1.0	58–63	—	—	1–1.7	0.5	—
617	N06617	0.15	20–24	3.0	1.0	Bal.	8–10	—	0.8–1.5	0.5	Co: 10–15
625	N06625	0.10	20–23	5.0	0.5	Bal.	8–10	—	0.40	1.0	Nb: 3.15–4.15
690	N06690	0.05	27–31	7–11	0.5	58.0 min	—	—	—	0.5	—
C-4	N06455	0.015	14–18	3.0	1.0	Bal.	14–17	—	—	0.5	—
C-22	N06022	0.01	20–24	3.0	0.5	Bal.	12–14	—	—	0.08	Co: 2.5, W: 3.0
C-276	N10276	0.02	14.5–16.5	4–7	1.0	Bal.	15–17	—	—	0.08	Co: 2.5
59	N06059	0.10	22–24	1.5	0.50	Bal.	15–16.5	—	0.4	0.08	—
230	N06230	0.05–0.15	20–24	3.0	0.30–1.0	Bal.	1–3	—	0.2–5	0.1	—
RA333	N06333	0.08	24–27	Bal.	2.0	44–47	2.5–4	—	—	—	—
HX	N06006	0.05–0.15	20.5–23.0	17–20	1.0	Bal.	8–10	—	0.1–0.5	0.75–1.5	W: 0.2–1.0
S	N06635	0.02	14.5–17	3.0	0.30–1.0	Bal.	14–16.5	—	—	0.2–0.75	—
W	N10004	0.12	5.0	6.0	1.0	63.0	24	—	—	—	—

		C	Cr	Fe	Mn	Ni	Mo	Ti	Al	Si	Other
X	N06002	0.05–0.15	20.5–23.0	17–20	1.0	Bal.	8–10	—	—	0.5	Co: 0.5–2.5, W: 0.2–1.0
686	N06686	0.10	19–23	5.0	0.75	Bal.	15–17	—	—	—	W: 3.0–4.4
Iron–nickel–chromium alloys											
HP	N08705	0.35–0.75	19–23	Bal.	2.00	35–37	—	—	0.15–0.6	2.5	—
800	N08800	0.10	19–23	Bal.	1.5	30–35	—	—	0.15–0.6	1.0	—
801	N08801	0.10	19–22	Bal.	1.5	30–34	—	—	—	1.0	—
802	N08802	0.2–0.5	19–23	Bal.	1.50	30–35	—	—	0.15–1.	0.75	—
800H	N08810	0.05–0.1	19–23	Bal.	1.50	30–35	—	—	0.15–0.6	1.0	—
800HT	N08811	0.06–0.1	19–23	39.5 min	1.50	30–35	—	0.85–1.2	0.15–0.6	1.0	—
825	N08825	0.05	19.5–23.5	Bal.	1.00	38–46	—	0.6–1.2	0.2	0.5	—
Iron–nickel low-expansion alloys											
36	K93601	0.10	0.5	Bal.	0.6	34–38	0.5	—	0.1	0.35	—
42	K94100	0.05	0.5	Bal	0.8	42.0nom	0.5	—	0.15	0.3	—
48	K94800	0.05	0.25	Bal.	0.8	48.0nom	—	—	0.1	0.05	—
902	N09902	0.06	4.9–5.75	Bal.	0.8	41–43.5	—	2.2–2.75	0.3–0.8	1.0	—
903	N19903	—	—	42.0	—	38.0	—	1.4	0.9	—	Co: 15.0, Nb: 3.0
907	N19907	—	—	42.0	—	38.0	—	1.5	0.03	0.15	Co: 13.0, Nb: 4.7
909	N19909	0.01	—	42.0	—	38.0	—	1.5	0.03	0.4	Co: 13.0, Nb: 4.7
Precipitation-strengthened alloys											
K500	N05500	0.25	—	2.00	1.50	63.0–70.0	—	0.35–0.85	2.3–3.15	—	Cu bal.
263	N07263	0.04–0.08	19.0–21.0	0.7	0.60	Bal.	5.6–6.1	1.9–2.4	0.3–0.6	—	Co: 19.0–21.0
713	N07713	0.08–0.20	12.0–14.0	2.50	0.25	Bal.	3.8–5.2	0.5–1.0	5.5–6.5	—	Nb: 1.8–2.8
718	N07718	0.08	17.0–21.0	Bal.	0.35	50.0–55.0	2.8–3.3	0.65–1.15	0.2–0.8	—	Nb: 4.75–5.50
Waspaloy	N07001	0.03–0.10	18.0–21.0	2.00	1.00	Bal.	3.5–5.0	2.75–3.25	1.2–1.6	—	Co: 12–15
R41	N07041	0.12	18.0–22.0	5.00	0.10	Bal.	9.0–10.5	3.0–3.3	1.4–1.8	—	Co: 10.0–12.0
214	N07214	0.05	15.0–17.0	2.0–4.0	0.5	Bal.	0.5	0.5	4.0–5.0	—	Co: 2.0
725	N07725	0.03	19.0–22.5	Bal.	0.35	55.0–59.0	7.0–9.5	1.0–1.7	0.35	—	Nb: 2.75–4.0
706	N09706	0.06	14.5–17.5	Bal.	0.35	39.0–44.0	—	1.5–2.0	0.40	—	Nb: 2.5–3.3
925	N09925	0.03	19.5–23.5	22.0Min	1.00	38.0–46.0	2.5–3.5	1.9–2.4	0.1–0.5	—	Cu: 1.5–3.00

Filler Metals for Ni-base Alloys

AWS classification	Alloy	UNS no.	Composition (wt%) (single values are maxima unless noted)							
			C	Cr	Fe	Mn	Ni	Mo	Si	Others
Coated welding electrodes										
ENi-1	WE141	W82141	0.10	—	0.75	0.75	92.0 min	—	1.25	Ti: 1.0–4.0
ENi-CI	WE99	W82001	2.00	—	8.00	2.50	85.0 min	—	4.00	Cu: 2.5
ENiCu-7	WE190	W84190	0.15	—	2.50	4.00	62.0–69.0	—	1.50	Cu bal., Al: 1.75, Ti: 1.0
ENiCrFe-1	WE132	W86132	0.08	13–17	11.0	3.50	62.0 min	—	0.75	Nb: 1.5–4.0
ENiCrFe-2	Weld A	W86133	0.10	13–17	12.0	1.00–3.5	62.0 min	0.5–2.5	0.75	Nb: 0.5–3.0
ENiCrFe-3	WE182	W86182	0.10	13–17	10.0	5.00–9.5	59.0 min	—	1.0	Nb: 1–2.5, Ti: 1.0
ENiCrFe-7	WE152	W86152	0.05	28.0–31.5	7–12	5.00	Bal.	—	0.75	Nb: 1–2.5
ENiMo-7	HAST B-2	W80665	0.02	1.0	2.25	1.75	Bal.	26–30	0.2	W: 1.0
ENiCrMo-3	WE112	W86112	0.10	20–23	7.0	1.0	55.0 min	8–10	0.75	Nb: 3.15–4.15
ENiCrMo-4	WE C-276	W80276	0.02	14.5–16.5	4.0–7.0	1.0	Bal.	15–17	0.2	W: 3.0–4.5
ENiCrMo-10	WE C-22	W86022	0.02	20–22.5	2.0–6.0	1.0	Bal.	12.4–14.5	0.2	W: 2.5–3.5, Co: 2.5
ENiCrMo-14	WE686	W86686	0.02	19–23	5.0	1.0	Bal.	15–17	0.25	W: 3.0–4.4
ENiCrCoMo-1	WE117	W86117	0.05–0.15	21–26	5.0	0.3–2.5	Bal.	8–10	0.75	Co: 9.0–15.0, Nb: 1.0
Bare welding electrodes and rods										
ERNi-1	FM61	N02061	0.15	—	1.00	1.0	93.0 min	—	0.75	Cu: 0.25, Ti: 2.5–3.5, Al: 1.5
ERNi-CI	FM99	N02215	1.00	—	4.00	2.5	90.0 min	—	0.75	Cu: 4.0
ERNiFeMn-CI	FM44	N02216	0.50	—	Bal.	10.0–14.0	35.0–45.0	—	1.0	Cu: 2.5
ERNiCu-7	FM60	N04060	0.15	—	2.50	4.0	62.0–69.0	—	1.25	Cu bal., Ti: 1.5–3.0, Al: 1.25
ERNiCu-8	FM64	N05504	0.25	—	2.0	1.5	63.0–70.0	—	—	Cu bal., Ti: 0.35–0.85, Al: 2.3–3.15
ERNiCr-3	FM82	N06082	0.10	18.0–22.0	3.0	2.5–3.5	67.0 min	—	—	Nb: 2.0–3.0
ERNiCr-4	FM72	N06072	0.01–0.10	42.0–46.0	0.50	0.2	Bal.	—	—	Ti: 0.3–1.0

ERNiCrFe-5	FM62	N06062	0.08	14.0–17.0	6.00–10.0	1.0	70.0 min	—	0.35	Nb: 1.5–3.0
ERNiCrFe-6	FM92	N07092	0.08	14.0–17.0	8.00	2.0–2.7	67.0 min	—	0.35	Ti: 2.5–3.5
ERNiCrFe-7	FM52	N06052	0.04	28.0–31.5	7.00–11.0	1.0	Bal.	—	0.50	Al: 1.10, Ti: 1.0
ERNiCrFe-7A	FM52M	N06054	0.04	28.0–31.5	7.00–11.0	1.0	Bal.	—	—	Al: 1.10, Ti: 1.0, Nb: 0.5–1.0
ERNiCrFe-8	FM69	N07069	0.08	14.0–17.0	5.0–9.0	1.0	70.0 min	—	—	Ti: 2.25–2.75, Al: 0.4–1.0, Nb: 0.70–1.2
ERNiFeCr-1	FM65	N08065	0.05	19.5–23.5	22.0 min	1.0	38.0–46.0	2.5–3.5	0.5	Cu: 1.50–3.0, Al: 0.20, Ti: 0.6–1.2
ERNiFeCr-2	FM718	N07718	0.08	17.0–21.0	Bal.	0.35	50.0–55.0	2.8–3.3	—	Ti: 0.65–1.15, Al: 0.2–0.8, Nb: 4.75–5.50
ERNiMo-3	HAST W	N10004	0.12	4.0–6.0	4.0–7.0	1.0	Bal.	23–26	1.0	W: 1.0, Co: 2.5
ERNiCrMo-3	625	N06625	0.10	20–23	5.0	0.5	58.0 min	8–10	0.5	Nb: 3.15–4.15, Al: 0.40, Ti: 0.4
ERNiCrMo-7	C-4	N06455	0.015	14–18	3.0	1.0	Bal.	14–18	0.08	W: 0.50, Ti: 0.70
ERNiCrMo-10	C-22	N06022	0.015	20–22.5	2–6	0.5	Bal.	12.5–14.5	0.08	W: 2.5–3.5, Co: 2.5
ERNiCrMo-13	59	N06059	0.01	22–24	1.5	0.5	Bal.	15–16.5	0.1	Al: 0.1–0.4
ERNiCrMo-14	FM686	N06686	0.01	19–23.	5.0	1.0	Bal.	15–17	0.08	W: 3.0–4.4, Al: 0.5, Ti: 0.25
ERNiCrMo-15	FM725	N07725	0.03	19.0–22.5	Bal.	0.35	55.0–59.0	7.0–9.5	—	Ti: 1.0–1.7, Al: 0.35, Nb: 2.75–4.0
ERNiCrMo-17	C-2000	N06200	0.01	22–24	3.0	0.5	Bal.	15–17	0.08	Al: 0.5, Cu: 1.3–1.9, Co: 2.0
ERNiCrWMo-1	230-W	N06231	0.05–0.15	20–24	3.0	0.3–1.0	Bal.	1–3	0.50	Al: 0.2–0.5, Co: 5.0, W: 13–15
ERNiCrCoMo-1	617	N06617	0.05–0.15	20–24	3.0	1.0	Bal.	8–10	1.0	Co: 10–15, Al: 0.8–1.5

APPENDIX D

ETCHING TECHNIQUES

The microstructure of welds can be revealed using a variety of etching techniques. The weld metal, the partially melted zone around the weld, and sometimes the HAZ are not homogeneous, so they tend to etch differently than base metals. Etching techniques can be divided into chemical methods, electrolytic methods, and staining methods. In general, chemical methods are simpler to apply and require less equipment, so they tend to be favored by the nonspecialist. Electrolytic methods tend to be favored by those who specialize in examination of corrosion-resistant alloys. Various staining methods help to bring out phases of interest at higher contrast (or with color) than possible with the other techniques.

This section provides some etching techniques that are commonly used for steels, stainless steels, and nickel-base alloys. The last section describes a cleaning technique that is very helpful in removing oxide scale from steel fracture surfaces. The etchants listed here are those with which the author has personal experience. A complete list of etching techniques can be found in the *Metals Handbook* and the *CRC Handbook of Metal Etchants* [1, 2].

A4.1 STEELS

The most common etchant for plain-carbon and low-alloy steels is nital. Nital is a dilute mixture of ethanol (or methanol) and concentrated nitric acid. Mixtures of 1–5% are the most common. The polished sample is simply immersed in the etchant until the desired

Welding Metallurgy and Weldability, First Edition. John C. Lippold.
© 2015 John Wiley & Sons, Inc. Published 2015 by John Wiley & Sons, Inc.

TABLE A4.1   Chemical etchants for steels

Etchant	Materials	Composition/use	Notes
Nital	Plain-carbon and low-alloy steels	1–5 ml HNO_3 in 100 ml of ethanol or methanol. Immerse sample for several seconds	General microstructure for most steel welds. Develops good contrast of ferrite at prior austenite grain boundaries
Picral	Plain-carbon and low-alloy steels	4% (saturated) solution of picric acid in ethanol or methanol. Immerse sample for several seconds	General microstructure for most steel welds. Provides better detail of fine structure than nital
Super picral	Alloy steels	Same solution as above, but with a few drops of HCl. Immerse sample for several seconds	This etchant is often used for heat-treated microstructures
Vilella's	Alloy steels	5 ml HCl, 1 g picric acid, and 100 ml ethanol or methanol. Immerse sample for several seconds	Used for alloys steels containing chromium. Reveals prior austenite grain boundaries in martensitic and bainitic microstructures
Macroetching	All steels	2–25% HCl in ethanol. Immerse for several minutes to reveal general structure	Reveals overall weld outline and is used for procedure development. Sample surface should be ground to an approximate 240 grit condition for best contrast

etching effect is achieved. As the chromium content of the steel increases, nital becomes less effective and more aggressive reagents must be used. For example, alloys such as A508, Grade 22 (2.25Cr–1Mo steel), and 9Cr steels are usually etched with Vilella's reagent. There are also a number of etchants that include picric acid. These etchants reveal the fine details of the microstructure, such as pearlite colonies and martensite lath structure. For macroetching of steel weldments to reveal weld penetration or general weld shape, a solution of 5–25% nitric acid in ethanol (or methanol) can be used. The sample is immersed in this solution until the desired effect is achieved. A list of some commonly used etchants for steels is provided in Table A4.1.

A4.2   STAINLESS STEELS

Because stainless steels contain a minimum 11% Cr, more aggressive etching solutions must be used to reveal the weld microstructure. Standard chemical etchants are provided in Table A4.2. Mixed acids and glyceregia are both general-purpose

TABLE A4.2 Chemical etchants for stainless steels

Etchant	Materials	Composition/use	Notes
Kalling's no. 1	Martensitic	1.5 g $CuCl_2$, 33 ml HCl, 33 ml ethanol, and 33 ml H_2O. Immerse or swab at room temperature	Darkens martensite, colors ferrite, and does not attack austenite
Vilella's	Martensitic	1 g picric acid, 5 ml HCl, and 100 ml ethanol. Immerse or swab at room temperature	Etches martensite, leaves carbides, sigma phase, and ferrite outlined
Kalling's no. 2	Austenitic Duplex	5 g $CuCl_2$, 100 ml HCl, and 100 ml ethanol. Immerse or swab at room temperature	Darkens ferrite in nominally austenitic and duplex stainless steel weld metal and leaves austenite light. A very versatile etch but should be used fresh because older etch tends to produce pitting
Mixed acids	Ferritic Austenitic Duplex	Equal parts HCl, HNO_3, and acetic acids. Use fresh. Swab at room temperature	A general attack etch will reveal ferrite and austenite, segregation patterns, precipitates, and grain boundaries. This etchant must be used within a few minutes of mixing and then discarded when it turns orange in color
Glyceregia	Martensitic Ferritic Austenitic Duplex	3 parts glycerol, 2–5 parts HCl, and 1 part HNO_3. Use fresh. Immerse or swab at room temperature	A general-purpose etch similar to mixed acids, but not so aggressive. It outlines ferrite and austenite and attacks martensite and sigma phase. It should also be used fresh and discarded when it becomes orange in color
Dilute aqua regia	Dissimilar welds	15 ml HCl, 5 ml HNO_3, and 100 ml H_2O at room temperature	Useful for etching stainless steel to carbon steel or low-alloy steel joints, for stainless filler metal used in high-carbon steels, and for cladding of stainless steel on carbon steel. It tends to reveal structure also in carbon steel and low-alloy steel heat-affected zones, though exposure time must be limited to avoid overetching the HAZ. It darkens austenite and outlines ferrite, sigma phase, and carbides
Murakami's	Duplex and superduplex	10 g $K_3Fe(CN)_6$, 10 g KOH or 7 g NaOH, and 100 mL H_2O, heated to 80–100°C. Immerse in fresh solution. Use under an exhaust hood as cyanide fumes are released	Should be used fresh, as evaporation changes its strength. It darkens ferrite, outlines sigma phase and colors it light blue, and leaves austenite unattacked

TABLE A4.3   Electrolytic etching techniques for stainless steels

Etchant	Materials	Composition/use	Notes
Oxalic acid: 10%	Ferritic Austenitic	10 g oxalic acid and 90 ml H_2O. Use at room temperature. Etch at 3–6 volts for 5–60 s	Very effective at revealing grain boundaries, particularly if there is carbide precipitation. This technique is often used to detect sensitization
Chromic acid: 10%	Austenitic	10 g CrO_3 (chromic acid) and 90 ml water. Use at room temperature. Etch at 3–6 V for 5–60 s	A good general etchant for revealing the microstructure of base metals and welds. Reveals segregation behavior and grain boundaries in weld metals For dissimilar welds with carbon or low-alloy steels, this etchant can be used after the steel has been etched with nital
NaOH: 40%	Duplex	40 g sodium hydroxide (NaOH) in 100 g of distilled water. Etch at 1–3 V for 5–60 s	Good contrast between ferrite and austenite. Can be used to distinguish sigma phase per ASTM A923
Ramirez's [5]	Duplex and superduplex	40 vol% HNO_3 in water. Step 1: apply 1–1.2 V for 2 min. Step 2: apply 0.75 V for 7 min	This technique reveals the various types of austenite that form without overetching the chromium nitrides

etchants that work well to reveal microstructure in most stainless steel alloys. These also serve well for macroetching.

A number of electrolytic etching techniques are provided in Table A4.3. The 10% oxalic acid technique is widely used, since it reveals carbide precipitation at grain boundaries in ferritic and austenitic alloys. It also provides good contrast of ferrite and austenite in weld metals and segregation effects in fully austenitic weld metals. The 40% NaOH electrolytic technique provides good ferrite–austenite phase contrast in duplex stainless steels and is also used to detect sigma-phase precipitation per ASTM A923.

To differentiate ferrite from austenite in austenitic and duplex grades, "staining" techniques have also been developed that give good contrast between the ferrite and austenite phases. A modified Murakami's technique that involves both electrolytic

TABLE A4.4   Staining techniques for stainless steels

Etchant	Materials	Composition/use	Notes
Modified Murakami's [6]	Duplex	Electrolytic 10% oxalic acid at 6 V for 10–20 s. Immersion in boiling Murakami's (10 g permanganate, 10 g KOH, 100 ml H_2O) for up to 60 min	The first two steps reveal the austenite and ferrite phases. Boiling Murakami's produces an interference film on the ferrite that reveals segregation patterns within the ferrite. Austenite remains white
Ferrofluid [7]	Austenitic Duplex	Apply colloidal suspension of "ferrofluid" (Fe_3O_4) in paraffin. Rinse from sample and dry	The Fe_3O_4 particles attach to the ferrite phase due to its remnant magnetism and color the ferrite. The austenite remains white. This technique provides very good contrast between ferrite and austenite

etching and immersion has been shown to be quite effective in staining the different phases to provide color contrast. This technique can also differentiate sigma phase. Another approach is to use the remnant ferromagnetism of the ferrite to provide contrast. Using a solution of Ferrofluid® (particles of Fe_3O_4) in paraffin, the solution is applied to the surface of a polished sample and then removed by alcohol rinsing. The Fe_3O_4 particles attach to the ferrite, but not the austenite, which is paramagnetic (Table A4.4).

For dissimilar welds between austenitic stainless steels and carbon steels, a two-step procedure is recommended. After polishing, the sample is etched with nital to attack the carbon steel without etching the stainless steel. The sample is then electrolytically etched with 10% chromic acid solution. The prior nital etching protects (passivates) the carbon steel and prevents any additional attack by the chromic acid solution.

A4.3   NICKEL-BASE ALLOYS

As with the stainless steels, the etching techniques described here can be divided into chemical methods and electrolytic methods. In general, chemical methods are simpler to apply and require less equipment. Tables A4.5–A4. 6 include etching techniques that the author has found useful in examining the microstructure of Ni-base alloy welds. This is not intended to be an exhaustive list. More extensive listings of etchants and etching methods for Ni-base alloys can be found in *ASM Metals Handbook* [1], *CRC Handbook of Metal Etchants* [2], and *Metallography of Superalloys* [3] (Table A4.7).

TABLE A4.5   Macroetchants for Ni-base alloys

Etchant	Composition/use	Notes
Lepito's	(a) 15 g $(NH_4)_2S_2O_8$ (ammonium persulfate) and 75 ml water, (b) 250 g $FeCl_3$ and 100 ml HCl, and (c) 30 ml HNO_3. Mix (a) and (b) and then add (c), immerse for 30–120 s at room temperature	Macroetch for general structure of base metals and welds. Good for determining weld penetration and measuring grain size. Not effective at revealing solidification structure
Mixed acids	Equal parts of HCl, HNO_3, and acetic acids. Use fresh. Swab at room temperature	A general attack etch that reveals both macro- and microstructure. This etchant must be used within a few minutes of mixing and then discarded when it turns orange in color
HCl and peroxide	Equal parts of HCl and H_2O_2. Immerse or swab at room temperature	Reveals general structure

TABLE A4.6   Microetchants (swab or immerse) for Ni-base alloys

Etchant	Composition/use	Notes
Mixed acids	Equal parts of HCl, HNO_3, and acetic acids. Use fresh. Swab at room temperature	A general attack etch will reveal segregation patterns, precipitates, and grain boundaries. This etchant must be used within a few minutes of mixing and then discarded when it turns orange in color
Glyceregia	15 ml glycerol, 10 ml HCl, and 5 ml HNO_3. Use fresh. Immerse or swab at room temperature	Can also use 15/25/5 and 20/10/5 ratios. A general-purpose etch similar to mixed acids, but not so aggressive. It outlines ferrite and austenite, attacks martensite and sigma phase, and reveals carbides and grain boundaries. Good for NiCrFe and NiFeCr alloys. It should also be used fresh and discarded when it becomes orange in color
Nitric–acetic	10 ml HNO_3 and 90 ml acetic acid. Immerse or swab at room temperature	Use for Ni and Ni–Cu alloys
Nitric–hydrofluoric	20 ml HNO_3 and 3 ml hydrofluoric acid (HF). Immerse or swab at room temperature	General microstructure etchant. Can also use 30/3 and 50/3 ratios
HCl/bromine	Immerse in concentrated HCl for 3 s and then rinse in alcohol. Immerse in mixture of 1 part bromine and 99 parts methanol for 10–20 s	Use for NiCrMo and NiFeCrMo alloys to reveal grain boundaries. **Caution: Do not breathe bromine fumes**

TABLE A4.7   Microetchants (electrolytic) for Ni-base alloys

Etchant	Composition/use	Notes
10% chromic	10 g CrO$_3$ (chromic acid) and 90 ml water. Use at room temperature. Etch at 3–6 volts for 5–60 s	A good general etchant for revealing the microstructure of base metals and welds. Reveals segregation behavior and grain boundaries in weld metals. May be used for dissimilar welds with carbon steels if the carbon steel is previously immersion etched with nital
Phosphoric	80 ml H$_3$PO$_4$ and 10 mL water. Use at room temperature. Etch at 3 V for 5–10 s	50/50 and 20/80 ratios may also be used. Reveals grain boundaries in NiFeCo and NiCrFe alloys. Tends to pit the sample if overetched
Hydrofluoric and glycerol	5 ml HF, 10 ml glycerol, and 85 ml ethanol. Use at room temperature. Etch at 6–12 V	Reveals gamma prime precipitation in Ni-base superalloys
Nital	5 ml HNO$_3$ and 95 ml methanol	Reveals grain boundaries in NiCr, NiFeCr, and NiCrFe alloys

A4.4   FRACTURE SURFACE CLEANING

As discussed in Section 8.3.4, steel samples exposed to the environment will readily oxidize and form an adherent "rust" on the surface. This can pose problems from a fracture analysis standpoint since the oxidation will obscure the underlying fracture surface. In some cases, water-based detergent cleaning can be used to remove surface debris and light oxidation. A detergent that is commonly used is Alconox®. It comes in powder form and can be purchased at most hardware stores. A solution is prepared by dissolving 15 g of Alconox in approximately 350 ml of water. This solution is then heated to approximately 95°C (205°F) and placed in an ultrasonic cleaner. The sample is immersed under ultrasonic agitation for up to 30 min, rinsed with alcohol, and dried.

For heavily oxidized surfaces, more aggressive solutions must be used. A number of these are described in *ASM Handbook, Volume 12: Fractography* [4]. A very effective method for removing tenacious oxide from the surface of steels and stainless steels uses a reagent solution consisting of concentrated hydrochloric acid (HCl) and 10 g/liter of 1,3-*n*-butyl-2 thiourea (known as DBT). This stock solution is then diluted 50/50 with distilled water prior to use [8]. The sample is immersed in this solution under ultrasonic agitation for a short period until the oxidation is removed. The advantage of this technique is that it is quite fast (usually only a few minutes of agitation) and it has been shown to do virtually no damage to the underlying fracture surface. It works particularly well with carbon and low-alloy steels but is also effective in the cleaning of stainless steels.

REFERENCES

[1] *ASM Handbook.* 9th ed. Volume 9, Materials Park, OH: ASM International; 1985. p 279–296.

[2] Walker P, Tarn WH, editors. *CRC Handbook of Metal Etchants.* Boca Raton, FL: CRC Press; 1991. p 1188–1199.

[3] Vander Voort GF. *Metallography of Superalloys.* Lake Bluff, IL: Buehler Ltd.; October 2003.

[4] *ASM Handbook. Volume 12: Fractography.* 9th ed. Materials Park, OH: ASM International; 1987. p 74–76.

[5] Ramirez AJ, Brandi SD, Lippold JC. Study of secondary austenite precipitation by scanning electron microscopy. Acta Microsc 2001;1 (Suppl. A):147.

[6] Varol I, Baeslack WA, Lippold JC. Characterization of weld solidification cracking in a duplex stainless steel. Metallography 1989;23:1–19.

[7] Ginn BJ. A technique for determining austenite to ferrite ratios in welded duplex stainless steels. Weld Inst Res Bull 1985;26:365–367.

[8] Kayafas I. Technical note: Corrosion product removal from steel fracture surfaces for metallographic examination. Corrosion 1980;36(8):443–445.

INDEX

Welding Metallurgy and Weldability, First Edition. John C. Lippold.
© 2015 John Wiley & Sons, Inc. Published 2015 by John Wiley & Sons, Inc.